不只做一个技术者，
更要做一个思考者！

谭勇德/Tom

咕泡学院 Java架构师成长丛书
gupaoedu.com

设计模式就该这样学

基于经典框架源码和真实业务场景

谭勇德（Tom）◎著

電子工業出版社
Publishing House of Electronics Industry
北京•BEIJING

内 容 简 介

本书从软件架构设计必备的通用技能UML开始，重点介绍常用的类图和时序图；然后介绍软件架构设计常用的七大原则；接着结合JDK、Spring、MyBatis、Tomcat、Netty等经典框架源码对GoF的23种设计模式展开分析，并结合作者多年“踩坑填坑”和“教学答疑”经验，用深刻、全面、通俗、生动、有趣、接地气的方式结合真实业务场景分析每种设计模式，治愈“设计模式选择困难症”；之后介绍4种常用的新设计模式；最后总结软件架构、设计模式与应用框架之间的区别。

如果你已经有编程经验，那么你一定要学一遍设计模式；如果你是资深工程师或者架构师，那么本书将颠覆你以前的认知；如果你觉得设计模式晦涩难懂，那么本书一定能让你醍醐灌顶；如果你看框架源码总是力不从心，那么本书就是“内功心法”。

图书在版编目（CIP）数据

设计模式就该这样学：基于经典框架源码和真实业务场景 / 谭勇德著. —北京：电子工业出版社，2020.8
（咕泡学院Java架构师成长丛书）
ISBN 978-7-121-39208-5

Ⅰ. ①设…　Ⅱ. ①谭…　Ⅲ. ①JAVA语言—程序设计　Ⅳ. ①TP312.8

中国版本图书馆CIP数据核字（2020）第116205号

责任编辑：董　英
印　　刷：北京天宇星印刷厂
装　　订：北京天宇星印刷厂
出版发行：电子工业出版社
　　　　　北京市海淀区万寿路173信箱　　　邮编：100036
开　　本：787×980　1/16　　印张：31.5　　字数：706千字
版　　次：2020年8月第1版
印　　次：2022年7月第5次印刷
印　　数：7001~7500册　　定价：118.00元

凡所购买电子工业出版社图书有缺损问题，请向购买书店调换。若书店售缺，请与本社发行部联系，联系及邮购电话：（010）88254888，88258888。

质量投诉请发邮件至zlts@phei.com.cn，盗版侵权举报请发邮件至dbqq@phei.com.cn。

本书咨询联系方式：（010）51260888-819，faq@phei.com.cn。

序　言

Design Patterns: Elements of Reusable Object-Oriented Software（以下简称《设计模式》），一书由 Erich Gamma、Richard Helm、Ralph Johnson 和 John Vlissides 合著（Addison-Wesley Professional，1994），这四位作者常被称为“四人组（Gang of Four，GoF）”，而这本书也就被称为“四人组（或 GoF）”书。他们首次给我们总结出一套在软件开发中可以反复使用的经验，帮助我们提高代码的可重用性、系统的可维护性等，解决软件开发中的复杂问题。

设计模式已诞生 20 多年，其间相继出版的关于设计模式的经典著作不计其数。如果说 GoF 的《设计模式》是设计模式领域的“圣经”，那么之后出版的各种关于设计模式的书籍可称为“圣经”的“批注版”或者“白话版”。本书正是基于 GoF 的《设计模式》来编写的。

本书可以作为笔者对“圣经”实践的精华总结，是一本真正能够落地的“设计模式”之书，也是目前唯一从框架源码如何落地“设计模式”这个角度来理解设计模式的书。本书会结合 JDK、Spring、MyBatis、Tomcat、Netty 等经典框架源码展开对设计模式的分析。当然，本书还会结合笔者多年的“踩坑填坑”经验和“教学答疑”经验，用比“圣经”更深刻、更全面、更通俗、更生动、更有趣、更接地气的方式结合真实业务场景分析每种设计模式的优缺点，治愈“设计模式选择困难症”。选设计模式就像相亲选对象，一旦做好了接受他或她的缺点的准备，那么他或她就一定属于你。所以，本书对于日常开发而言更具有指导意义。

书中部分观点若有不妥之处，恳请纠正，共同进步！

关于本书

适用对象	如果你已经有编程经验，那么你一定要学一遍设计模式； 如果你是资深工程师或者架构师，那么本书将颠覆你以前的认知； 如果你觉得设计模式晦涩难懂，那么本书一定能让你醍醐灌顶； 如果你看框架源码总是力不从心，那么本书就是“内功心法”
JDK版本	1.8及以上
源码版本	Spring 5.0.2.RELEASE MyBatis 3.5.0 Netty 4.1.6.Final Tomcat 8.5.49 JSF 1.2 JUnit 4.12 SLF4J 1.7.2 Guava 18.0
IDE版本	IntelliJ IDEA 2017.1.4
Maven版本	3.5.0及以上

随书源码会在 https://github.com/gupaoedu-tom/design-samples 中持续更新。

读者服务

微信扫码回复：39208

· 获取博文视点学院 20 元付费内容抵扣券

· 获取免费增值资源

· 加入读者交流群，与本书作者互动

· 获取精选书单推荐

关于我

为什么都叫我"文艺汤"

我自幼爱好书法和美术，长了一双能书会画的手，而且手指又长又白，因此以前的艺名叫"玉手藝人"。中学期间，我曾获市级书法竞赛一等奖，校园美术竞赛一等奖，校园征文比赛二等奖，担任过学生会宣传部长，负责校园黑板报，以及校园刊物的编辑、排版和设计。

2008 年参加工作后，我做过家具建模、平面设计等工作，亲自设计了咕泡学院的 Logo。做讲师之后，我给自己起了一个跟姓氏谐音的英文名字"Tom"，江湖人称"编程界写字写得最好的、书法界编程最牛的文艺汤"。

我的技术生涯

我的 IT 技术生涯应该算是从 2009 年开始的，在此之前做过 UI 设计，做过前端网页，到 2009 年才真正开始参与 Java 后台开发。在这里要感谢所有帮助我入门编程的同事和老师。从 2010 年至 2014 年担任过项目组长、项目经理、架构师、技术总监，对很多开源框架都建立了自己的独特见解。我会习惯性地用形象思维来理解抽象世界。譬如：看到二进制数 0 和 1，我会想到《周易》中的两仪——阴和阳；看到颜色值用 RGB 表示，我会想到美术理论中的太阳光折射三原色；下班回家看到炒菜流程，我会想到模板方法模式；坐公交车看到学生卡、老人卡、爱心卡，我会想到策略模式；等等。大家看到的这本书，很多地方都融入了这种形象思维。

众多图书之下为什么写此书

首先，自《Spring 5 核心原理与 30 个类手写实战》和《Netty 4 核心原理与手写 RPC 框架实战》出版以来，各位"汤粉"给我带来了非常多惊喜，这些惊喜让我觉得肩上的责任更大了。

非常感谢各位“汤粉”的大力支持和认可，感谢大家喜欢我的“形象思维”。

其次，我传播知识的宗旨是：将抽象的知识形象化，将枯燥的知识趣味化，将难懂的知识通俗化。而设计模式又是大家公认的最为枯燥的知识，也是很难落地的。这一次，我将压箱底多年的干货都掏出来，以此奉献社会，回馈社会。我总结多年实战经验，并结合经典框架源码分析设计模式，让设计模式真正能够为大家所用，真正可以用来解决实际问题。

最后，再次感谢各位“汤粉”的支持，感谢为本书手稿提出宝贵修改意见的学员，也感谢电子工业出版社有限公司董英团队的辛勤付出。

谭勇德（Tom）

2020 年 6 月 于长沙

目　录

第 1 篇　软件设计前奏篇

第 2 篇　创建型设计模式

第 3 篇 结构型设计模式

第5篇　设计模式总结篇

第6篇 架构设计扩展篇

第 1 篇

软件设计前奏篇

1

第 1 章 重新认识 UML

1.1 UML 的定义

统一建模语言（Unified Modeling Language，UML）是一种为面向对象系统的产品进行说明、可视化和编制文档的标准语言，是非专利的第三代建模和规约语言。UML 是一种面向对象设计的建模工具，是在开发阶段说明、可视化、构建和书写一个面向对象软件密集系统的制品的开放方法，但独立于任何具体的程序设计语言。

1.2 UML 应用场景

UML 最佳的应用是工程实践，在对大规模、复杂系统进行建模方面，特别是在软件架构层次，已经被验证有效。UML 模型大多以图表的方式表现出来。一份典型的建模图表通常包含几个块或框、连接线和作为模型附加信息的文本。这些虽简单却非常重要，在 UML 规则中相互联系和扩展。

UML 的目标是以面向对象图的方式来描述任何类型的系统，具有很宽的应用领域。其中最常用的是建立软件系统的模型，但它同样可以用于描述非软件领域的系统，如机械系统、企业机构或业务过程，以及处理复杂数据的信息系统、具有实时要求的工业系统或工业过程等。总之，UML

可以对任何具有静态结构和动态行为的系统进行建模，而且适用于从需求规格描述直至系统完成后的测试和维护等系统开发的各个阶段。

1.3　UML 基本构件简介

UML 建模的核心是模型，模型是现实的简化、真实系统的抽象。UML 提供了系统的设计蓝图。当给软件系统建模时，需要采用通用的符号语言，这种描述模型所使用的语言被称为建模语言。在 UML 中，所有的描述由事物、关系和图这些构件组成。下图完整地描述了所有构件的关系。

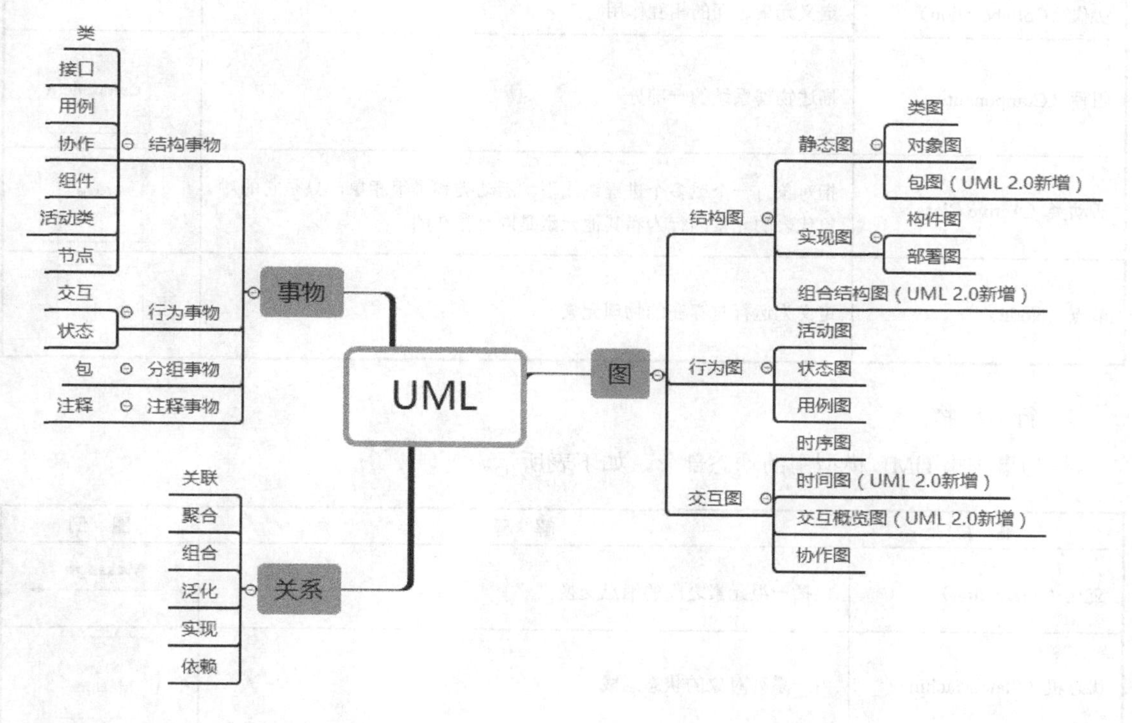

1.3.1　事物

事物是抽象化的最终结果，分为结构事物、行为事物、分组事物和注释事物。

1. 结构事物

结构事物是模型中的静态部分，用以呈现概念或实体的表现元素，如下表所示。

事　物	解　释	图　例
类（Class）	具有相同属性、方法、关系和语义的对象集合	Class / attributes / operations
接口（Interface）	指一个类或构件的一个服务的操作集合。它仅仅定义了一组操作的规范，并没有给出这组操作的具体实现	Interface
用例（User Case）	指对一组动作序列的描述，系统执行这些动作将产生一个对特定的参与者（Actor）有价值且可观察的结果	User Case
协作（Collaboration）	定义元素之间的相互作用	
组件（Component）	描述物理系统的一部分	Component
活动类（Active Class）	指对象有一个或多个进程或线程。活动类和类很相象，只是它的对象代表的元素的行为和其他元素是同时存在的	Active Class
节点（Node）	定义为运行时存在的物理元素	Node

2. 行为事物

行为事物指 UML 模型中的动态部分，如下表所示。

事　物	解　释	图　例
交互（Interaction）	包括一组元素之间的消息交换	Message
状态机（State Machine）	由一系列对象的状态组成	State Machine

3. 分组事物

目前只有一种分组事物，即包。包纯粹是概念上的，只存在于开发阶段，结构事物、行为事物甚至分组事物都有可能放在一个包中，如下表所示。

事　物	解　释	图　例
包（Package）	UML中唯一的组织机制	Package attributes

4. 注释事物

注释事物是解释 UML 模型元素的部分，如下表所示。

事　物	解　释	图　例
注释（Note）	用于解析说明UML元素	Note

1.3.2　关系

UML 将事物之间的联系归纳为 6 种，并用对应的图形类表示，如下表所示。

事物关系	说　明	图　例
关联（Association）	【定义描述】表示一种拥有的关系，具有方向性。如果一个类单方向地访问另一个类，则称为单向关联；如果两个类对象可以互相访问，则称为双向关联；一个对象能访问关联对象的数目叫作“多重性” 【图例解释】用带普通箭头的实线表示，或者用不带箭头的实线表示 【箭头方向】箭头指向被拥有者	引用　被引用
聚合（Aggregate）	【定义描述】表示整体与部分的关系。当某个实体聚合成另一个实体时，该实体还可以是另一个实体的部分 【图例解释】用带空心菱形的实线表示 【箭头方向】菱形“指向”整体，箭头（一般省略箭头）指向个体	个体　整体
组合（Combination）	【定义描述】表示整体与部分的关系，组合比聚合更加严格。当某个实体组合成另一个实体时，二者具有相同的生命周期，例如手臂和人之间是组合关系 【图例解释】用带实心菱形的实线表示 【箭头方向】菱形“指向”整体，箭头（一般省略箭头）指向个体	个体　整体
泛化（Generalization）	【定义描述】表示一个更泛化的元素与一个更具体的元素之间的关系，与继承是同一个概念 【图例解释】用带三角箭头的实线表示 【箭头方向】箭头指向父类	子类　父类

续表

事物关系	说　明	图　例
实现（Realization）	【定义描述】表示类与接口的关系，类实现接口 【图例解释】用带三角箭头的虚线表示 【箭头方向】箭头指向父接口	实现　接口
依赖（Dependency）	【定义描述】如果一个类的改动会影响另一个类，则两个类之间存在依赖关系，一般而言，依赖是单向的 【图例解释】用带普通箭头的虚线表示 【箭头方向】箭头指向被依赖者	引用　被依赖

1.3.3　图

UML 2.0 一共有 13 种图（UML 1.5 定义了 9 种，UML 2.0 增加了 4 种），分别是类图、对象图、构件图、部署图、活动图、状态图、用例图、时序图、协作图 9 种，以及包图、组合结构图、时间图、交互概览图 4 种。

图 名 称	解　释
类图（Class Diagrams）	用于定义系统中的类
对象图（Object Diagrams）	类图的一个实例，描述了系统在具体时间点上所包含的对象及各个对象之间的关系
构件图（Component Diagrams）	一种特殊的UML图，描述系统的静态实现视图
部署图（Deployment Diagrams）	定义系统中软硬件的物理体系结构
活动图（Activity Diagrams）	用来描述满足用例要求所要进行的活动及活动间的约束关系
状态图（State Chart Diagrams）	用来描述类的对象的所有可能的状态和时间发生时，状态的转移条件
用例图（Usecase Diagrams）	用来描述用户的需求，从用户的角度描述系统的功能，并指出各功能的执行者，强调谁在使用系统、系统为执行者完成哪些功能
时序图（Sequence Diagrams）	描述对象之间的交互顺序，着重体现对象间消息传递的时间顺序，强调对象之间消息的发送顺序，同时显示对象之间的交互过程
协作图（Collaboration Diagrams）	描述对象之间的合作关系，更侧重向用户对象说明哪些对象有消息的传递
包图（Package Diagrams）	对构成系统的模型元素进行分组整理的图
组合结构图（Composite Structure Diagrams）	表示类或者构建内部结构的图
时间图（Timing Diagrams）	用来显示随时间变化，一个或多个元素的值或状态的更改，也显示时间控制事件之间的交互及管理它们的时间和期限约束
交互概览图（Interaction Overview Diagrams）	用活动图来表示多个交互之间的控制关系的图

第 2 章 设计模式常用的 UML 图

2.1 类图

在 UML 2.0 的 13 种图中，类图（Class Diagrams）是使用频率最高的 UML 图之一。类图描述系统中的类，以及各个类之间的关系的静态视图，能够让我们在正确编写代码之前对系统有一个全面的认识。类图是一种模型类型，确切地说，是一种静态模型类型。类图表示类、接口和它们之间的协作关系，用于系统设计阶段。

类图用 3 个矩形拼接表示，最上面的部分标识类的名称，中间的部分标识类的属性，最下面的部分标识类的方法，如下图所示。

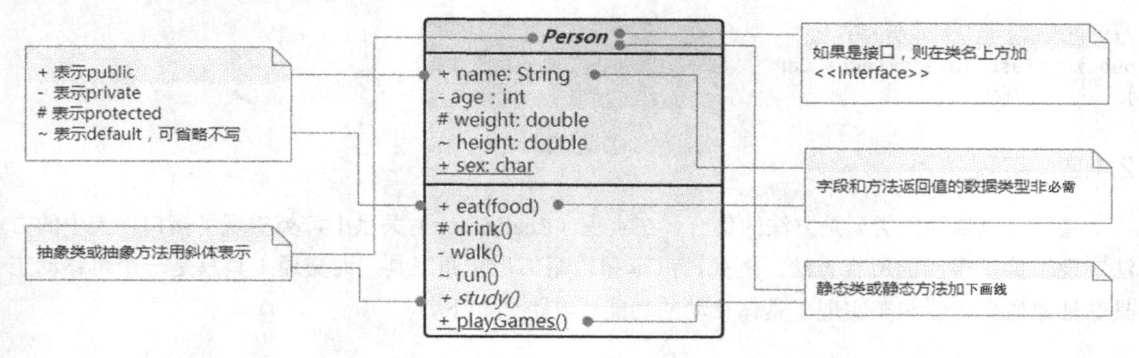

类与类之间的关系（即事物关系）有继承（泛化）关系、实现关系、组合关系、聚合关系、关联关系和依赖关系 6 种。下面我们来详细分析类关系的具体内容。

2.1.1 继承关系

在继承（Generalization，又叫作泛化）关系中，子类继承父类的所有功能，父类所具有的属性、方法，子类都应该有。除了与父类一致的信息，子类中还包括额外的信息。例如，公交车、出租车和小轿车都是汽车，它们都有名称，并且都能在路上行驶。其类图如下。

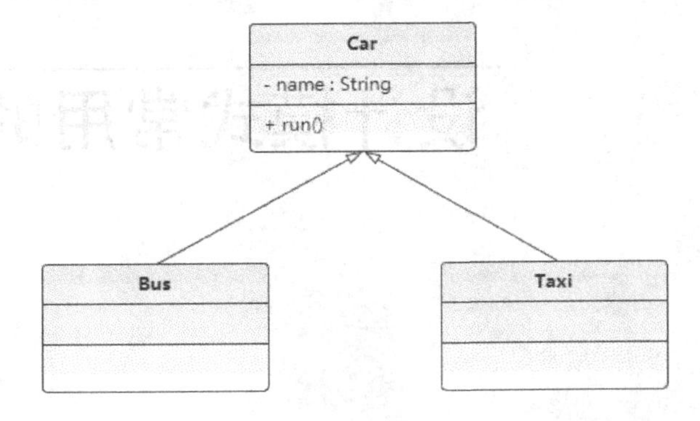

其代码结构如下。

```
//汽车类
public class Car {
    private String name;
    public void run(){}
}

//公共汽车类是汽车类的子类
public class Bus extends Car {
}

//出租汽车类也是汽车类的子类
public class Taxi extends Car {
}
```

2.1.2 实现关系

接口（包括抽象类）是方法的集合，在实现（Realization）关系中，类实现了接口，类中的方法实现了接口声明的所有方法。例如，汽车和轮船都是交通工具，而交通工具只是一个可移动工具的抽象概念，船和车实现了具体移动的功能。其类图如下。

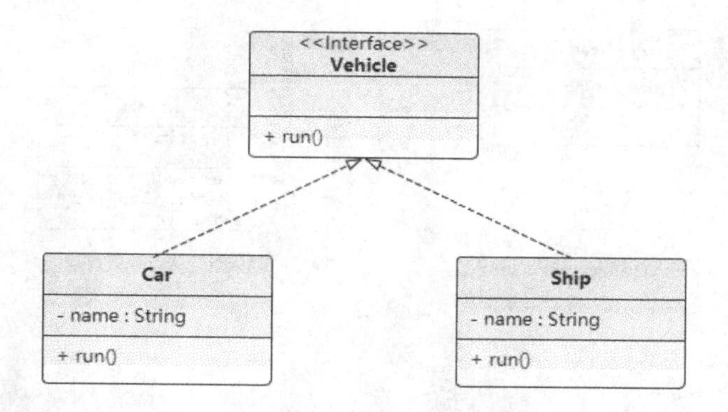

其代码结构如下。

```
//交通工具接口
public interface Vehicle {
    //只要是交通工具都能跑
    void run();
}

//汽车实现交通工具的接口
public class Car implements Vehicle {
    private String name;
    public void run() {
    }
}

//轮船实现交通工具的接口
public class Ship implements Vehicle {
    private String name;
    public void run() {
    }
}
```

2.1.3　组合关系

组合（Combination）关系表示类之间整体与部分的关系，整体与部分有一致的生存期。一旦整体对象不存在，部分对象也将不存在，整体和部分是同生共死的关系。例如，人由头部和身体组成，两者不可分割，共同存在。其类图如下。

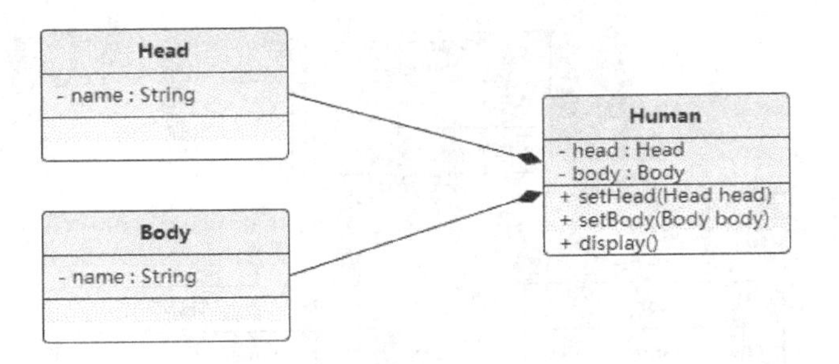

其代码结构如下。

```
//头部类
public class Head {
    private String name;
}

//身体类
public class Body {
    private String name;
}

//人类
public class Human {
    private Head head; //头部属于人的一部分，和人具有相同的生命周期
    private Body body; //身体属于人的一部分，和人具有相同的生命周期

    public void setHead(Head head) {
        this.head = head;
    }

    public void setBody(Body body) {
        this.body = body;
    }

    public void display(){
    }
}
```

2.1.4 聚合关系

聚合（Aggregate）关系也表示类之间整体与部分的关系，成员对象是整体对象的一部分，但是成员对象可以脱离整体对象独立存在。例如，公交车司机和工作服、工作帽是整体与部分的关系，但是可以分开，没有共同的生命周期。工作服、工作帽可以穿、戴在别的司机身上，公交车司机也可以换别人的工作服、工作帽。其类图如下。

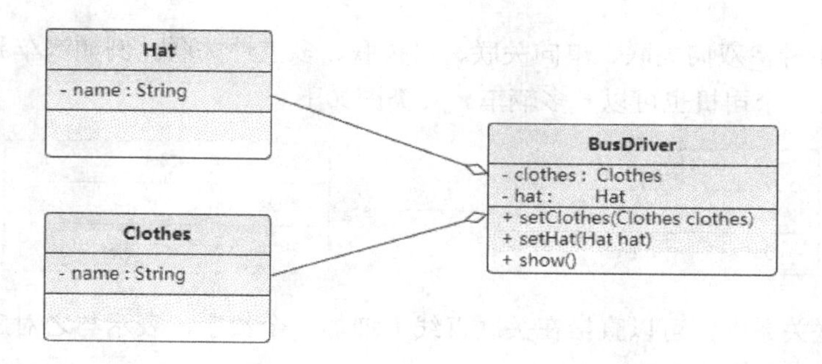

其代码结构如下。

```
//工作帽类
public class Hat {
    private String name;
}

//工作服类
public class Clothes {
    private String name;
}

//公交车司机类
public class BusDriver {
    private Clothes clothes;
    private Hat hat;

    //公交车司机可以更换工作服
    public void setClothes(Clothes clothes) {
        this.clothes = clothes;
    }

    //公交车司机可以更换工作帽
    public void setHat(Hat hat) {
        this.hat = hat;
    }

    public void show(){
    }
}
```

2.1.5　关联关系

关联（Association）关系是类与类之间最常用的一种关系，表示一类对象与另一类对象之间有联系。组合、聚合也属于关联关系，只是关联关系的类间关系比其他两种关系要弱。

关联关系有 4 种：双向关联、单向关联、自关联、多重性关联。例如汽车和司机，一辆汽车对应特定的司机，一个司机也可以开多辆车。其类图如下。

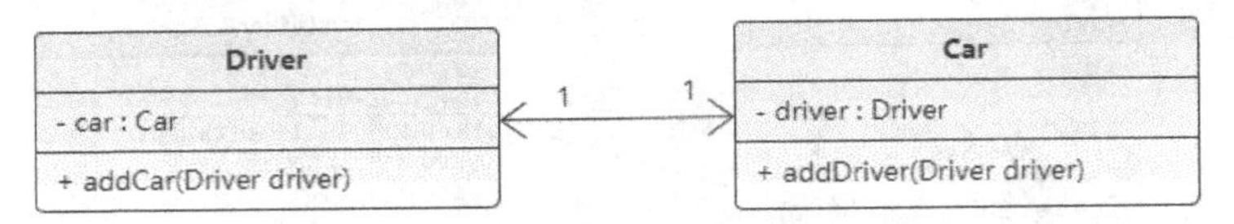

在多重性关联关系中，可以直接在关联直线上增加一个数字，表示与之对应的另一个类的对象的个数，具体含义如下表所示。

表示方式	含 义
1..1	仅一个
0..*	零个或多个
1..*	一个或多个
0..1	没有或只有一个
m..n	最少m、最多n个（$m<=n$）

其代码结构如下。

```
//司机类
public class Driver {
    private Car car; //司机需要持有汽车对象的引用
    //司机驾驶哪辆汽车，要在启动前才赋值
    public void addCar(Car car){
        this.car = car;
    }
}

//汽车类
public class Car {
    private Driver driver;  //汽车需要持有司机的引用
    //汽车需要司机才能行驶，具体的司机要在汽车启动前才赋值
    public void addDriver(Driver driver){
        this.driver = driver;
    }
}
```

2.1.6 依赖关系

依赖（Dependency）关系是一种“使用”关系，特定事物的改变有可能会影响到使用该事物的其他事物，当需要表示一个事物使用另一个事物时，使用依赖关系。在大多数情况下，依赖关

系体现在某个类的方法使用另一个类的对象作为参数。例如，汽车依赖汽油，如果没有汽油，则汽车将无法行驶。其类图如下。

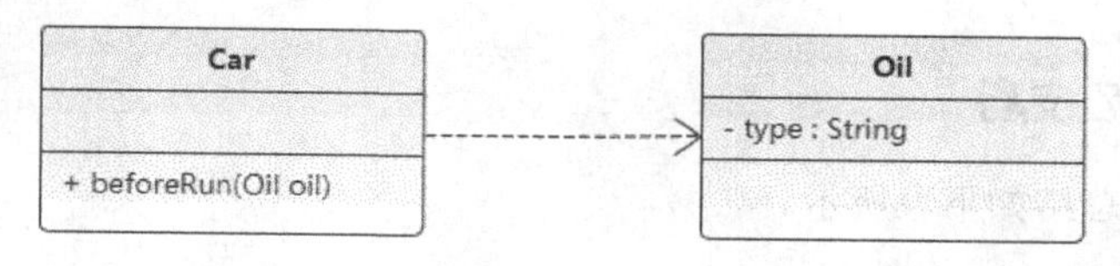

其代码结构如下。

```
//汽油类
public class Oil {
    private String type;
}

//汽车类
public class Car {
    //汽车开动前需要先加油，因此汽车依赖汽油
    public void beforeRun(Oil oil){
    }
}
```

在这 6 种类关系中，组合、聚合和关联的代码结构一样，可以从关系的强弱来理解，各类关系从强到弱依次是：继承 > 实现 > 组合 > 聚合 > 关联 > 依赖。下面我们用一张完整的类图，将前面描述的所有类与类之间的关系串联起来。

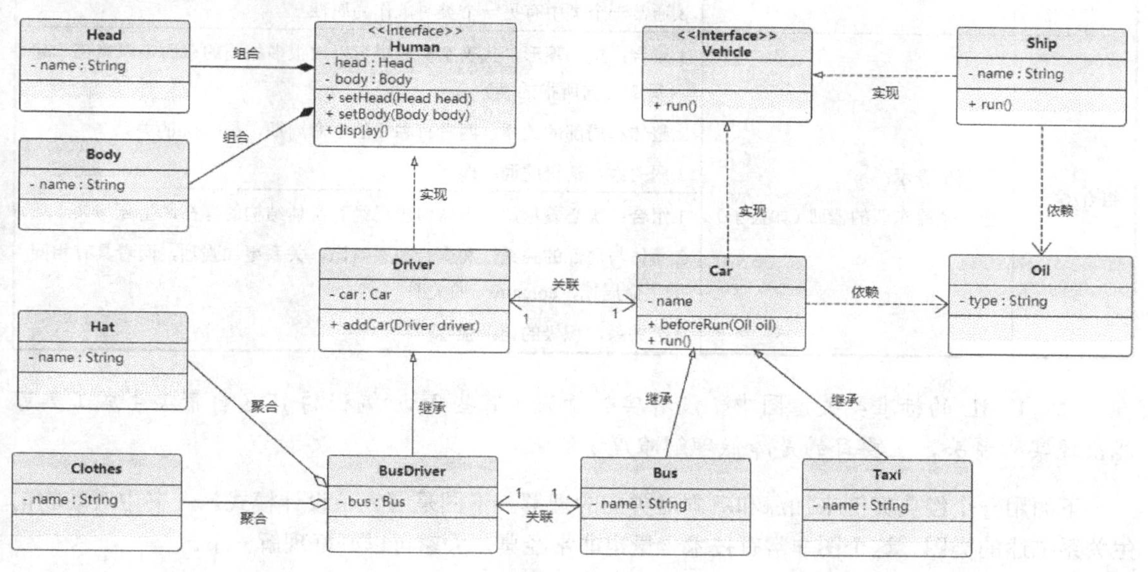

UML 类图是面向对象设计的辅助工具，但并非是必须工具，所以我们把它作为架构师软技能来讲解。

2.1.7 类关系记忆技巧

类关系记忆技巧总结如下表所示。

<table>
<tr><th>分　类</th><th>箭头特征</th><th>记忆技巧</th></tr>
<tr><td>箭头方向</td><td>从子类指向父类</td><td>1.定义子类需要通过extends关键字指定父类
2.子类一定是知道父类定义的，但父类并不知道子类的定义
3.只有知道对方信息时才能指向对方
4.箭头的方向是从子类指向父类</td></tr>
<tr><td rowspan="2">继承/实现</td><td rowspan="2">用线条连接两个类；
空心三角箭头表示继承或实现</td><td>实线表示继承，是is-a的关系，表示扩展，不虚，很结实</td></tr>
<tr><td>虚线表示实现，虚线代表“虚”无实体</td></tr>
<tr><td rowspan="2">关联/依赖</td><td rowspan="2">用线条连接两个类；
普通箭头表示关联或依赖</td><td>1.虚线表示依赖关系：临时用一下，若即若离，虚无缥缈，若有若无
2.表示一种使用关系，一个类需要借助另一个类来实现功能
3.一般一个类将另一个类作为参数使用，或作为返回值</td></tr>
<tr><td>1.实线表示关联关系：关系稳定，实打实的关系，“铁哥们”
2.表示一个类对象和另一个类对象有关联
3.通常一个类中有另一个类对象作为属性</td></tr>
<tr><td rowspan="2">组合/聚合</td><td rowspan="2">用菱形表示；
像一个盛东西的器皿（如盘子）</td><td>1.聚合：空心菱形，代表空器皿里可以放很多相同的东西，聚集在一起（箭头方向所指的类）
2.整体和局部的关系，两者有独立的生命周期，是has-a的关系
3.弱关系，消极的词：弱-空</td></tr>
<tr><td>1.组合：实心菱形，代表器皿里已经有实体结构的存在，生死与共
2.整体与局部的关系，和聚合关系对比，关系更加强烈，两者具有相同的生命周期，contains-a的关系
3.强关系，积极的词：强-满</td></tr>
</table>

注：UML 的标准类关系图中，没有实心箭头（有些 Java 编程的 IDE 自带类生成工具可能出现实心箭头，主要目的是降低理解难度）。

下面用一个经典案例来加深和巩固对类图的理解。下图是《大话设计模式》一书中对动物衍生关系描述的类图。这个图非常有技术含量也非常经典，大家可以好好理解一下。

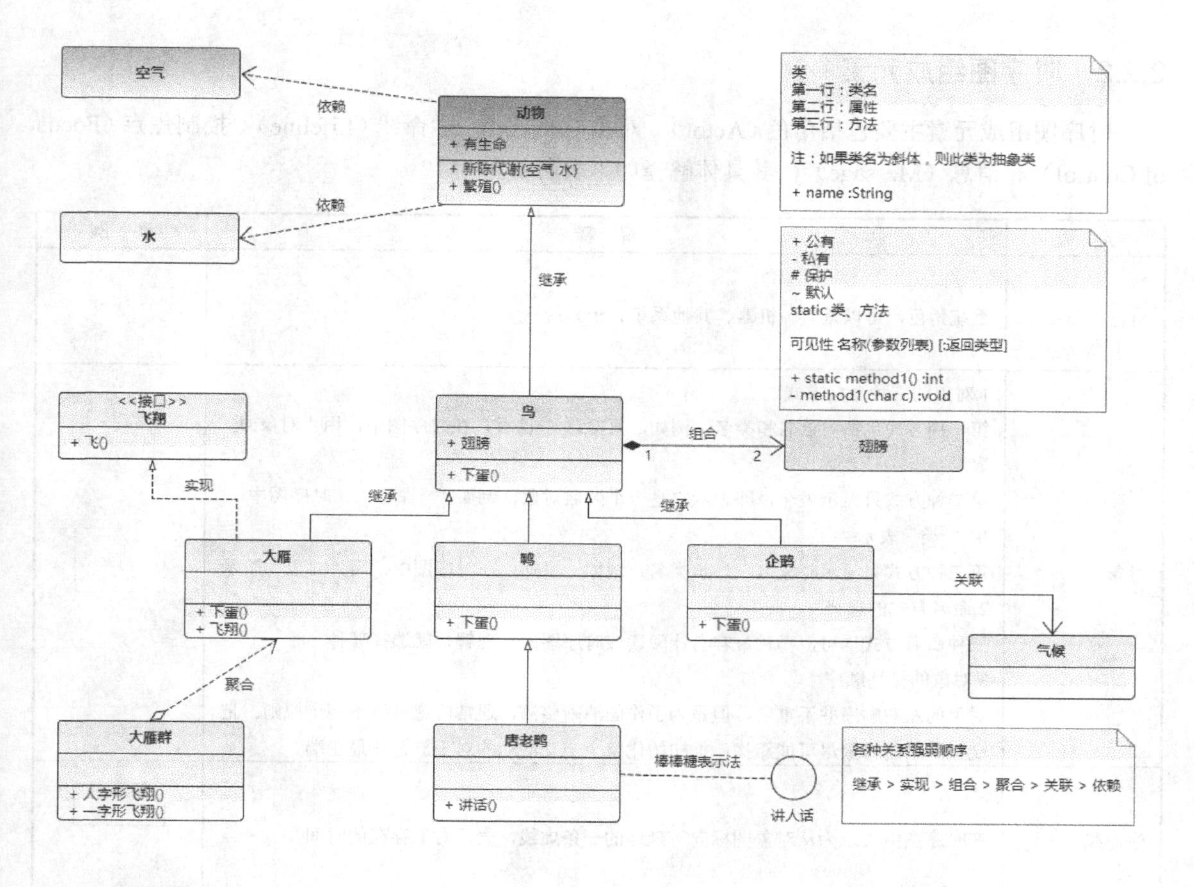

2.2　时序图

时序图（Sequence Diagrams）描述对象之间消息的发送顺序，强调时间顺序。时序图是一个二维图，横轴表示对象，纵轴表示时间，消息在各对象之间横向传递，按照时间顺序纵向排列。用箭头表示消息，用竖虚线表示对象生命线。

2.2.1　时序图的作用

（1）展示对象之间交互的顺序。将交互行为建模为消息传递，通过描述消息如何在对象间发送和接收来动态展示对象之间的交互。

（2）相对于其他 UML 图，时序图更强调交互的时间顺序。

（3）可以直观地描述并发进程。

2.2.2 时序图组成元素

时序图组成元素主要包括角色（Actor）、对象（Object）、生命线（Lifeline）、控制焦点（Focus of Control）和消息（Message），其具体解释如下表所示。

元　素	解　释	图　例
角色	系统角色，可以是人、机器、其他系统、子系统	
对象	1.对象的三种命名方式 第一种方式包括对象名和类名，例如，直播课时:课时，在时序图中，用“对象:类”表示 第二种方式只显示类名，即表示它是一个匿名对象，例如， :课程；在时序图中，用“:类”表示 第三种方式只显示对象名不显示类名，例如，讲师；在时序图中，用“对象”表示 2.命名方式的选择 三种命名方式均可，哪种最容易让阅读该时序图的人理解，就选择哪种 3.对象的排列顺序 对象的左右顺序并不重要，但是为了作图清晰整洁，通常应遵循以下两个原则：把交互频繁的对象尽可能靠拢；把初始化整个交互活动的对象放置在最左端	Object
生命线	在时序图中表示为从对象图标向下延伸的一条虚线，表示对象存在的时间	
控制焦点	又被称为激活期，表示时间段的符号，表示在这个时间段内对象将执行相应的操作。可以理解为Java中一对大括号中的内容	
消息	消息一般分为同步消息（Synchronous Message）、异步消息（Asynchronous Message）和返回消息（Return Message） 1. 消息发送者把控制传递给消息接收者，然后停止活动，等待消息接收者放弃或者返回控制。用来表示同步的意义 2. 消息发送者通过消息把信号传递给消息接收者，然后继续自己的活动，不等待接收者返回消息或者控制。异步消息的接收者和发送者是并发工作的 3. 返回消息表示从过程调用返回	Message

2.2.3 时序图组合片段

组合片段（Combined Fragments）用来解决交互执行的条件和方式，它允许在时序图中直接表示逻辑组件，用于通过指定条件或子进程的应用区域，为任何生命线的任何部分定义特殊条件和子进程。组合片段共有 13 种，名称及含义如下表所示。

类　型	名　称	说　明
Alt	抉择	包含一个片段列表，这些片段包含备选消息序列。在任何场合下只发生一个序列。可以在每个片段中都设置一个临界来指示该片段可以运行的条件。else 的临界指示其他任何临界都不为true时应运行的片段。如果所有临界都为false并且没有else，则不执行任何片段
Opt	选项	包含一个可能发生或可能不发生的序列。可以在临界中指定序列发生的条件
Loop	循环	片段重复一定次数。可以在临界中指示片段重复的条件。Loop组合片段具有Min和Max属性，它们指示片段可以重复的最小和最大次数。默认值是无限制
Break	中断	如果执行此片段，则放弃序列的其余部分。可以使用临界来指示发生中断的条件
Par	并行	并行处理。片段中的事件可以交错
Critical	关键	用在Par或Seq片段中。指示此片段中的消息不得与其他消息交错
Seq	弱顺序	有两个或更多操作数片段。涉及同一生命线的消息必须按片段的顺序发生。如果消息涉及的生命线不同，则来自不同片段的消息可能会并行交错
Strict	强顺序	有两个或更多操作数片段。这些片段必须按给定顺序发生
Consider	考虑	指定此片段描述的消息列表。其他消息可发生在运行的系统中，但对此描述来说意义不大。在Messages属性中键入该列表
Ignore	忽略	指定此片段未描述的消息列表。这些消息可发生在运行的系统中，但对此描述来说意义不大。在Messages属性中键入该列表
Assert	断言	操作数片段指定唯一有效的序列。通常用在Consider或Ignore片段中
Neg	否定	此片段中显示的序列不得发生。通常用在 Consider或Ignore片段中

常用组合片段举例如下。

1. 抉择（Alt）

抉择用来指明在两个或更多消息序列之间的互斥的选择，相当于经典的 if...else。抉择在任何场合下只发生一个序列。可以在每个片段中都设置一个临界来指示该片段可以运行的条件。else 的临界指示其他任何临界都不为 true 时应运行的片段。如果所有临界都为 false 并且没有 else，则不执行任何片段，如下图所示。

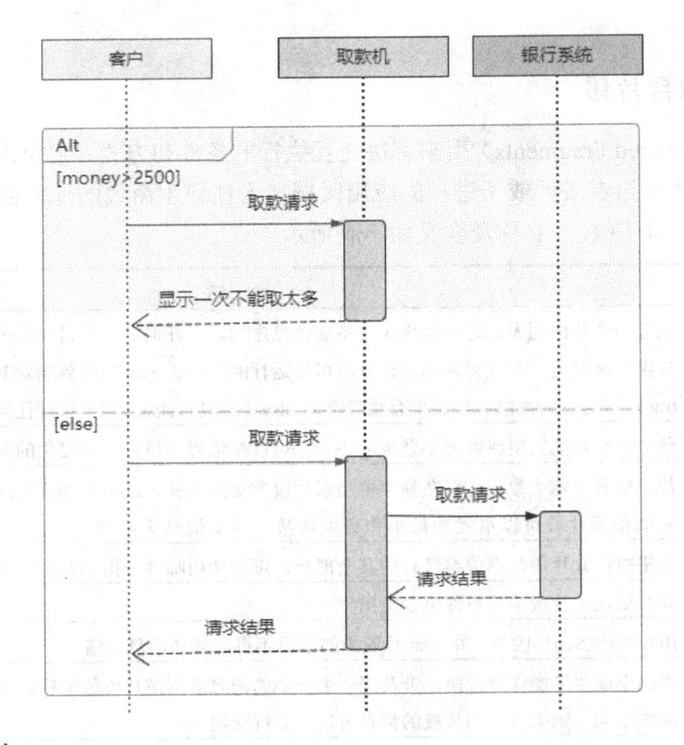

2. 选项（Opt）

包含一个可能发生或不发生的序列，如下图所示。

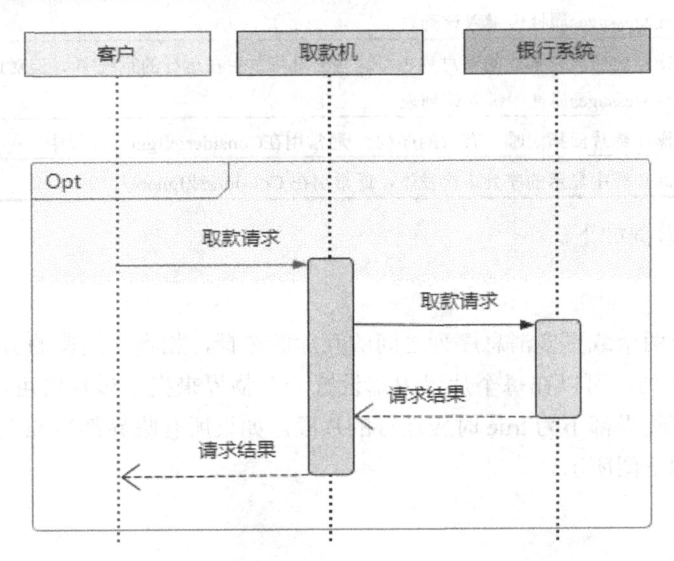

3. 循环（Loop）

片段重复一定次数，可以在临界中指示片段重复的条件，如下图所示。

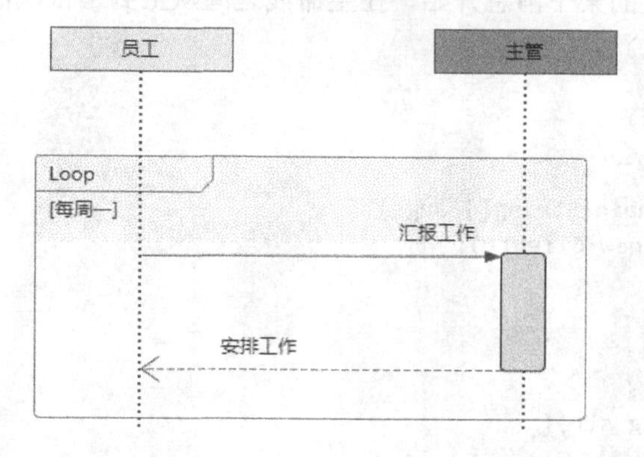

4. 并行（Par）

并行处理，片段中的事件可以并行交错，Par 相当于多线程，如下图所示。

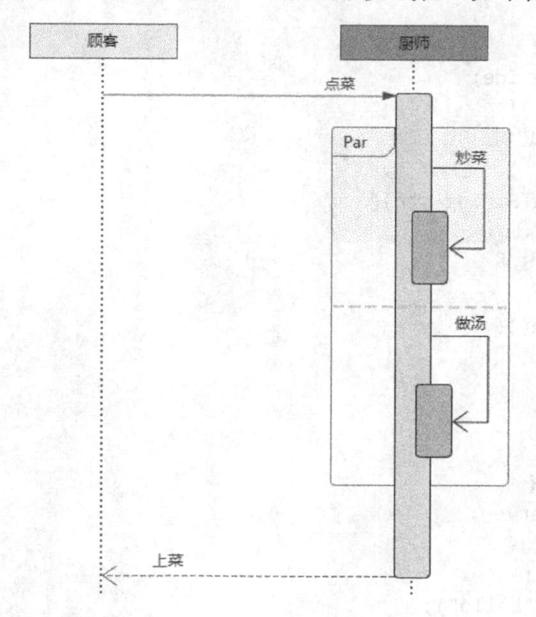

2.2.4　时序图画法及应用实践

时序图的绘制步骤可简单总结如下。

（1）划清边界，识别交互的语境。

（2）将所要绘制的交互场景中的角色及对象梳理出来。

（3）从触发整个交互的某个消息开始，在生命线之间从上到下依次画出所有消息，并注明每个消息的特性（如参数等）。

假设有如下一段代码。

```
public class Test {
    public static void main(String[] args) {
        Client client = new Client();
        client.work();
    }

    static class Device {
        void write(String str){
            //此处省略处理逻辑
         }
    }

    static class Server {
        private Device device;
        public void open(){
            //此处省略处理逻辑
        }
        public void print(String str){
            device.write(str);
            //此处省略处理逻辑
        }
        public void close(){
            //此处省略处理逻辑
        }
    }

    static class Client {
        private Server server;
        public void work(){
            server.open();
            server.print("hello");
            server.close();
        }
    }
}
```

上面代码执行对应的时序图如下。

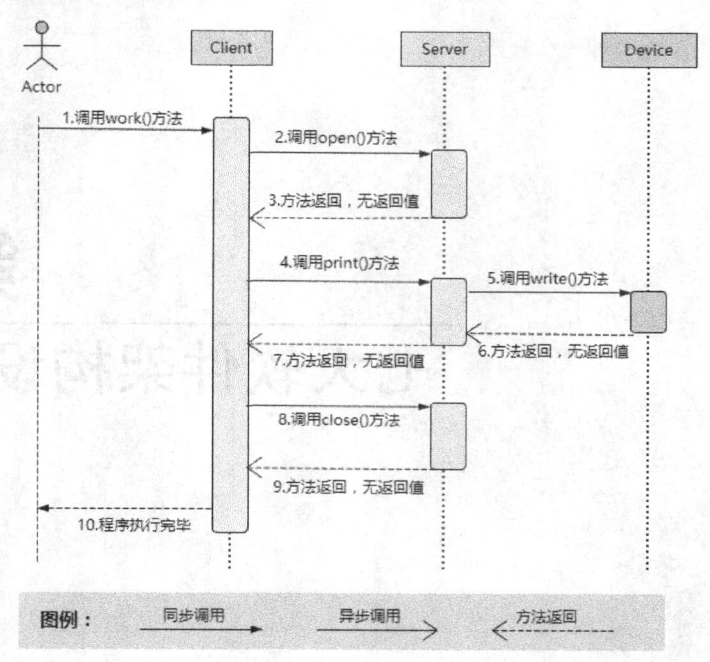

3

第 3 章 七大软件架构设计原则

3.1 开闭原则

3.1.1 开闭原则的定义

开闭原则（Open-Closed Principle，OCP）指一个软件实体如类、模块和函数应该对扩展开放，对修改关闭。所谓开闭，也正是对扩展和修改两个行为的一个原则。强调的是用抽象构建框架，用实现扩展细节，可以提高软件系统的可复用性及可维护性。开闭原则是面向对象设计中最基础的设计原则。它指导我们如何建立稳定灵活的系统，例如版本更新，我们尽可能不修改源码，但是可以增加新功能。

在现实生活中，开闭原则也有体现。比如，很多互联网公司都实行弹性制作息时间，规定每天工作 8 小时。意思就是，对于每天工作 8 小时这个规定是关闭的，但是什么时候来、什么时候走是开放的。早来早走，晚来晚走。

实现开闭原则的核心思想就是面向抽象编程。

3.1.2 使用开闭原则解决实际问题

我们来看一段代码，以咕泡学院的课程体系为例，首先创建一个课程接口 ICourse。

```
public interface ICourse {
    Integer getId();
    String getName();
    Double getPrice();
}
```

整个课程生态有 Java 架构、大数据、人工智能、前端、软件测试等。我们创建一个 Java 架构课程的类 JavaCourse。

```
public class JavaCourse implements ICourse{
    private Integer Id;
    private String name;
    private Double price;
    public JavaCourse(Integer id, String name, Double price) {
        this.Id = id;
        this.name = name;
        this.price = price;
    }
    public Integer getId() {
        return this.Id;
    }
    public String getName() {
        return this.name;
    }
    public Double getPrice() {
        return this.price;
    }
}
```

现在要给 Java 架构课程做活动，价格优惠。如果修改 JavaCourse 中的 getPrice()方法，则会存在一定风险，可能影响其他地方的调用结果。如何在不修改原有代码的前提下，实现价格优惠这个功能呢？我们再写一个处理优惠逻辑的类——JavaDiscountCourse 类（可以思考一下为什么要叫 JavaDiscountCourse，而不叫 DiscountCourse）。

```
public class JavaDiscountCourse extends JavaCourse {
    public JavaDiscountCourse(Integer id, String name, Double price) {
        super(id, name, price);
    }
    public Double getOriginPrice(){
        return super.getPrice();
    }
```

```
    public Double getPrice(){
        return super.getPrice() * 0.61;
    }
}
```

简单回顾一下类图，如下图所示。

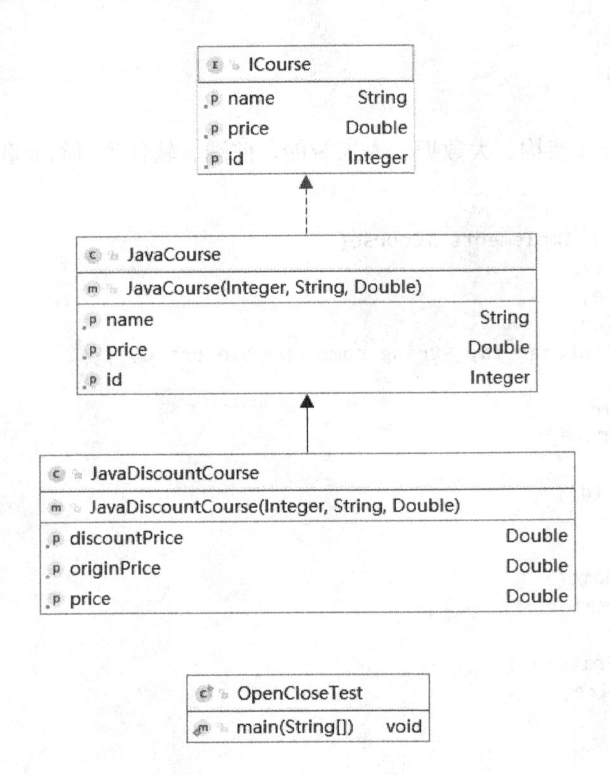

从类图中可以看出，JavaDiscountCourse 中保留了覆盖 JavaCourse 的 getPrice()方法，而不是直接修改 JavaCourse 类。

3.2 依赖倒置原则

3.2.1 依赖倒置原则的定义

依赖倒置原则（Dependence Inversion Principle，DIP）指设计代码结构时，高层模块不应该依赖底层模块，二者都应该依赖其抽象。抽象不应该依赖细节，细节应该依赖抽象。通过依赖倒置，可以降低类与类之间的耦合性，提高系统的稳定性，提高代码的可读性和可维护性，并降低修改程序带来的风险。

3.2.2　使用依赖倒置原则解决实际问题

我们来看一个案例，还是以课程为例，首先创建一个类 Tom。

```
public class Tom {
    public void studyJavaCourse(){
        System.out.println("Tom 在学习 Java 的课程");
    }

    public void studyPythonCourse(){
        System.out.println("Tom 在学习 Python 的课程");
    }
}
```

然后编写客户端测试代码并调用。

```
public static void main(String[] args) {
    Tom tom = new Tom();
    tom.studyJavaCourse();
    tom.studyPythonCourse();
}
```

Tom 热爱学习，目前正在学习 Java 课程和 Python 课程。大家都知道，学习是会上瘾的。随着学习兴趣的暴涨，Tom 还想学习 AI 课程。这个时候，需要业务扩展，代码要从底层到高层（调用层）一次修改代码。在 Tom 类中增加 studyAICourse()的方法，在高层也要追加调用。如此一来，在系统发布以后，实际上是非常不稳定的，在修改代码的同时会带来意想不到的风险。因此我们优化代码，首先创建一个课程的抽象接口 ICourse。

```
public interface ICourse {
    void study();
}
```

然后写 JavaCourse 类。

```
public class JavaCourse implements ICourse {
    @Override
    public void study() {
        System.out.println("Tom 在学习 Java 课程");
    }
}
```

再实现 PythonCourse 类。

```
public class PythonCourse implements ICourse {
    @Override
    public void study() {
        System.out.println("Tom 在学习 Python 课程");
    }
}
```

最后修改 Tom 类。

```
public class Tom {
    public void study(ICourse course){
        course.study();
    }
}
```

来看客户端测试代码。

```
public static void main(String[] args) {
    Tom tom = new Tom();
    tom.study(new JavaCourse());
    tom.study(new PythonCourse());
}
```

这时候再看代码，Tom 的兴趣无论怎么暴涨，对于新的课程，只需要新建一个类，通过传参的方式告诉 Tom，而不需要修改底层代码。实际上，这是一种大家非常熟悉的方式，叫作依赖注入。注入的方式还有构造器注入方式和 Setter 注入方式。下面来看构造器注入方式。

```
public class Tom {

    private ICourse course;

    public Tom(ICourse course){
        this.course = course;
    }

    public void study(){
        course.study();
    }
}
```

来看客户端代码，将 JavaCourse 对象作为 Tom 对象的构造参数注入。

```
public static void main(String[] args) {
    Tom tom = new Tom(new JavaCourse());
    tom.study();
}
```

根据构造器注入方式，当调用时，每次都要创建实例。那么，如果 Tom 是全局单例，则只能选择 Setter 注入方式，继续修改 Tom 类的代码。

```
public class Tom {
    private ICourse course;
    public void setCourse(ICourse course) {
        this.course = course;
    }
    public void study(){
        course.study();
    }
}
```

来看客户端代码，调用 Tom 对象的 setCourse()方法，将 JavaCourse 对象作为参数。

```
public static void main(String[] args) {
    Tom tom = new Tom();
    tom.setCourse(new JavaCourse());
    tom.study();

    tom.setCourse(new PythonCourse());
    tom.study();
}
```

最终得到的类图如下。

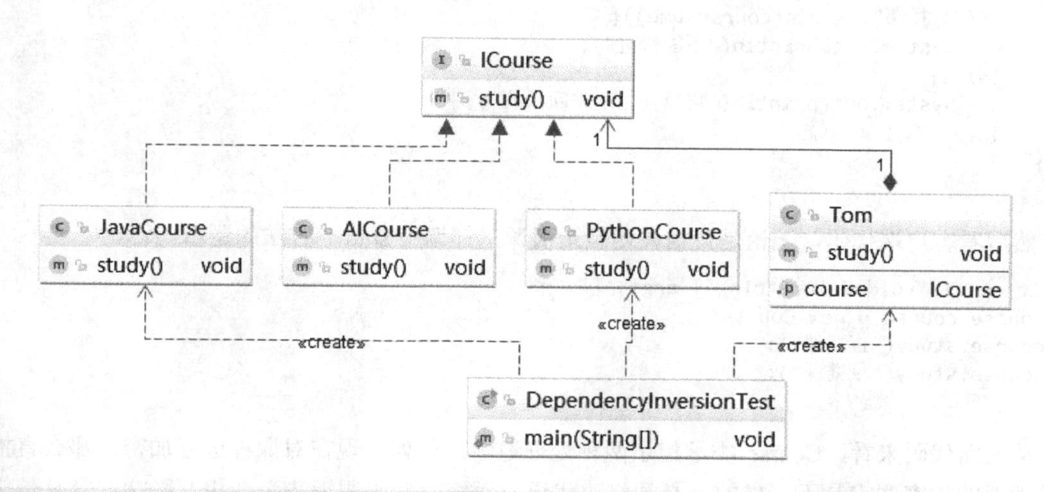

注：以抽象为基准比以细节为基准搭建起来的架构要稳定得多，因此大家在拿到需求后，要面向接口编程，按照先顶层再细节的顺序设计代码结构。

3.3 单一职责原则

3.3.1 单一职责原则的定义

单一职责原则（Simple Responsibility Principle，SRP）指不要存在一个以上导致类变更的原因。假设有一个 Class 负责两个职责，一旦发生需求变更，修改其中一个职责的逻辑代码，有可能会导致另一个职责的功能发生故障。这样一来，这个 Class 就存在两个导致类变更的原因。如何解决这个问题呢？我们就要分别用两个 Class 来实现两个职责，进行解耦。后期需求变更维护互不影响。这样的设计，可以降低类的复杂度，提高类的可读性，提高系统的可维护性，降低变更引起的风险。总体来说就是一个 Class、Interface、Method 只负责一项职责。

3.3.2 使用单一职责原则解决实际问题

我们来看代码实例，还是用课程举例，我们的课程有直播课和录播课。直播课不能快进和快退，录播课可以任意地反复观看，功能职责不一样。首先创建一个 Course 类。

```
public class Course {
    public  void study(String courseName){
        if("直播课".equals(courseName)){
            System.out.println("不能快进");
        }else{
            System.out.println("可以任意地来回播放");
        }
    }
}
```

然后看客户端代码，无论是直播课还是录播课，都调用 study()方法的逻辑。

```
public static void main(String[] args) {
    Course course = new Course();
    course.study("直播课");
    course.study("录播课");
}
```

从上面代码来看，Course 类承担了两种处理逻辑。假如，现在对课程进行加密，那么直播课和录播课的加密逻辑是不一样的，必须修改代码。而修改代码逻辑势必会相互影响，容易带来不可控的风险。我们对职责进行分离解耦，分别创建两个类 LiveCourse 和 ReplayCourse。

LiveCourse 类的代码如下。

```
public class LiveCourse {
```

```
    public void study(String courseName){
        System.out.println(courseName + "不能快进看");
    }
}
```

ReplayCourse 类的代码如下。

```
public class ReplayCourse {
    public  void  study(String courseName){
        System.out.println("可以任意地来回播放");
    }
}
```

客户端代码如下，将直播课的处理逻辑调用 LiveCourse 类，录播课的处理逻辑调用 ReplayCourse 类。

```
public static void main(String[] args) {
    LiveCourse liveCourse = new LiveCourse();
    liveCourse.study("直播课");

    ReplayCourse replayCourse = new ReplayCourse();
    replayCourse.study("录播课");
}
```

当业务继续发展时，要对课程做权限。没有付费的学员可以获得课程的基本信息，已经付费的学员可以获得视频流，即学习权限。那么对于控制课程层面，至少有两个职责。我们可以把展示职责和管理职责分离开，都实现同一个抽象依赖。设计一个顶层接口，创建 ICourse 接口。

```
public interface ICourse {

    //获得课程的基本信息
    String getCourseName();

    //获得视频流
    byte[] getCourseVideo();

    //学习课程
    void studyCourse();
    //退款
    void refundCourse();
}
```

可以把这个接口拆成两个接口，创建一个接口 ICourseInfo 和 ICourseManager。

ICourseInfo 接口的代码如下。

```
public interface ICourseInfo {
    String getCourseName();
```

```
    byte[] getCourseVideo();
}
```

ICourseManager 接口的代码如下。

```
public interface ICourseManager {
    void studyCourse();
    void refundCourse();
}
```

来看一下类图，如下图所示。

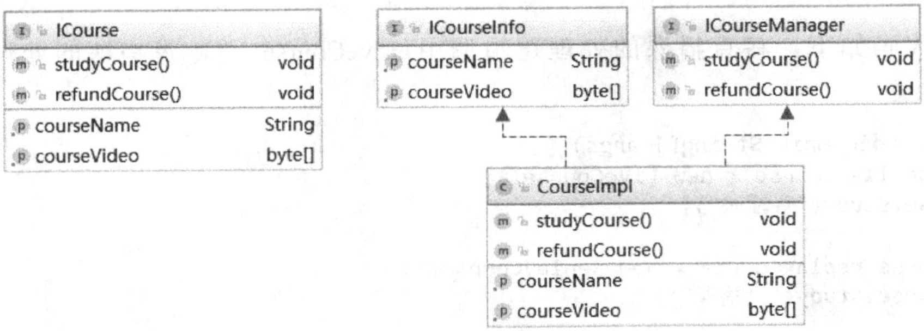

下面来看方法层面的单一职责设计。有时候，为了偷懒，通常会把一个方法写成下面这样。

```
private void modifyUserInfo(String userName,String address){
    userName = "Tom";
    address = "Changsha";
}
```

还可能写成这样。

```
private void modifyUserInfo(String userName,String... fields){
        userName = "Tom";
}
private void modifyUserInfo(String userName,String address,boolean bool){
    if(bool){

    }else{

    }

    userName = "Tom";
    address = "Changsha";
}
```

显然，上面两种写法的 modifyUserInfo()方法都承担了多个职责，既可以修改 userName，也可以修改 address，甚至更多，明显不符合单一职责原则。那么我们做如下修改，把这个方法拆成两个。

```
private void modifyUserName(String userName){
    userName = "Tom";
}
private void modifyAddress(String address){
    address = "Changsha";
}
```

代码在修改之后，开发起来简单，维护起来也容易。在实际项目中，代码会存在依赖、组合、聚合关系，在项目开发过程中还受到项目的规模、周期、技术人员水平、对进度把控的影响，导致很多类都不能满足单一职责原则。但是，我们在编写代码的过程中，尽可能地让接口和方法保持单一职责，对项目后期的维护是有很大帮助的。

3.4　接口隔离原则

3.4.1　接口隔离原则的定义

接口隔离原则（Interface Segregation Principle，ISP）指用多个专门的接口，而不使用单一的总接口，客户端不应该依赖它不需要的接口。这个原则指导我们在设计接口时，应当注意以下几点。

（1）一个类对另一个类的依赖应该建立在最小接口上。

（2）建立单一接口，不要建立庞大臃肿的接口。

（3）尽量细化接口，接口中的方法尽量少（不是越少越好，一定要适度）。

接口隔离原则符合“高聚合、低耦合”的设计思想，使得类具有很好的可读性、可扩展性和可维护性。在设计接口的时候，要多花时间思考，要考虑业务模型，包括还要对以后可能发生变更的地方做一些预判。所以，在实际开发中，我们对抽象、业务模型的理解是非常重要的。

3.4.2　使用接口隔离原则解决实际问题

我们来写一个动物行为的抽象。

IAnimal 接口的代码如下。

```
public interface IAnimal {
    void eat();
    void fly();
```

```
    void swim();
}
```

Bird 类实现的代码如下。

```
public class Bird implements IAnimal {
    @Override
    public void eat() {}
    @Override
    public void fly() {}
    @Override
    public void swim() {}
}
```

Dog 类实现的代码如下。

```
public class Dog implements IAnimal {
    @Override
    public void eat() {}
    @Override
    public void fly() {}
    @Override
    public void swim() {}
}
```

由上面代码可以看出，Bird 类的 swim()方法可能只能空着，Dog 类的 fly()方法显然是不可能的。这时候，我们针对不同动物的行为来设计不同的接口，分别设计 IEatAnimal、IFlyAnimal 和 ISwimAnimal 接口。

IEatAnimal 接口的代码如下。

```
public interface IEatAnimal {
    void eat();
}
```

IFlyAnimal 接口的代码如下。

```
public interface IFlyAnimal {
    void fly();
}
```

ISwimAnimal 接口的代码如下。

```
public interface ISwimAnimal {
    void swim();
}
```

Dog 只实现 IEatAnimal 和 ISwimAnimal 接口。

```
public class Dog implements ISwimAnimal,IEatAnimal {
```

```
    @Override
    public void eat() {}
    @Override
    public void swim() {}
}
```

我们来看两种类图的对比，如下图所示，还是非常清晰明了的。

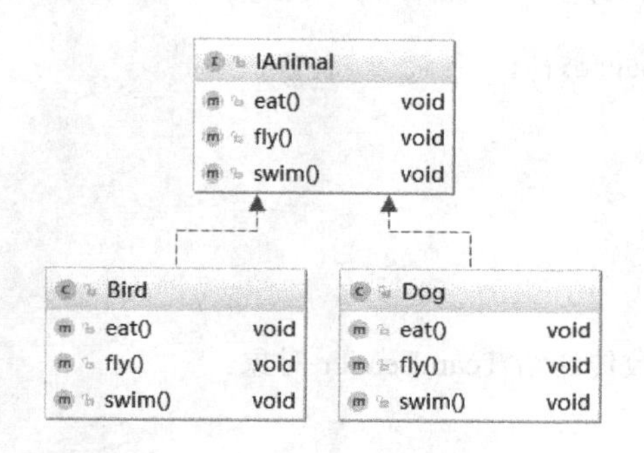

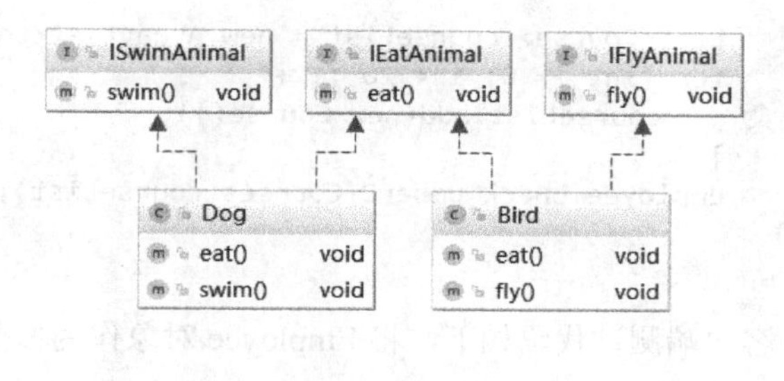

3.5　迪米特法则

3.5.1　迪米特法则的定义

迪米特法则（Law of Demeter，LoD）又叫作最少知道原则（Least Knowledge Principle，LKP），指一个对象应该对其他对象保持最少的了解，尽量降低类与类之间的耦合。迪米特法则主要强调只和朋友交流，不和陌生人说话。出现在成员变量、方法的输入和输出参数中的类都可以被称为成员朋友类，而出现在方法体内部的类不属于朋友类。

3.5.2　使用迪米特法则解决实际问题

我们来设计一个权限系统，TeamLeader 需要查看目前发布到线上的课程数量。这时候，TeamLeader 要让 Employee 去进行统计，Employee 再把统计结果告诉 TeamLeader，来看代码。

Course 类的代码如下。

```
public class Course {
}
```

Employee 类的代码如下。

```
public class Employee{
    public void checkNumberOfCourses(List<Course> courseList){
```

```
        System.out.println("目前已发布的课程数量是：" + courseList.size());
    }
}
```

TeamLeader 类的代码如下。

```
public class TeamLeader{
    public void commandCheckNumber(Employee employee){

        List<Course> courseList = new ArrayList<Course>();
        for (int i= 0; i < 20 ;i ++){
            courseList.add(new Course());
        }
        employee.checkNumberOfCourses(courseList);
    }
}
```

客户端测试代码如下，将 Employee 对象作为参数传送给 TeamLeader 对象。

```
public static void main(String[] args) {
    TeamLeader teamLeader = new TeamLeader();
    Employee employee = new Employee();
    teamLeader.commandCheckNumber(employee);
}
```

写到这里，其实功能都已经实现，代码看上去也没什么问题。根据迪米特法则，TeamLeader 只想要结果，不需要跟 Course 产生直接交流。而 Employee 统计需要引用 Course 对象，TeamLeader 和 Course 并不是朋友，从如下图所示的类图就可以看出来。

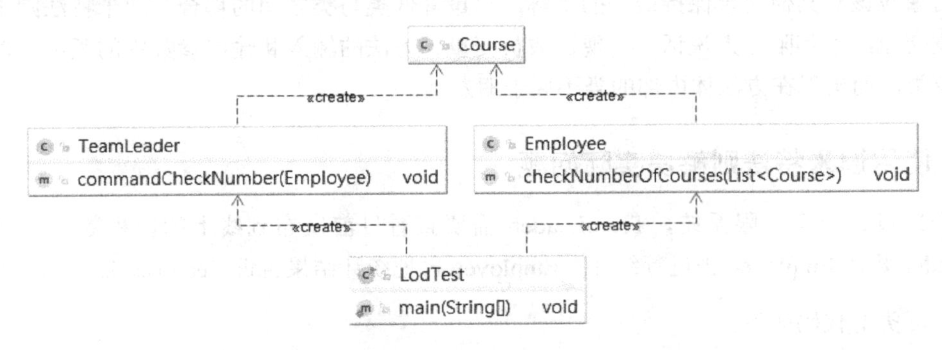

下面对代码进行改造。

Employee 类的代码如下。

```
public class Employee {

    public void checkNumberOfCourses(){
```

```
        List<Course> courseList = new ArrayList<Course>();
        for (int i= 0; i < 20 ;i ++){
            courseList.add(new Course());
        }
        System.out.println("目前已发布的课程数量是："+courseList.size());
    }
}
```

TeamLeader 类的代码如下。

```
public class TeamLeader {

    public void commandCheckNumber(Employee employee){
        employee.checkNumberOfCourses();
    }
}
```

再来看如下图所示的类图，Course 和 TeamLeader 已经没有关联了。

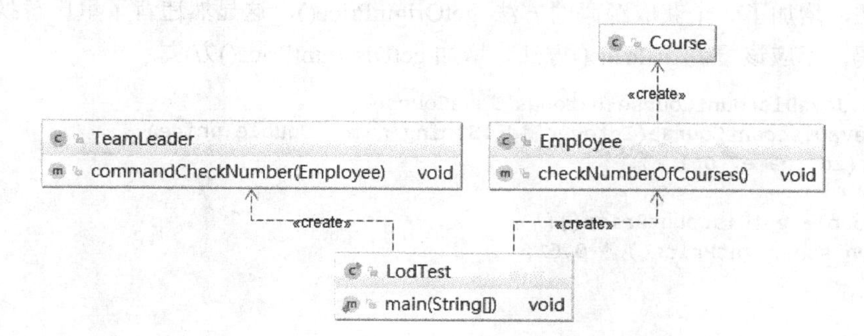

学习软件设计原则，千万不能形成强迫症。当碰到业务复杂的场景时，需要随机应变。

3.6 里氏替换原则

3.6.1 里氏替换原则的定义

里氏替换原则（Liskov Substitution Principle，LSP）指如果对每一个类型为 T_1 的对象 O_1，都有类型为 T_2 的对象 O_2，使得以 T_1 定义的所有程序 P 在所有对象 O_1 都替换成 O_2 时，程序 P 的行为没有发生变化，那么类型 T_2 是类型 T_1 的子类型。

定义看上去比较抽象，我们重新解释一下，可以理解为一个软件实体如果适用于一个父类，则一定适用于其子类，所有引用父类的地方必须能透明地使用其子类的对象，子类对象能够替换父类对象，而程序逻辑不变。也可以理解为，子类可以扩展父类的功能，但不能改变父类原有的功能。根据这个理解，我们对里氏替换原则的定义总结如下。

（1）子类可以实现父类的抽象方法，但不能覆盖父类的非抽象方法。

（2）子类中可以增加自己特有的方法。

（3）当子类的方法重载父类的方法时，方法的前置条件（即方法的输入参数）要比父类的方法更宽松。

（4）当子类的方法实现父类的方法时（重写/重载或实现抽象方法），方法的后置条件（即方法的输出/返回值）要比父类的方法更严格或相等。

3.6.2 使用里氏替换原则解决实际问题

在讲开闭原则的时候，我们埋下了一个伏笔。我们在获取折扣价格后重写覆盖了父类的getPrice()方法，增加了一个获取源码的方法 getOriginPrice()，这显然违背了里氏替换原则。我们修改一下代码，不应该覆盖 getPrice()方法，增加 getDiscountPrice()方法。

```
public class JavaDiscountCourse extends JavaCourse {
    public JavaDiscountCourse(Integer id, String name, Double price) {
        super(id, name, price);
    }
    public Double getDiscountPrice(){
        return super.getPrice() * 0.61;
    }
}
```

使用里氏替换原则有以下优点。

（1）约束继承泛滥，是开闭原则的一种体现。

（2）加强程序的健壮性，同时变更时可以做到非常好的兼容性，提高程序的维护性、可扩展性，降低需求变更时引入的风险。

现在来描述一个经典的业务场景，用正方形、矩形和四边形的关系说明里氏替换原则，我们都知道正方形是一个特殊的长方形，那么可以创建一个长方形的父类 Rectangle 类，代码如下。

```
public class Rectangle {
    private long height;
    private long width;

    public long getHeight() {
        return height;
    }
```

```
    public void setHeight(long height) {
        this.height = height;
    }

    public long getWidth() {
        return width;
    }

    public void setWidth(long width) {
        this.width = width;
    }
}
```

创建正方形 Square 类继承长方形，代码如下。

```
public class Square extends Rectangle {
    private long length;

    public long getLength() {
        return length;
    }

    public void setLength(long length) {
        this.length = length;
    }

    @Override
    public long getHeight() {
        return getLength();
    }

    @Override
    public void setHeight(long height) {
        setLength(height);
    }

    @Override
    public long getWidth() {
        return getLength();
    }

    @Override
    public void setWidth(long width) {
        setLength(width);
    }
}
```

在测试类中，创建 resize()方法。根据逻辑，长方形的宽应该大于等于高，我们让高一直自增，直到高等于宽变成正方形，代码如下。

```
    public static void resize(Rectangle rectangle){
        while (rectangle.getWidth() >= rectangle.getHeight()){
            rectangle.setHeight(rectangle.getHeight() + 1);
            System.out.println("Width:" +rectangle.getWidth() +
                          ",Height:" + rectangle.getHeight());
        }
        System.out.println("Resize End,Width:" +rectangle.getWidth() +
                      ",Height:" + rectangle.getHeight());
    }
```

客户端测试代码如下。

```
public static void main(String[] args) {
    Rectangle rectangle = new Rectangle();
    rectangle.setWidth(20);
    rectangle.setHeight(10);
    resize(rectangle);
}
```

运行结果如下所示。

```
Width:20,Height:11
Width:20,Height:12
Width:20,Height:13
Width:20,Height:14
Width:20,Height:15
Width:20,Height:16
Width:20,Height:17
Width:20,Height:18
Width:20,Height:19
Width:20,Height:20
Width:20,Height:21
Resize End,Width:20,Height:21
```

由运行结果可知，高比宽还大，这在长方形中是一种非常正常的情况。再来看下面的代码，把长方形替换成它的子类正方形，修改客户端测试代码如下。

```
public static void main(String[] args) {
    Square square = new Square();
    square.setLength(10);
    resize(square);
}
```

此时，运行出现了死循环，违背了里氏替换原则，在将父类替换为子类后，程序运行结果没有达到预期。因此，代码设计是存在一定风险的。里氏替换原则只存在于父类与子类之间，约束继承泛滥。再来创建一个基于长方形与正方形共同的抽象——四边形 QuardRangle 接口，代码如下。

```
public interface QuardRangle {
    long getWidth();
    long getHeight();
}
```

修改长方形 Rectangle 类的代码如下。

```
public class Rectangle implements QuardRangle {
    private long height;
    private long width;

    public long getHeight() {
        return height;
    }

    public void setHeight(long height) {
        this.height = height;
    }

    public long getWidth() {
        return width;
    }

    public void setWidth(long width) {
        this.width = width;
    }
}
```

修改正方形 Square 类的代码如下。

```
public class Square implements QuardRangle {
    private long length;

    public long getLength() {
        return length;
    }

    public void setLength(long length) {
        this.length = length;
    }

    public long getWidth() {
        return length;
    }
```

```
    public long getHeight() {
        return length;
    }
}
```

此时，如果把 resize()方法的参数换成四边形 QuardRangle 类，方法内部就会报错。因为正方形已经没有了 setWidth()和 setHeight()方法，所以，为了约束继承泛滥，resize()方法的参数只能用长方形 Rectangle 类。

3.7 合成复用原则

3.7.1 合成复用原则的定义

合成复用原则（Composite/Aggregate Reuse Principle，CARP）指尽量使用对象组合（has-a）或对象聚合（contanis-a）的方式实现代码复用，而不是用继承关系达到代码复用的目的。合成复用原则可以使系统更加灵活，降低类与类之间的耦合度，一个类的变化对其他类造成的影响相对较小。

继承，又被称为白箱复用，相当于把所有实现细节暴露给子类。组合/聚合又被称为黑箱复用，对类以外的对象是无法获取实现细节的。我们要根据具体的业务场景来做代码设计，其实也都需要遵循面向对象编程（Object Oriented Programming，OOP）模型。

3.7.2 使用合成复用原则解决实际问题

还是以数据库操作为例，首先创建 DBConnection 类。

```
public class DBConnection {
    public String getConnection(){
        return "MySQL 数据库连接";
    }
}
```

创建 ProductDao 类。

```
public class ProductDao{
    private DBConnection dbConnection;
    public void setDbConnection(DBConnection dbConnection) {
        this.dbConnection = dbConnection;
    }
    public void addProduct(){
        String conn = dbConnection.getConnection();
        System.out.println("使用"+conn+"增加产品");
    }
}
```

这是一种非常典型的合成复用原则应用场景。但是，对于目前的设计来说，DBConnection 还不是一种抽象，不便于系统扩展。目前的系统支持 MySQL 数据库连接，假设业务发生变化，数据库操作层要支持 Oracle 数据库。当然，我们可以在 DBConnection 中增加对 Oracle 数据库支持的方法，但是这违背了开闭原则。其实，可以不必修改 Dao 的代码，将 DBConnection 修改为 abstract，代码如下。

```
public abstract class DBConnection {
    public abstract String getConnection();
}
```

然后将 MySQL 的逻辑抽离。

```
public class MySQLConnection extends DBConnection {
    @Override
    public String getConnection() {
        return "MySQL 数据库连接";
    }
}
```

再创建 Oracle 支持的逻辑。

```
public class OracleConnection extends DBConnection {
    @Override
    public String getConnection() {
        return "Oracle 数据库连接";
    }
}
```

具体选择交给应用层，来看如下图所示的类图。

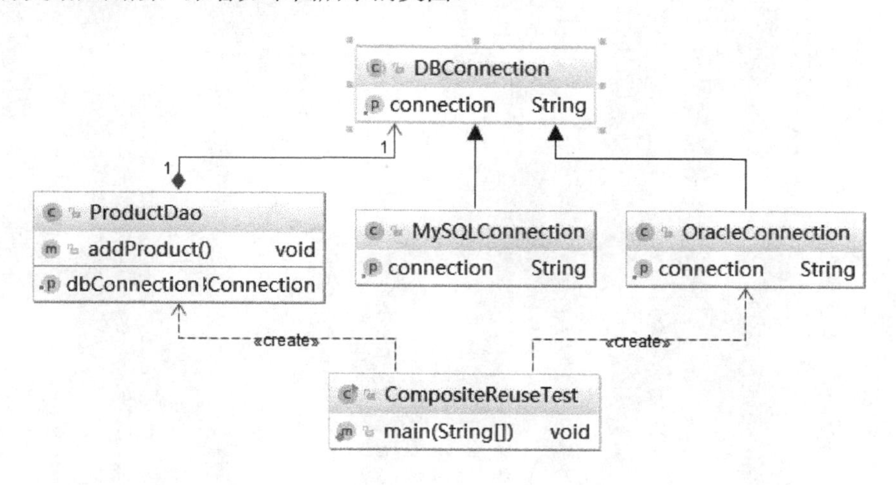

3.8 软件架构设计原则小结

学习设计原则是学习设计模式的基础。在实际开发过程中，并不是一定要求所有代码都遵循设计原则，而是要综合考虑人力、时间、成本、质量，不刻意追求完美，要在适当的场景遵循设计原则。这体现的是一种平衡取舍，可以帮助我们设计出更加优雅的代码结构。

分别用一句话归纳总结软件设计七大原则，如下表所示。

设计原则	一句话归纳	目　的
开闭原则	对扩展开放，对修改关闭	降低维护带来的新风险
依赖倒置原则	高层不应该依赖低层	更利于代码结构的升级扩展
单一职责原则	一个类只干一件事	便于理解，提高代码的可读性
接口隔离原则	一个接口只干一件事	功能解耦，高聚合、低耦合
迪米特法则	不该知道的不要知道	只和朋友交流，不和陌生人说话，减少代码臃肿
里氏替换原则	子类重写方法功能发生改变，不应该影响父类方法的含义	防止继承泛滥
合成复用原则	尽量使用组合实现代码复用，而不使用继承	降低代码耦合

第4章 关于设计模式的那些事儿

4.1 本书与 GoF 的《设计模式》的关系

《设计模式》这本书最大的功能是目录，该目录列举并描述了 23 种设计模式。近年来，这一清单又增加了一些类别，最重要的是使涵盖范围扩展到更具体的问题类型。例如，Mark Grand 在 *Patterns in Java: A Catalog of Reusable Design Patterns Illustrated with UML*（以下简称《模式 Java 版》）一书中增加了解决涉及诸如并发等问题的模式，而由 Deepak Alur、John Crupi 和 Dan Malks 合著的 *Core J2EE Patterns: Best Practices and Design Strategies* 一书中主要关注使用 Java 2 企业技术的多层应用程序上的模式。

很多人并没有注意到这点，学完 Java 基础语言就直接学习 J2EE，有的甚至赶鸭子上架，直接使用 Weblogic 等具体 J2EE 软件，一段时间后，发现 J2EE 也不过如此，挺简单好用，但是你真正理解 J2EE 了吗？你在具体案例中的应用是否也在延伸 J2EE 的思想？对软件设计模式的研究造就了一本可能是面向对象设计方面最有影响的书籍——《设计模式》。

由此可见，设计模式和 J2EE 在思想上和动机上是一脉相承的，笔者总结了以下几个原因。

（1）设计模式更抽象。J2EE 是具体的产品代码，可以接触得到，而设计模式在针对每个应用时才产生具体代码。

（2）设计模式是比 J2EE 等框架软件更小的体系结构，J2EE 中许多具体程序都是应用设计模式来完成的，当你深入 J2EE 的内部代码研究时，这点尤其明显。因此，如果不具备设计模式的基础知识（GoF 的《设计模式》），则很难快速地理解 J2EE。不能理解 J2EE，又如何能灵活应用？

（3）J2EE 只是适合企业计算应用的框架软件，但是 GoF 的《设计模式》几乎可以用于任何应用！因此，GoF 的《设计模式》应该是 J2EE 的重要理论基础之一。

所以说，GoF 的《设计模式》是 Java 基础知识和 J2EE 框架知识之间一座隐性的“桥”。

设计模式其实也是一门艺术。设计模式源于生活，不要为了套用设计模式而使用设计模式。设计模式是在我们遇到问题没有头绪时提供的一种解决问题的方案，或者说用好设计模式可以防患于未然。自古以来，当人生迷茫时，我们往往都会寻求帮助，或上门咨询，或查阅资料。在几千年前，孔子就教给我们怎样做人。中国人都知道，从初生襁褓、十有五而志于学，到二十弱冠、三十而立、四十而不惑，再到五十而知天命、六十而耳顺，七十而从心所欲，不逾矩……我们的人生也在套用模板，当然，有些人不会选择这套模板。

《设计模式》总结的是经验之谈，千万不要死记硬背，生搬硬套。下面来总体预览一下设计模式的分类和总结，如下表所示。

分　类	解　释	举　例
创建型设计模式（Creational）	这类设计模式提供了一种在创建对象的同时隐藏创建逻辑的方式，而不是使用新的运算符直接实例化对象，这使得程序在判断针对某个给定实例需要创建哪些对象时更加灵活	工厂方法模式（Factory Method Pattern）、抽象工厂模式（Abstract Factory Pattern）、单例模式（Singleton Pattern）、原型模式（Prototype Pattern）、建造者模式（Builder Pattern）
结构型设计模式（Structural）	这类设计模式关注类和对象的组合。继承的概念被用来组合接口和定义组合对象获得新功能的方式	代理模式（Proxy Pattern）、门面模式（Facade Pattern）、装饰器模式（Decorator Pattern）、享元模式（Flyweight Pattern）、组合模式（Composite Pattern）、适配器模式（Adapter Pattern）、桥接模式（Bridge Pattern）
行为型设计模式（Behavioral）	这类设计模式特别关注对象之间的通信	模板方法模式（Template Method Pattern）、策略模式（Strategy Pattern）、责任链模式（Chain of Responsibility Pattern）、迭代器模式（Iterator Pattern）、命令模式（Command Pattern）、状态模式（State Pattern）、备忘录模式（Memento Pattern）、中介者模式（Mediator Pattern）、解释器模式（Interpreter Pattern）、观察者模式（Observer Pattern）、访问者模式（Visitor Pattern）

4.2　为什么一定要学习设计模式

先来看一个生活案例，当我们开心时，也许会寻求享乐。在学习设计模式之前，你可能会这样感叹：

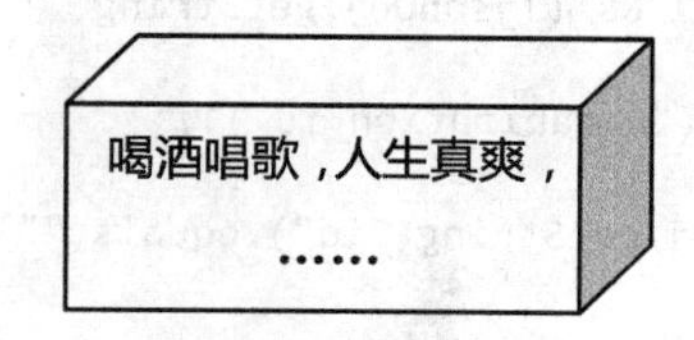

学完设计模式之后，你可能会这样感叹：

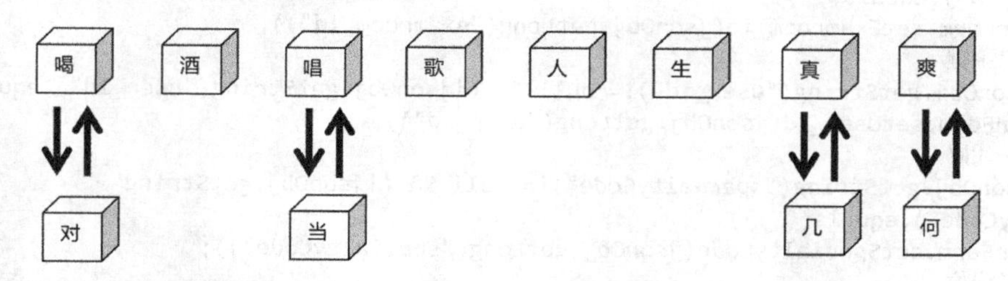

大家对比一下前后的区别，有何感受？

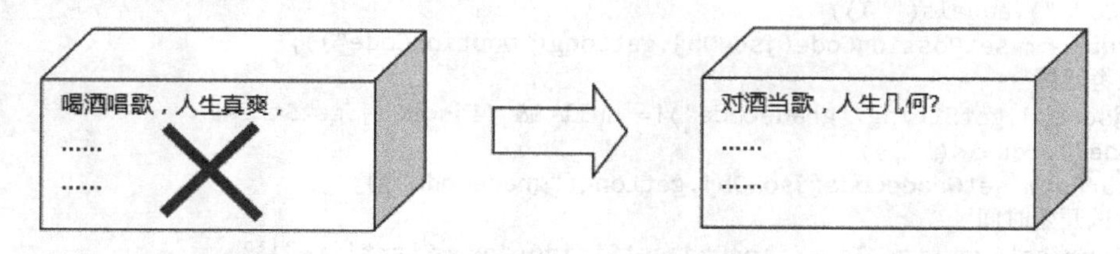

回到代码中，我们来思考一下，设计模式能解决哪些问题？

4.2.1　写出优雅的代码

先来看一段笔者很多年前写的代码。

```
public void setExammingForm(ExammingForm curForm,String parameters)throws BaseException {

    ...

    JSONObject jsonObj = new JSONObject(parameters);
    //试卷主键
```

```
    if(jsonObj.getString("examinationPaper_id")!= null && (!jsonObj.getString
("examinationPaper_id").equals("")))
        curForm.setExaminationPaper_id(jsonObj.getLong("examinationPaper_id"));
    //剩余时间
    if(jsonObj.getString("leavTime") != null && (!jsonObj.getString("leavTime").equals("")))
        curForm.setLeavTime(jsonObj.getInt("leavTime"));
    //单位主键
    if(jsonObj.getString("organization_id")!= null && (!jsonObj.getString
("organization_id").equals("")))
        curForm.setOrganization_id(jsonObj.getLong("organization_id"));
    //考试主键
    if(jsonObj.getString("id")!= null && (!jsonObj.getString("id").equals("")))
        curForm.setId(jsonObj.getLong("id"));
    //考场主键
    if(jsonObj.getString("examroom_id")!= null && (!jsonObj.getString
("examroom_id").equals("")))
        curForm.setExamroom_id(jsonObj.getLong("examroom_id"));
    //用户主键
    if(jsonObj.getString("user_id")!= null && (!jsonObj.getString("user_id").equals("")))
        curForm.setUser_id(jsonObj.getLong("user_id"));
    //专业代码
    if(jsonObj.getString("specialtyCode")!= null && (!jsonObj.getString
("specialtyCode").equals("")))
        curForm.setSpecialtyCode(jsonObj.getLong("specialtyCode"));
    //报考岗位
    if(jsonObj.getString("postionCode")!= null && (!jsonObj.getString
("postionCode").equals("")))
        curForm.setPostionCode(jsonObj.getLong("postionCode"));
    //报考等级
    if(jsonObj.getString("gradeCode")!= null && (!jsonObj.getString
("gradeCode").equals("")))
        curForm.setGradeCode(jsonObj.getLong("gradeCode"));
    //考试开始时间
    curForm.setExamStartTime(jsonObj.getString("examStartTime"));
    //考试结束时间
    curForm.setExamEndTime(jsonObj.getString("examEndTime"));

    ...
}
```

优化之后的代码如下。

```
public class ExammingFormVo extends ExammingForm{

   private String examinationPaperId;     //试卷主键
   private String leavTime;               //剩余时间
   private String organizationId;         //单位主键
```

```
    private String id;                    //考试主键
    private String examRoomId;            //考场主键
    private String userId;                //用户主键
    private String specialtyCode;         //专业代码
    private String postionCode;           //报考岗位
    private String gradeCode;             //报考等级
    private String examStartTime;         //考试开始时间
    private String examEndTime;           //考试结束时间

    ...
}

public void setExammingForm(ExammingForm form,String parameters)throws BaseException {
    try {
        JSONObject json = new JSONObject(parameters);
        ExammingFormVo  vo = JSONObject.parseObject(json,ExammingFormVo.class);

        form = vo;

    }catch (Exception e){
        e.printStackTrace();
    }

}
```

4.2.2　更好地重构项目

平时我们写的代码虽然满足了需求，但往往不利于项目的开发与维护，以下面的 JDBC 代码为例。

```
public void save(Student stu){
    String sql = "INSERT INTO t_student(name,age) VALUES(?,?)";
    Connection conn = null;
    Statement st = null;
    try{
        //1. 加载注册驱动
        Class.forName("com.mysql.jdbc.Driver");
        //2. 获取数据库连接
        conn=DriverManager.getConnection("jdbc:mysql:///jdbc_demo","root","root");
        //3. 创建语句对象
        PreparedStatement ps=conn.prepareStatement(sql);
        ps.setObject(1,stu.getName());
        ps.setObject(2,stu.getAge());
        //4. 执行 SQL 语句
        ps.executeUpdate();
        //5. 释放资源
```

```
    }catch(Exception e){
        e.printStackTrace();
    }finally{
        try{
            if(st != null)
                st.close();
        }catch(SQLException e){
            e.printStackTrace();
        }finally{
            try{
                if(conn != null)
                    conn.close();
            }catch(SQLException e){
                e.printStackTrace();
            }
        }
    }
}

//删除学生信息
public void delete(Long id){
    String sql = "DELETE  FROM t_student WHERE id=?";
    Connection conn = null;
    Statement st = null;
    try{
        //1. 加载注册驱动
        Class.forName("com.mysql.jdbc.Driver");
        //2. 获取数据库连接
        conn=DriverManager.getConnection("jdbc:mysql:///jdbc_demo","root","root");
        //3. 创建语句对象
        PreparedStatement ps = conn.prepareStatement(sql);
        ps.setObject(1,id);
        //4. 执行 SQL 语句
        ps.executeUpdate();
        //5. 释放资源
    }catch(Exception e){
        e.printStackTrace();
    }finally{
        try{
            if(st != null)
                st.close();
        }catch(SQLException e){
            e.printStackTrace();
        }finally{
            try{
                if(conn != null)
                    conn.close();
```

```
            }catch(SQLException e){
                e.printStackTrace();
            }
        }
    }
}

//修改学生信息
public void update(Student stu){
    String sql = "UPDATE t_student SET name=?,age=? WHERE id=?";
    Connection conn = null;
    Statement st = null;
    try{
        //1. 加载注册驱动
        Class.forName("com.mysql.jdbc.Driver");
        //2. 获取数据库连接
        conn=DriverManager.getConnection("jdbc:mysql:///jdbc_demo","root","root");
        //3. 创建语句对象
        PreparedStatement ps = conn.prepareStatement(sql);
        ps.setObject(1,stu.getName());
        ps.setObject(2,stu.getAge());
        ps.setObject(3,stu.getId());
        //4. 执行 SQL 语句
        ps.executeUpdate();
        //5. 释放资源
    }catch(Exception e){
        e.printStackTrace();
    }finally{
        try{
            if(st != null)
                st.close();
        }catch(SQLException e){
            e.printStackTrace();
        }finally{
            try{
                if(conn != null)
                    conn.close();
            }catch(SQLException e){
                e.printStackTrace();
            }
        }
    }
}
```

上述代码的功能没问题，但是代码重复得太多，因此可以进行抽取，把重复代码放到一个工具类 JdbcUtil 里。

```
//工具类
public class JdbcUtil {
    private JdbcUtil() { }
    static {
        //1. 加载注册驱动
        try {
            Class.forName("com.mysql.jdbc.Driver");
        } catch (Exception e) {
            e.printStackTrace();
        }
    }

    public static Connection getConnection() {
        try {
            //2. 获取数据库连接
            return DriverManager.getConnection("jdbc:mysql:///jdbc_demo", "root", "root");
        } catch (Exception e) {
            e.printStackTrace();
        }
        return null;
    }

    //释放资源
    public static void close(ResultSet rs, Statement st, Connection conn) {
        try {
            if (rs != null)
                rs.close();
        } catch (SQLException e) {
            e.printStackTrace();
        } finally {
            try {
                if (st != null)
                    st.close();
            } catch (SQLException e) {
                e.printStackTrace();
            } finally {
                try {
                    if (conn != null)
                        conn.close();
                } catch (SQLException e) {
                    e.printStackTrace();
                }
            }
        }
    }
}
```

只需要在实现类中直接调用工具类 JdbcUtil 中的方法即可。

```
//增加学生信息
public void save(Student stu) {
    String sql = "INSERT INTO t_student(name,age) VALUES(?,?)";
    Connection conn = null;
    PreparedStatement ps=null;
    try {
        conn = JDBCUtil.getConnection();
        //3. 创建语句对象
        ps = conn.prepareStatement(sql);
        ps.setObject(1, stu.getName());
        ps.setObject(2, stu.getAge());
        //4. 执行 SQL 语句
        ps.executeUpdate();
        //5. 释放资源
    } catch (Exception e) {
        e.printStackTrace();
    } finally {
        JDBCUtil.close(null, ps, conn);
    }

}

//删除学生信息
public void delete(Long id) {
    String sql = "DELETE  FROM t_student WHERE id=?";
    Connection conn = null;
    PreparedStatement ps = null;
    try {
        conn=JDBCUtil.getConnection();
        //3. 创建语句对象
        ps = conn.prepareStatement(sql);
        ps.setObject(1, id);
        //4. 执行 SQL 语句
        ps.executeUpdate();
        //5. 释放资源
    } catch (Exception e) {
        e.printStackTrace();
    } finally {
        JDBCUtil.close(null, ps, conn);
    }

}

//修改学生信息
public void update(Student stu) {
```

```
    String sql = "UPDATE t_student SET name=?,age=? WHERE id=?";
    Connection conn = null;
    PreparedStatement ps = null;
    try {
        conn=JDBCUtil.getConnection();
        //3. 创建语句对象
        ps = conn.prepareStatement(sql);
        ps.setObject(1, stu.getName());
        ps.setObject(2, stu.getAge());
        ps.setObject(3, stu.getId());
        //4. 执行 SQL 语句
        ps.executeUpdate();
        //5. 释放资源
    } catch (Exception e) {
        e.printStackTrace();
    } finally {
        JDBCUtil.close(null, ps, conn);
    }

}

public Student get(Long id) {
    String sql = "SELECT * FROM t_student WHERE id=?";
    Connection conn = null;
    Statement st = null;
    ResultSet rs = null;
    PreparedStatement ps=null;
    try {
        conn = JDBCUtil.getConnection();
        //3. 创建语句对象
        ps = conn.prepareStatement(sql);
        ps.setObject(1, id);
        //4. 执行 SQL 语句
        rs = ps.executeQuery();
        if (rs.next()) {
            String name = rs.getString("name");
            int age = rs.getInt("age");
            Student stu = new Student(id, name, age);
            return stu;
        }
        //5. 释放资源
    } catch (Exception e) {
        e.printStackTrace();
    } finally {
        JDBCUtil.close(rs, ps, conn);
    }
    return null;
```

```
}

public List<Student> list() {
    List<Student> list = new ArrayList<>();
    String sql = "SELECT * FROM t_student ";
    Connection conn = null;
    Statement st = null;
    ResultSet rs = null;
    PreparedStatement ps=null;
    try {
        conn=JDBCUtil.getConnection();
        //3. 创建语句对象
        ps = conn.prepareStatement(sql);
        //4. 执行 SQL 语句
        rs = ps.executeQuery();
        while (rs.next()) {
            long id = rs.getLong("id");
            String name = rs.getString("name");
            int age = rs.getInt("age");
            Student stu = new Student(id, name, age);
            list.add(stu);
        }
        //5. 释放资源
    } catch (Exception e) {
        e.printStackTrace();
    } finally {
        JDBCUtil.close(rs, ps, conn);
    }
    return list;
}
```

虽然完成了重复代码的抽取，但数据库中的账号、密码等直接显示在代码中，不利于后期账户密码改动的维护。可以建立一个 db.propertise 文件，用来存储这些信息。

```
driverClassName=com.mysql.jdbc.Driver
url=jdbc:mysql:///jdbcdemo
username=root
password=root
```

只需要在工具类 JdbcUtil 中获取里面的信息即可。

```
static {
    //1. 加载注册驱动
    try {
        ClassLoader loader = Thread.currentThread().getContextClassLoader();
        InputStream inputStream = loader.getResourceAsStream("db.properties");
        p = new Properties();
        p.load(inputStream);
```

```
        Class.forName(p.getProperty("driverClassName"));
    } catch (Exception e) {
        e.printStackTrace();
    }
}

public static Connection getConnection() {
    try {
        //2. 获取数据库连接
        return DriverManager.getConnection(p.getProperty("url"), p.getProperty("username"),
        p.getProperty("password"));
    } catch (Exception e) {
        e.printStackTrace();
    }
    return null;
}
```

代码抽取到这里，貌似已经完成，但在实现类中，依然存在部分重复代码，在 DML 操作中，除了 SQL 和设置值的不同，其他都相同，把相同的部分抽取出来，把不同的部分通过参数传递进来，无法直接放在工具类中。此时，可以创建一个模板类 JdbcTemplate，创建一个 DML 和 DQL 的模板来对代码进行重构。

```
//查询统一模板
public static List<Student> query(String sql,Object...params){
    List<Student> list=new ArrayList<>();
    Connection conn = null;
    PreparedStatement ps=null;
    ResultSet rs = null;
    try {
        conn=JDBCUtil.getConnection();
        ps=conn.prepareStatement(sql);
        //设置值
        for (int i = 0; i < params.length; i++) {
            ps.setObject(i+1, params[i]);
        }
        rs = ps.executeQuery();
        while (rs.next()) {
            long id = rs.getLong("id");
            String name = rs.getString("name");
            int age = rs.getInt("age");
            Student stu = new Student(id, name, age);
            list.add(stu);
        }
        //5. 释放资源
    } catch (Exception e) {
        e.printStackTrace();
```

```
    } finally {
        JDBCUtil.close(rs, ps, conn);
    }
    return list;
}
```

实现类直接调用方法即可。

```
//增加学生信息
public void save(Student stu) {
    String sql = "INSERT INTO t_student(name,age) VALUES(?,?)";
    Object[] params=new Object[]{stu.getName(),stu.getAge()};
    JdbcTemplate.update(sql, params);
}

//删除学生信息
public void delete(Long id) {
    String sql = "DELETE FROM t_student WHERE id = ?";
    JdbcTemplate.update(sql, id);
}

//修改学生信息
public void update(Student stu) {
    String sql = "UPDATE t_student SET name = ?,age = ? WHERE id = ?";
    Object[] params=new Object[]{stu.getName(),stu.getAge(),stu.getId()};
    JdbcTemplate.update(sql, params);
}

public Student get(Long id) {
    String sql = "SELECT * FROM t_student WHERE id=?";
    List<Student> list = JDBCTemplate.query(sql, id);
    return list.size()>0? list.get(0):null;
}

public List<Student> list() {
    String sql = "SELECT * FROM t_student ";
    return JDBCTemplate.query(sql);
}
```

这样重复的代码基本就解决了，但有一个很严重的问题，就是这个程序 DQL 操作中只能处理 Student 类和 t_student 表的相关数据，无法处理其他类，比如 Teacher 类和 t_teacher 表。不同的表（不同的对象）应该有不同的列，不同列处理结果集的代码就应该不一样，处理结果集的操作只有 DAO 自己最清楚。也就是说，处理结果的方法根本就不应该放在模板方法中，应该由每个 DAO 自己来处理。因此，可以创建一个 IRowMapper 接口来处理结果集。

```
public interface IRowMapper {
```

```
    //处理结果集
    List rowMapper(ResultSet rs) throws Exception;
}
```

DQL 模板类中调用 IRowMapper 接口中的 handle 方法，提醒实现类自己去实现 mapping 方法。

```
public static List<Student> query(String sql,IRowMapper rsh, Object...params){
    List<Student> list = new ArrayList<>();
    Connection conn = null;
    PreparedStatement ps=null;
    ResultSet rs = null;
    try {
        conn = JdbcUtil.getConnection();
        ps = conn.prepareStatement(sql);
        //设置值
        for (int i = 0; i < params.length; i++) {
            ps.setObject(i+1, params[i]);
        }
        rs = ps.executeQuery();
        return rsh.mapping(rs);
        //5. 释放资源
    } catch (Exception e) {
        e.printStackTrace();
    } finally {
        JdbcUtil.close(rs, ps, conn);
    }
    return list ;
}
```

实现类自己去实现 IRowMapper 接口的 mapping 方法，想要处理什么类型的数据在里面定义即可。

```
public Student get(Long id) {
    String sql = "SELECT * FROM t_student WHERE id = ?";
    List<Student> list = JdbcTemplate.query(sql,new StudentRowMapper(), id);
    return list.size()>0? list.get(0):null;
}
public List<Student> list() {
    String sql = "SELECT * FROM t_student ";
    return JdbcTemplate.query(sql,new StudentRowMapper());
}
class StudentRowMapper implements IRowMapper{
    public List mapping(ResultSet rs) throws Exception {
        List<Student> list=new ArrayList<>();
        while(rs.next()){
            long id = rs.getLong("id");
            String name = rs.getString("name");
            int age = rs.getInt("age");
```

```
            Student stu=new Student(id, name, age);
            list.add(stu);
        }
        return list;
    }
}
```

到这里为止，实现 ORM 的关键代码已经大功告成，但是 DQL 查询不单单要查询学生信息（List 类型），还要查询学生数量，这时就要通过泛型来完成。

```
public interface IRowMapper<T> {
    //处理结果集
    T mapping(ResultSet rs) throws Exception;
}

public static <T> T query(String sql,IRowMapper<T> rsh, Object...params){
    Connection conn = null;
    PreparedStatement ps=null;
    ResultSet rs = null;
    try {
        conn = JdbcUtil.getConnection();
        ps = conn.prepareStatement(sql);
        //设置值
        for (int i = 0; i < params.length; i++) {
            ps.setObject(i+1, params[i]);
        }
        rs = ps.executeQuery();
        return rsh.mapping(rs);
        //5. 释放资源
    } catch (Exception e) {
        e.printStackTrace();
    } finally {
        JdbcUtil.close(rs, ps, conn);
    }
    return null;
}
```

StudentRowMapper 类的代码如下。

```
class StudentRowMapper implements IRowMapper<List<Student>>{
    public List<Student> mapping(ResultSet rs) throws Exception {
        List<Student> list=new ArrayList<>();
        while(rs.next()){
            long id = rs.getLong("id");
            String name = rs.getString("name");
            int age = rs.getInt("age");
            Student stu=new Student(id, name, age);
            list.add(stu);
```

```
        }
        return list;
    }
}
```

这样，不仅可以查询 List，还可以查询学生数量。

```
public Long getCount(){
    String sql = "SELECT COUNT(*) total FROM t_student";
    Long totalCount = (Long) JdbcTemplate.query(sql,
            new IRowMapper<Long>() {
                public Long mapping(ResultSet rs) throws Exception {
                    Long totalCount = null;
                    if(rs.next()){
                        totalCount = rs.getLong("total");
                    }
                    return totalCount;
                }
            });
    return totalCount;
}
```

这样，重构设计就已经完成，好的代码能让我们以后维护更方便，因此学会对代码重构是非常重要的。

4.2.3 经典框架都在用设计模式解决问题

比如，Spring 就是一个把设计模式用得淋漓尽致的经典框架。本书会结合 JDK、Spring、MyBatis、Netty、Tomcat、Dubbo 等经典框架的源码对设计模式展开分析，帮助大家更好、更深入地理解设计模式在框架源码中的落地。

2

第 2 篇

创建型设计模式

第 5 章 简单工厂模式

5.1 工厂模式的历史由来

我们都知道在现实生活中，原始社会自给自足（没有工厂），农耕社会小作坊（简单工厂，民间酒坊），工业革命流水线（工厂方法，自产自销），现代产业链代工厂（抽象工厂，富士康）。我们的项目代码同样是由简至繁一步一步迭代而来的，但对于调用者来说，却越来越简单。在日常开发中，凡是需要生成复杂对象的地方，都可以尝试考虑使用工厂模式来代替。

> **注：**上述复杂对象指的是类的构造函数参数过多等对类的构造有影响的情况，因为类的构造过于复杂，如果直接在其他业务类内使用，则两者的耦合过重，后续业务更改，就需要在任何引用该类的源代码内进行更改，光是查找所有依赖就很消耗时间了，更别说要一个一个修改了。

工厂模式，按照实际业务场景进行划分，有 3 种不同的实现方式，分别是简单工厂模式、工厂方法模式和抽象工厂模式。

5.2　简单工厂模式概述

5.2.1　简单工厂模式的定义

简单工厂模式（Simple Factory Pattern）又叫作静态工厂方法模式（Static Factory Method Pattern），简单来说，简单工厂模式有一个具体的工厂类，可以生成多个不同的产品，属于创建型设计模式。简单工厂模式不在 GoF 23 种设计模式之列。

5.2.2　简单工厂模式的应用场景

对于产品种类相对较少的情况，考虑使用简单工厂模式可以很方便地创建所需产品。使用简单工厂模式的客户端只需要传入工厂类的参数，不需要关心如何创建对象的逻辑。

5.2.3　简单工厂模式的 UML 类图

简单工厂模式的 UML 类图如下。

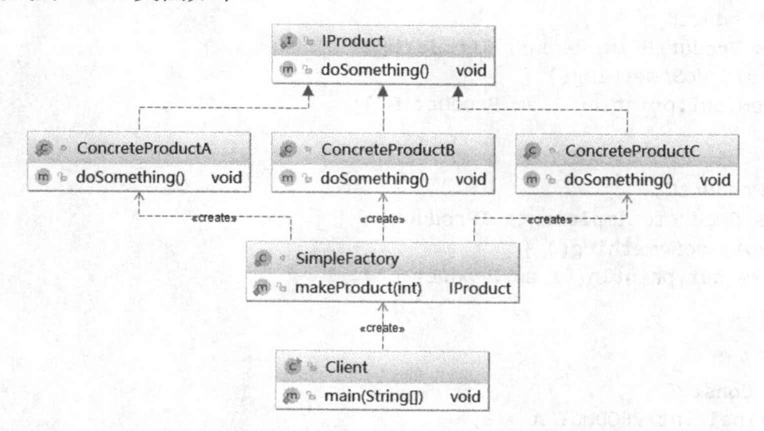

由上图可以看到，简单工厂模式主要包含 3 个角色。

（1）简单工厂（SimpleFactory）：是简单工厂模式的核心，负责实现创建所有实例的内部逻辑。工厂类的创建产品类的方法可以被外界直接调用，创建所需的产品对象。

（2）抽象产品（IProduct）：是简单工厂创建的所有对象的父类，负责描述所有实例共有的公共接口。

（3）具体产品（ConcreteProduct）：是简单工厂模式的创建目标。

5.2.4 简单工厂模式的通用写法

以下是简单工厂模式的通用写法。

```
public class Client {

    public static void main(String[] args) {

        new SimpleFactory().makeProduct(1);
    }

    //抽象产品
    public interface IProduct {
        void doSomething();
    }
    //具体产品：ProductA
    static class ProductA implements IProduct{
        public void doSomething() {
            System.out.println("I am Product A");
        }
    }
    //具体产品：ProductB
    static class ProductB implements IProduct{
        public void doSomething() {
            System.out.println("I am Product B");
        }
    }
    //具体产品：ProductC
    static class ProductC implements IProduct{
        public void doSomething() {
            System.out.println("I am Product C");
        }
    }

    final class Const {
        static final int PRODUCT_A = 0;
        static final int PRODUCT_B = 1;
        static final int PRODUCT_C = 2;
    }

    static class SimpleFactory {
        public static IProduct makeProduct(int kind) {
            switch (kind) {
                case Const.PRODUCT_A:
                    return new ProductA();
                case Const.PRODUCT_B:
```

```
                    return new ProductB();
                case Const.PRODUCT_C:
                    return new ProductC();
            }
            return null;
        }
    }
}
```

5.3 使用简单工厂模式封装产品创建细节

接下来看代码，还是以课程为例。咕泡学院目前开设有 Java 架构、大数据、人工智能等课程，已经形成了一个生态。我们可以定义一个课程标准 ICourse 接口。

```
public interface ICourse {
    /** 录制视频 */
    public void record();
}
```

创建一个 Java 课程的实现类 JavaCourse。

```
public class JavaCourse implements ICourse {
    public void record() {
        System.out.println("录制 Java 课程");
    }
}
```

客户端调用代码如下。

```
public static void main(String[] args) {
    ICourse course = new JavaCourse();
    course.record();
}
```

由上面代码可知，父类 ICourse 指向子类 JavaCourse 的引用，应用层代码需要依赖 JavaCourse。如果业务扩展，则继续增加 PythonCourse，甚至更多，那么客户端的依赖会变得越来越臃肿。因此，我们要想办法把这种依赖减弱，把创建细节隐藏。虽然在目前的代码中，创建对象的过程并不复杂，但从代码设计角度来讲不易于扩展。因此，用简单工厂模式对代码进行优化。首先增加课程 PythonCourse 类。

```
public class PythonCourse implements ICourse {
    public void record() {
        System.out.println("录制 Python 课程");
    }
}
```

然后创建 CourseFactory 工厂类。

```
public class CourseFactory {
    public ICourse create(String name){
        if("java".equals(name)){
            return new JavaCourse();
        }else if("python".equals(name)){
            return new PythonCourse();
        }else {
            return null;
        }
    }
}
```

最后修改客户端调用代码。

```
public class SimpleFactoryTest {
    public static void main(String[] args) {
        CourseFactory factory = new CourseFactory();
        factory.create("java");
    }
}
```

当然，为了调用方便，可将 CourseFactory 的 create()方法改为静态方法，其类图如下。

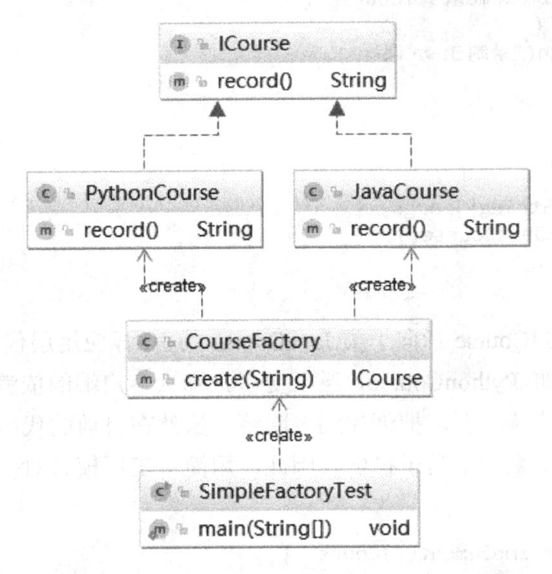

客户端调用虽然简单了，但如果业务继续扩展，要增加前端课程，则工厂中的 create()方法就要随着产品链的丰富每次都要修改代码逻辑，这不符合开闭原则。因此，我们可以采用反射技术

继续对简单工厂模式进行优化，代码如下。

```
public class CourseFactory {
    public ICourse create(String className){
        try {
            if (!(null == className || "".equals(className))) {
                return (ICourse) Class.forName(className).newInstance();
            }
        }catch (Exception e){
            e.printStackTrace();
        }
        return null;
    }
}
```

客户端调用代码修改如下。

```
public static void main(String[] args) {
        CourseFactory factory = new CourseFactory();
        ICourse                                 course                                  =
factory.create("com.gupaoedu.vip.pattern.factory.simplefactory.JavaCourse");
        course.record();
}
```

优化之后，产品不断丰富，不需要修改 CourseFactory 中的代码。但问题是，方法参数是字符串，可控性有待提升，而且还需要强制转型。继续修改代码。

```
public ICourse create(Class<? extends ICourse> clazz){
    try {
        if (null != clazz) {
            return clazz.newInstance();
        }
    }catch (Exception e){
        e.printStackTrace();
    }
    return null;
}
```

优化客户端测试代码。

```
public static void main(String[] args) {
    CourseFactory factory = new CourseFactory();
    ICourse course = factory.create(JavaCourse.class);
    course.record();
}
```

最后来看如下图所示的类图。

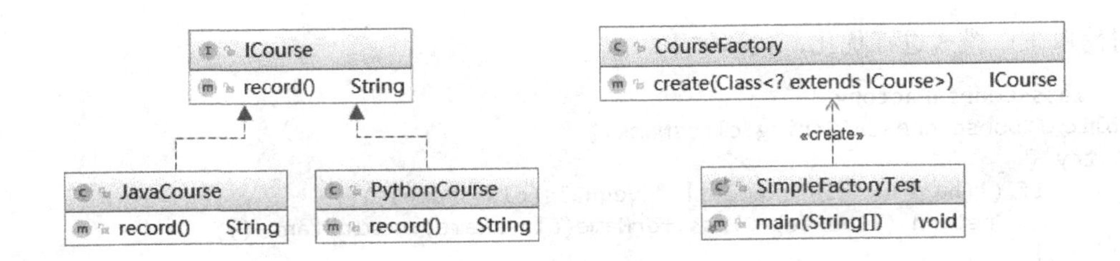

5.4 简单工厂模式在框架源码中的应用

5.4.1 简单工厂模式在 JDK 源码中的应用

简单工厂模式在 JDK 源码中无处不在，例如 Calendar 类，看 Calendar.getInstance()方法。下面打开的是 Calendar 的具体创建类。

```
private static Calendar createCalendar(TimeZone zone, Locale aLocale) {
    CalendarProvider provider =
        LocaleProviderAdapter.getAdapter(CalendarProvider.class, aLocale)
                             .getCalendarProvider();
    if (provider != null) {
        try {
            return provider.getInstance(zone, aLocale);
        } catch (IllegalArgumentException iae) {
        }
    }

    Calendar cal = null;

    if (aLocale.hasExtensions()) {
        String caltype = aLocale.getUnicodeLocaleType("ca");
        if (caltype != null) {
            switch (caltype) {
            case "buddhist":
            cal = new BuddhistCalendar(zone, aLocale);
                break;
            case "japanese":
                cal = new JapaneseImperialCalendar(zone, aLocale);
                break;
            case "gregory":
                cal = new GregorianCalendar(zone, aLocale);
                break;
            }
        }
    }
```

```
    if (cal == null) {

        if (aLocale.getLanguage() == "th" && aLocale.getCountry() == "TH") {
            cal = new BuddhistCalendar(zone, aLocale);
        } else if (aLocale.getVariant() == "JP" && aLocale.getLanguage() == "ja"
                && aLocale.getCountry() == "JP") {
            cal = new JapaneseImperialCalendar(zone, aLocale);
        } else {
            cal = new GregorianCalendar(zone, aLocale);
        }
    }
    return cal;
}
```

5.4.2　简单工厂模式在 Logback 源码中的应用

在大家经常使用的 Logback 中，可以看到 LoggerFactory 中有多个重载的方法 getLogger()。

```
public static Logger getLogger(String name) {
    ILoggerFactory iLoggerFactory = getILoggerFactory();
    return iLoggerFactory.getLogger(name);
}

public static Logger getLogger(Class clazz) {
    return getLogger(clazz.getName());
}
```

5.5　简单工厂模式扩展

5.5.1　简单工厂模式的优点

简单工厂模式的结构简单，调用方便。对于外界给定的信息，可以很方便地创建出相应的产品。工厂和产品的职责区分明确。

5.5.2　简单工厂模式的缺点

简单工厂模式的工厂类单一，负责所有产品的创建，但当产品基数增多时，工厂类代码会非常臃肿，违背高聚合原则。

6

第 6 章 工厂方法模式

6.1 工厂方法模式概述

6.1.1 工厂方法模式的定义

工厂方法模式（Factory Method Pattern）又叫作多态性工厂模式，指定义一个创建对象的接口，但由实现这个接口的类来决定实例化哪个类，工厂方法把类的实例化推迟到子类中进行。

原文：Define an interface for creating an object, but let subclasses decide which class to instantiate. Factory Method lets a class defer instantiation to subclasses.

在工厂方法模式中，不再由单一的工厂类生产产品，而是由工厂类的子类实现具体产品的创建。因此，当增加一个产品时，只需增加一个相应的工厂类的子类，实现生产这种产品，便可以解决简单工厂生产太多产品导致其内部代码臃肿（switch ... case 分支过多）的问题，也符合开闭原则。

6.1.2　工厂方法模式的应用场景

工厂方法模式主要适用于以下应用场景。

（1）创建对象需要大量重复的代码。

（2）客户端（应用层）不依赖产品类实例如何被创建、实现等细节。

（3）一个类通过其子类来指定创建哪个对象。

6.1.3　工厂方法模式的 UML 类图

工厂方法模式的 UML 类图如下。

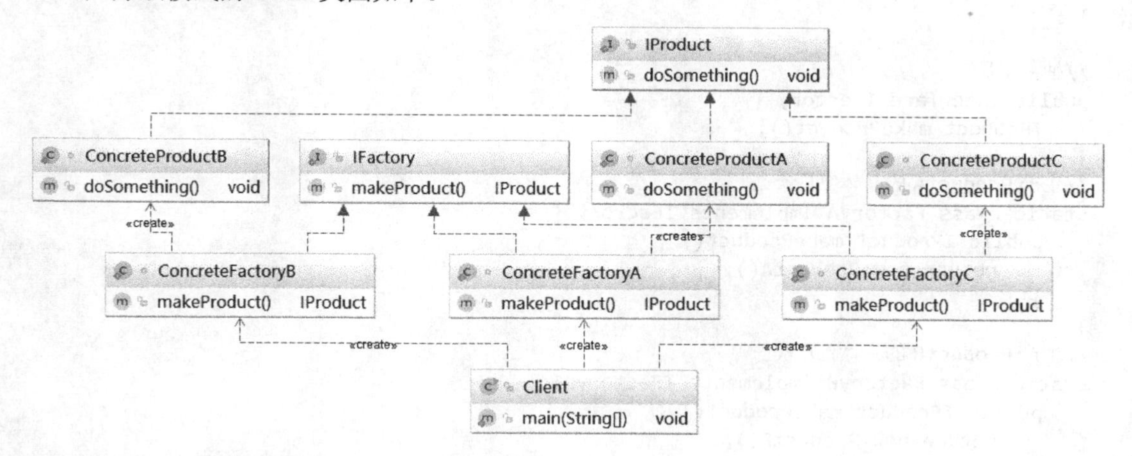

由上图可以看到，抽象工厂模式主要包含 4 个角色。

（1）抽象工厂（IFactory）：是工厂方法模式的核心，与应用程序无关。任何在模式中创建的对象的工厂类必须实现这个接口。

（2）具体工厂（ConcreteFactory）：是实现抽象工厂接口的具体工厂类，包含与应用程序密切相关的逻辑，并且被应用程序调用以创建产品对象。

（3）抽象产品（IProduct）：是工厂方法模式所创建的对象的超类型，也就是产品对象的共同父类或共同拥有的接口。

（4）具体产品（ConcreteProduct）：这个角色实现了抽象产品角色所定义的接口。某具体产品由专门的具体工厂创建，它们之间往往一一对应。

6.1.4 工厂方法模式的通用写法

以下是工厂方法模式的通用写法。

```
public class Client {
    public static void main(String[] args) {
        IFactory factory = new FactoryA();
        factory.makeProduct().doSomething();

        factory = new FactoryB();
        factory.makeProduct().doSomething();

        factory = new FactoryC();
        factory.makeProduct().doSomething();
    }

    //抽象工厂
    public interface IFactory {
        IProduct makeProduct();
    }
    //生产ProductA的具体工厂类
    static class FactoryA implements IFactory {
        public IProduct makeProduct() {
            return new ProductA();
        }
    }
    //生产ProductB的具体工厂类
    static class FactoryB implements IFactory {
        public IProduct makeProduct() {
            return new ProductB();
        }
    }

    //生产ProductC的具体工厂类
    static class FactoryC implements IFactory {
        public IProduct makeProduct() {
            return new ProductC();
        }
    }


    //抽象产品
    public interface IProduct {
        void doSomething();
    }
```

```
    //具体产品：ProductA
    static class ProductA implements IProduct {
        public void doSomething() {
            System.out.println("I am Product A");
        }
    }

    //具体产品：ProductB
    static class ProductB extends FactoryB implements IProduct {
        public void doSomething() {
            System.out.println("I am Product B");
        }
    }

    //具体产品：ProductC
    static class ProductC implements IProduct {
        public void doSomething() {
            System.out.println("I am Product C");
        }
    }
}
```

6.2　使用工厂方法模式实现产品扩展

工厂方法模式主要解决产品扩展的问题，在简单工厂中，随着产品链的丰富，如果每个课程的创建逻辑都有区别，则工厂的职责会变得越来越多，有点像万能工厂，并不便于维护。根据单一职责原则，我们将职能继续拆分，专人干专事。Java 课程由 Java 工厂创建，Python 课程由 Python 工厂创建，对工厂本身也做抽象。首先创建 ICourseFactory 接口。

```
public interface ICourseFactory {
    ICourse create();
}
```

然后分别创建子工厂，JavaCourseFactory 类的代码如下。

```
import com.gupaoedu.vip.pattern.factory.ICourse;
import com.gupaoedu.vip.pattern.factory.JavaCourse;

public class JavaCourseFactory implements ICourseFactory {
    public ICourse create() {
        return new JavaCourse();
    }
}
```

PythonCourseFactory 类的代码如下。

```
import com.gupaoedu.vip.pattern.factory.ICourse;
import com.gupaoedu.vip.pattern.factory.PythonCourse;

public class PythonCourseFactory implements ICourseFactory {
    public ICourse create() {
        return new PythonCourse();
    }
}
```

客户端测试代码如下。

```
public static void main(String[] args) {
    ICourseFactory factory = new PythonCourseFactory();
    ICourse course = factory.create();
    course.record();

    factory = new JavaCourseFactory();
    course = factory.create();
    course.record();
}
```

最后看如下图所示的类图。

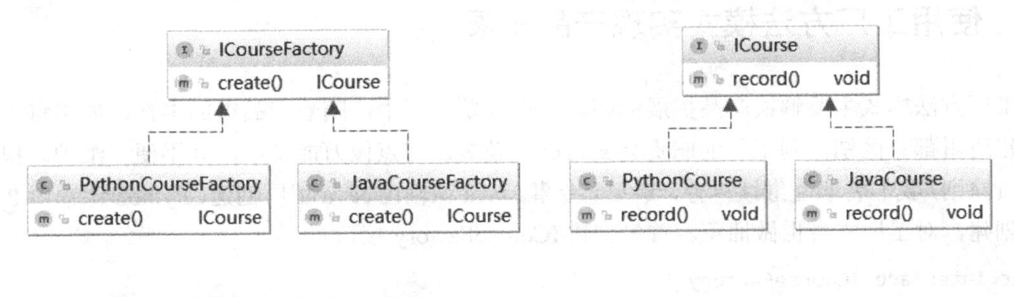

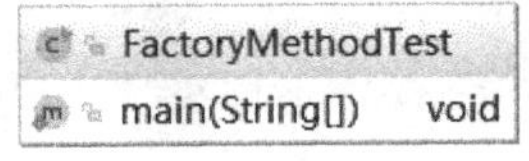

6.3 工厂方法模式在 Logback 源码中的应用

来看 Logback 中工厂方法模式的应用，其类图如下。

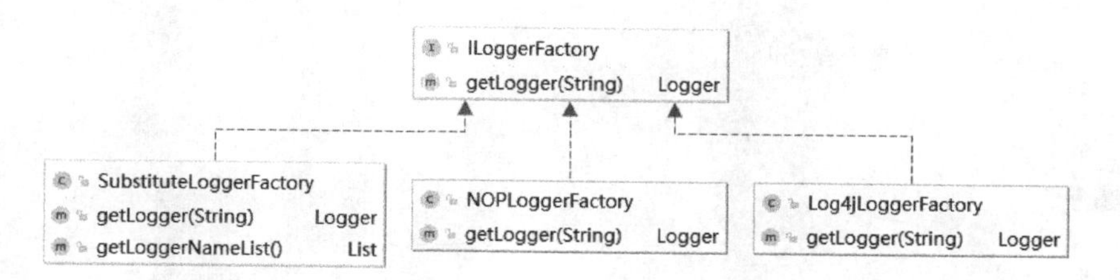

由上图可以看到，已经分离出不同工厂负责创建不同日志框架，如 Substitute 日志框架、NOP 日志框架、Log4J 日志框架，那么对应的 Logger 产品体系也是如此，如下图所示。

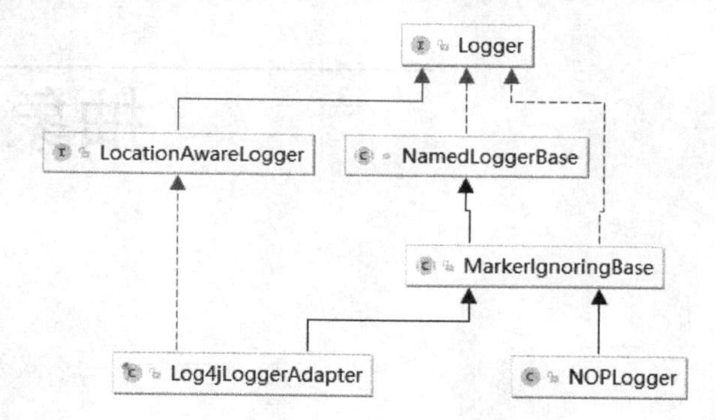

6.4　工厂方法模式扩展

6.4.1　工厂方法模式的优点

（1）灵活性增强，对于新产品的创建，只需多写一个相应的工厂类。

（2）典型的解耦框架。高层模块只需要知道产品的抽象类，无须关心其他实现类，满足迪米特法则、依赖倒置原则和里氏替换原则。

6.4.2　工厂方法模式的缺点

（1）类的个数容易过多，增加复杂度。

（2）增加了系统的抽象性和理解难度。

（3）抽象产品只能生产一种产品，此弊端可使用抽象工厂模式解决。

7

第 7 章 抽象工厂模式

7.1 抽象工厂模式概述

7.1.1 抽象工厂模式的定义

抽象工厂模式（Abstract Factory Pattern）指提供一个创建一系列相关或相互依赖对象的接口，无须指定它们具体的类。意思是客户端不必指定产品的具体类型，创建多个产品族中的产品对象。

原文： Provide an interface for creating families of related or dependent objects without specifying their concrete classes.

在抽象工厂模式中，客户端（应用层）不依赖产品类实例如何被创建、实现等细节，强调一系列相关的产品对象（属于同一产品族）一起创建对象，需要大量重复的代码。

需要提供一个产品类的库，所有产品以同样的接口出现，从而使客户端不依赖具体实现。

7.1.2　关于产品等级结构和产品族

在讲解抽象工厂之前，我们要了解两个概念：产品等级结构和产品族，如下图所示。

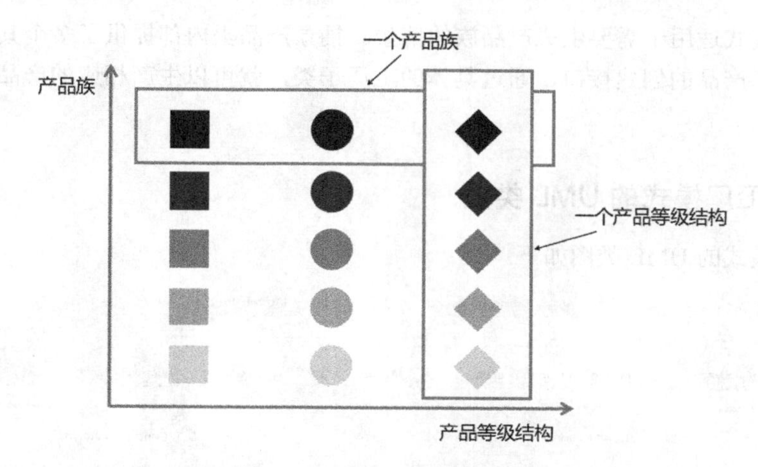

上图中有正方形、圆形和菱形 3 种图形，相同颜色、相同深浅的代表同一个产品族，相同形状的代表同一个产品等级结构。同样可以从生活中来举例，比如，美的电器生产多种家用电器，那么上图中，颜色最深的正方形就代表美的洗衣机，颜色最深的圆形代表美的空调，颜色最深的菱形代表美的热水器，颜色最深的一排都属于美的品牌，都属于美的电器这个产品族。再看最右侧的菱形，颜色最深的被指定了代表美的热水器，那么第二排颜色稍微浅一点的菱形代表海信热水器。同理，同一产品族下还有格力洗衣机、格力空调、格力热水器。

再看下图，最左侧的小房子被认为是具体的工厂，有美的工厂、海信工厂、格力工厂。每个品牌的工厂都生产洗衣机、空调和热水器。

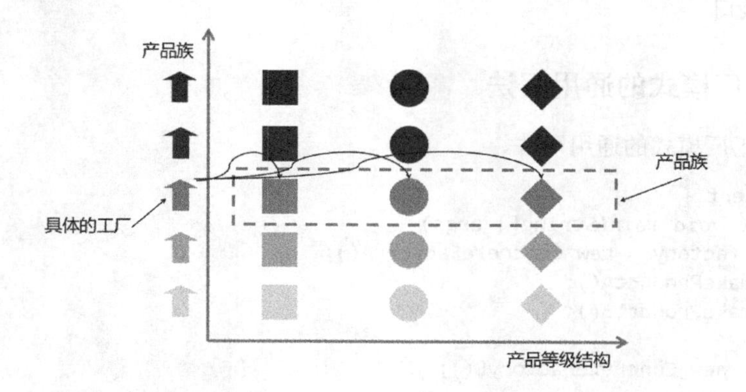

通过上面两张图的对比理解，相信大家对抽象工厂有了非常形象的理解。

7.1.3 抽象工厂模式的应用场景

抽象工厂模式适用于需要生成产品族的情景。抽象产品类内部提供了多个其他抽象产品，抽象工厂类定义了产品的创建接口，通过具体的工厂子类，就可以生产相应的产品族对象，供用户端使用。

7.1.4 抽象工厂模式的 UML 类图

抽象工厂模式的 UML 类图如下。

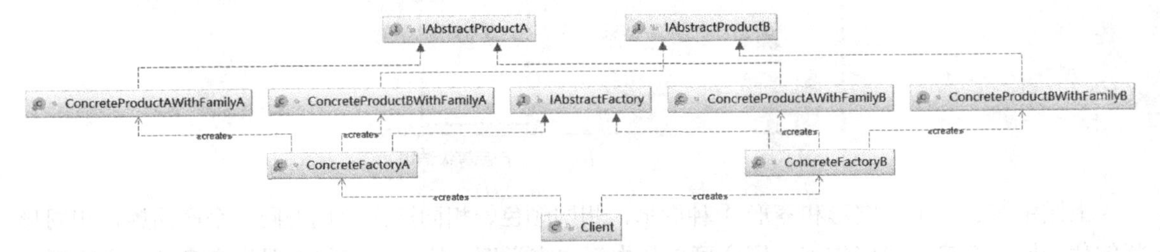

由上图可以看到，抽象工厂模式主要包含 4 个角色。

（1）抽象工厂（IAbstractFactory）：声明创建抽象产品对象的一个操作接口。

（2）具体工厂（ConcreteFactory）：实现创建具体产品对象的操作。

（3）抽象产品（IAbstractProduct）：为一类产品对象声明一个接口。

（4）具体产品（ConcreteProduct）：定义一个将被相应的具体工厂创建的产品对象，实现 AbstractProduct 接口。

7.1.5 抽象工厂模式的通用写法

以下是抽象工厂模式的通用写法。

```
public class Client {
    public static void main(String[] args) {
        IFactory factory = new ConcreteFactoryA();
        factory.makeProductA();
        factory.makeProductB();

        factory = new ConcreteFactoryB();
```

```
        factory.makeProductA();
        factory.makeProductB();
    }

    //抽象工厂
    public interface IFactory {
        IProductA makeProductA();

        IProductB makeProductB();
    }

    //产品 A 抽象
    public interface IProductA {
        void doA();
    }

    //产品 B 抽象
    public interface IProductB {
        void doB();
    }

    //产品族 A 的具体产品 A
    static class ConcreteProductAWithFamilyA implements IProductA{
        public void doA() {
            System.out.println("The ProductA be part of FamilyA");
        }
    }

    //产品族 A 的具体产品 B
    static class ConcreteProductBWithFamilyA implements IProductB{
        public void doB() {
            System.out.println("The ProductB be part of FamilyA");
        }
    }

    //产品族 B 的具体产品 A
    static class ConcreteProductAWithFamilyB implements IProductA{
        public void doA() {
            System.out.println("The ProductA be part of FamilyB");
        }
    }

    //产品族 B 的具体产品 B
    static class ConcreteProductBWithFamilyB implements IProductB{
        public void doB() {
            System.out.println("The ProductB be part of FamilyB");
        }
```

```
    }

    //具体工厂类A
    static class ConcreteFactoryA implements IFactory{
        public IProductA makeProductA() {
            return new ConcreteProductAWithFamilyA();
        }

        public IProductB makeProductB() {
            return new ConcreteProductBWithFamilyA();
        }
    }

    //具体工厂类B
    static class ConcreteFactoryB implements IFactory{
        public IProductA makeProductA() {
            return new ConcreteProductAWithFamilyB();
        }

        public IProductB makeProductB() {
            return new ConcreteProductBWithFamilyB();
        }
    }
}
```

7.2 使用抽象工厂模式解决实际问题

7.2.1 使用抽象工厂模式支持产品扩展

我们来看一个具体的业务场景，并且用代码来实现。还是以课程为例，咕泡学院的课程研发会有标准，每个课程不仅要提供课程的录播视频，还要提供老师的课堂笔记。相当于现在的业务变更为同一个课程不单纯包含一个课程信息，要同时包含录播视频、课堂笔记，甚至要提供源码才能构成一个完整的课程。首先在产品等级中增加两个产品：录播视频 IVideo 和课堂笔记 INote。

IVideo 接口的代码如下。

```
public interface IVideo {
    void record();
}
```

INote 接口的代码如下。

```
public interface INote {
    void edit();
}
```

然后创建一个抽象工厂 CourseFactory 类。

```
import com.gupaoedu.vip.pattern.factory.INote;
import com.gupaoedu.vip.pattern.factory.IVideo;

/**
 * 抽象工厂是用户的主入口
 * 在 Spring 中应用得最为广泛的一种设计模式
 * 易于扩展
 */
public abstract class CourseFactory {

    public void init(){
        System.out.println("初始化基础数据");
    }

    protected abstract INote createNote();

    protected abstract IVideo createVideo();

}
```

接下来创建 Java 产品族，Java 视频 JavaVideo 类的代码如下。

```
public class JavaVideo implements IVideo {
    public void record() {
        System.out.println("录制 Java 视频");
    }
}
```

扩展产品等级 Java 课堂笔记 JavaNote 类。

```
public class JavaNote implements INote {
    public void edit() {
        System.out.println("编写 Java 笔记");
    }
}
```

创建 Java 产品族的具体工厂 JavaCourseFactory。

```
public class JavaCourseFactory extends CourseFactory {

    public INote createNote() {
        super.init();
        return new JavaNote();
    }

    public IVideo createVideo() {
```

```
        super.init();
        return new JavaVideo();
    }
}
```

随后创建 Python 产品族，Python 视频 PythonVideo 类的代码如下。

```
public class PythonVideo implements IVideo {
    public void record() {
        System.out.println("录制 Python 视频");
    }
}
```

扩展产品等级 Python 课堂笔记 PythonNote 类。

```
public class PythonNote implements INote {
    public void edit() {
        System.out.println("编写 Python 笔记");
    }
}
```

创建 Python 产品族的具体工厂 PythonCourseFactory。

```
public class PythonCourseFactory implements CourseFactory {
    public INote createNote() {
        return new PythonNote();
    }
    public IVideo createVideo() {
        return new PythonVideo();
    }
}
```

最后来看客户端调用代码。

```
public static void main(String[] args) {
    JavaCourseFactory factory = new JavaCourseFactory();
    factory.createNote().edit();
    factory.createVideo().record();
}
```

上面代码完整地描述了 Java 课程和 Python 课程两个产品族，也描述了视频和笔记两个产品等级。抽象工厂非常完美、清晰地描述了这样一层复杂的关系。但是，不知道大家有没有发现，如果再继续扩展产品等级，将源码 Source 也加入课程中，则代码从抽象工厂到具体工厂要全部调整，这显然不符合开闭原则。

7.2.2 使用抽象工厂模式重构数据库连接池

还是演示课堂开始的 JDBC 操作案例，我们每次操作都需要重新创建数据库连接。其实每次创建都非常耗费性能，消耗业务调用时间。我们使用抽象工厂模式，将数据库连接预先创建好，放到容器中缓存着，当业务调用时就只需现取现用。我们来看代码。

Pool 抽象类的代码如下。

```
/**
 * 自定义连接池 getInstance()返回 POOL 唯一实例，第一次调用时将执行构造函数
 * 构造函数 Pool()调用驱动装载 loadDrivers()函数
 * 连接池创建 createPool()函数，loadDrivers()装载驱动
 * createPool()创建连接池，getConnection()返回一个连接实例
 * getConnection(long time)添加时间限制
 * freeConnection(Connection con)将 con 连接实例返回连接池，getnum()返回空闲连接数
 * getnumActive()返回当前使用的连接数
 */

public abstract class Pool {
   public String propertiesName = "connection-INF.properties";

   private static Pool instance = null;  //定义唯一实例

   /**
    * 最大连接数
    */
   protected int maxConnect = 100;        //最大连接数

   /**
    * 保持连接数
    */
   protected int normalConnect = 10;     //保持连接数

   /**
    * 驱动字符串
    */
   protected String driverName = null;   //驱动字符串

   /**
    * 驱动类
    */
   protected Driver driver = null;       //驱动变量


   /**
    * 私有构造函数，不允许外界访问
```

```
    */
    protected Pool() {
       try
       {
          init();
          loadDrivers(driverName);
       }catch(Exception e)
       {
          e.printStackTrace();
       }
    }

    /**
     * 初始化所有从配置文件中读取的成员变量
     */
    private void init() throws IOException {
       InputStream is = Pool.class.getResourceAsStream(propertiesName);
       Properties p = new Properties();
       p.load(is);
       this.driverName = p.getProperty("driverName");
       this.maxConnect = Integer.parseInt(p.getProperty("maxConnect"));
       this.normalConnect = Integer.parseInt(p.getProperty("normalConnect"));
    }

    /**
     * 装载和注册所有 JDBC 驱动程序
     * @param dri  接收驱动字符串
     */
    protected void loadDrivers(String dri) {
       String driverClassName = dri;
       try {
          driver = (Driver) Class.forName(driverClassName).newInstance();
          DriverManager.registerDriver(driver);
          System.out.println("成功注册 JDBC 驱动程序" + driverClassName);
       } catch (Exception e) {
          System.out.println("无法注册 JDBC 驱动程序:" + driverClassName + ",错误:" + e);
       }
    }

    /**
     * 创建连接池
     */
    public abstract void createPool();

    /**
     *
     *（单例模式）返回数据库连接池 Pool 的实例
```

```
 *
 * @param driverName 数据库驱动字符串
 * @return
 * @throws IOException
 * @throws ClassNotFoundException
 * @throws IllegalAccessException
 * @throws InstantiationException
 */
public static synchronized Pool getInstance() throws IOException,
      InstantiationException, IllegalAccessException,
      ClassNotFoundException {

   if (instance == null) {
      instance = (Pool) Class.forName("org.e_book.sqlhelp.Pool").newInstance();
   }
   return instance;
}

/**
 * 获得一个可用的连接，如果没有，则创建一个连接，并且小于最大连接限制
 * @return
 */
public abstract Connection getConnection();

/**
 * 获得一个连接，有时间限制
 * @param time 设置该连接的持续时间（以毫秒为单位）
 * @return
 */
public abstract Connection getConnection(long time);

/**
 * 将连接对象返回连接池
 * @param con 获得连接对象
 */
public abstract void freeConnection(Connection con);

/**
 * 返回当前空闲的连接数
 * @return
 */
public abstract int getnum();

/**
 * 返回当前工作的连接数
 * @return
 */
```

```
  public abstract int getnumActive();

  /**
   * 关闭所有连接，撤销驱动注册（此方法为单例方法）
   */
  protected synchronized void release() {
     //撤销驱动
     try {
        DriverManager.deregisterDriver(driver);
        System.out.println("撤销 JDBC 驱动程序 " + driver.getClass().getName());
     } catch (SQLException e) {
        System.out
              .println("无法撤销 JDBC 驱动程序的注册:" + driver.getClass().getName());
     }
  }
}
```

DBConnectionPool 数据库连接池的代码如下。

```
/**
 * 数据库连接池管理类
 */
public final class DBConnectionPool extends Pool {
  private int checkedOut;                          //正在使用的连接数
  /**
   * 存放产生的连接对象容器
   */
  private Vector<Connection> freeConnections = new Vector<Connection>();
                                                   //存放产生的连接对象容器

  private String passWord = null;                  //密码

  private String url = null;                       //连接字符串

  private String userName = null;                  //用户名

  private static int num = 0;                      //空闲连接数

  private static int numActive = 0;                //当前可用的连接数

  private static DBConnectionPool pool = null;     //连接池实例变量

  /**
   * 产生数据连接池
   * @return
   */
  public static synchronized DBConnectionPool getInstance()
```

```
{
   if(pool == null)
   {
      pool = new DBConnectionPool();
   }
   return pool;
}

/**
 * 获得一个数据库连接池的实例
 */
private DBConnectionPool() {
   try
   {
      init();
      for (int i = 0; i < normalConnect; i++) {      //初始 normalConn 个连接
         Connection c = newConnection();
         if (c != null) {
            freeConnections.addElement(c);           //往容器中添加一个连接对象
            num++; //记录总连接数
         }
      }
   }catch(Exception e)
   {
      e.printStackTrace();
   }
}
/**
 * 初始化
 * @throws IOException
 */
private void init() throws IOException
{
   InputStream is = DBConnectionPool.class.getResourceAsStream(propertiesName);
   Properties p = new Properties();
   p.load(is);
   this.userName = p.getProperty("userName");
   this.passWord = p.getProperty("passWord");
   this.driverName = p.getProperty("driverName");
   this.url = p.getProperty("url");
   this.driverName = p.getProperty("driverName");
   this.maxConnect = Integer.parseInt(p.getProperty("maxConnect"));
   this.normalConnect = Integer.parseInt(p.getProperty("normalConnect"));
}
/**
 * 如果不再使用某个连接对象，则可调此方法将该对象释放到连接池
 * @param con
```

```
 */
public synchronized void freeConnection(Connection con) {
   freeConnections.addElement(con);
   num++;
   checkedOut--;
   numActive--;
   notifyAll(); //解锁
}

/**
 * 创建一个新连接
 * @return
 */
private Connection newConnection() {
   Connection con = null;
   try {
      if (userName == null) { //用户、密码都为空
         con = DriverManager.getConnection(url);
      } else {
         con = DriverManager.getConnection(url, userName, passWord);
      }
      System.out.println("连接池创建一个新的连接");
   } catch (SQLException e) {
      System.out.println("无法创建这个URL的连接" + url);
      return null;
   }
   return con;
}

/**
 * 返回当前空闲的连接数
 * @return
 */
public int getnum() {
   return num;
}

/**
 * 返回当前可用的连接数
 * @return
 */
public int getnumActive() {
   return numActive;
}


/**
```

```
 * （单例模式）获取一个可用连接
 * @return
 */
public synchronized Connection getConnection() {
   Connection con = null;
   if (freeConnections.size() > 0) { //还有空闲的连接
      num--;
      con = (Connection) freeConnections.firstElement();
      freeConnections.removeElementAt(0);
      try {
         if (con.isClosed()) {
            System.out.println("从连接池删除一个无效连接");
            con = getConnection();
         }
      } catch (SQLException e) {
         System.out.println("从连接池删除一个无效连接");
         con = getConnection();
      }
      //没有空闲连接且当前连接小于最大允许值，若最大值为0，则不限制
   } else if (maxConnect == 0 || checkedOut < maxConnect) {
      con = newConnection();
   }
   if (con != null) { //当前连接数加1
      checkedOut++;
   }
   numActive++;
   return con;
}

/**
 * 获取一个连接，并加上等待时间限制，时间为毫秒
 * @param timeout  接受等待时间（以毫秒为单位）
 * @return
 */
public synchronized Connection getConnection(long timeout) {
   long startTime = new Date().getTime();
   Connection con;
   while ((con = getConnection()) == null) {
      try {
         wait(timeout); //线程等待
      } catch (InterruptedException e) {
      }
      if ((new Date().getTime() - startTime) >= timeout) {
         return null; //如果超时，则返回
      }
   }
   return con;
```

```
    }

    /**
     * 关闭所有连接
     */
    public synchronized void release() {
        try {
            //将当前连接赋值到枚举中
            Enumeration allConnections = freeConnections.elements();
            //使用循环关闭连接池中的所用连接
            while (allConnections.hasMoreElements()) {
                //如果此枚举对象至少还有一个可提供的元素，则返回此枚举的下一个元素
                Connection con = (Connection) allConnections.nextElement();
                try {
                    con.close();
                    num--;
                } catch (SQLException e) {
                    System.out.println("无法关闭连接池中的连接");
                }
            }
            freeConnections.removeAllElements();
            numActive = 0;
        } finally {
            super.release();
        }
    }

    /**
     * 建立连接池
     */
    public void createPool() {

        pool = new DBConnectionPool();
        if (pool != null) {
            System.out.println("创建连接池成功");
        } else {
            System.out.println("创建连接池失败");
        }
    }
}
```

7.3 抽象工厂模式在 Spring 源码中的应用

在 Spring 中，所有工厂都是 BeanFactory 的子类。通过对 BeanFactory 的实现，我们可以从 Spring

的容器访问 Bean。根据不同的策略调用 getBean()方法，从而获得具体对象。

```
public interface BeanFactory {
    String FACTORY_BEAN_PREFIX = "&";

    Object getBean(String name) throws BeansException;

    <T> T getBean(String name, @Nullable Class<T> requiredType) throws BeansException;

    Object getBean(String name, Object... args) throws BeansException;

    <T> T getBean(Class<T> requiredType) throws BeansException;

    <T> T getBean(Class<T> requiredType, Object... args) throws BeansException;

    boolean containsBean(String name);

    boolean isSingleton(String name) throws NoSuchBeanDefinitionException;

    boolean isPrototype(String name) throws NoSuchBeanDefinitionException;

    boolean isTypeMatch(String name, ResolvableType typeToMatch) throws NoSuchBeanDefinitionException;

    boolean isTypeMatch(String name, @Nullable Class<?> typeToMatch) throws NoSuchBean
    DefinitionException;

    @Nullable
    Class<?> getType(String name) throws NoSuchBeanDefinitionException;
    String[] getAliases(String name);

}
```

BeanFactory 的子类主要有 ClassPathXmlApplicationContext、XmlWebApplicationContext、StaticWebApplicationContext、StaticPortletApplicationContext、GenericApplicationContext 和 StaticApplicationContext。在 Spring 中，DefaultListableBeanFactory 实现了所有工厂的公共逻辑。

7.4　抽象工厂模式扩展

7.4.1　抽象工厂模式的优点

（1）当需要产品族时，抽象工厂可以保证客户端始终只使用同一个产品的产品族。

（2）抽象工厂增强了程序的可扩展性，对于新产品族的增加，只需实现一个新的具体工厂即可，不需要对已有代码进行修改，符合开闭原则。

7.4.2 抽象工厂模式的缺点

（1）规定了所有可能被创建的产品集合，产品族中扩展新的产品困难，需要修改抽象工厂的接口。

（2）增加了系统的抽象性和理解难度。

8

第 8 章 单例模式

8.1 单例模式概述

8.1.1 单例模式的定义

单例模式（Singleton Pattern）指确保一个类在任何情况下都绝对只有一个实例，并提供一个全局访问点，属于创建型设计模式。

原文：Ensure a class has only one instance, and provide a global point of access to it.

8.1.2 单例模式的应用场景

单例模式在现实生活中的应用非常广泛，例如公司 CEO、部门经理等都属于单例模型。J2EE 标准中的 ServletContext 和 ServletContextConfig、Spring 框架应用中的 ApplicationContext、数据库中的连接池等也都是单例模式。对于 Java 来说，单例模式可以保证在一个 JVM 中只存在单一实例。单例模式的应用场景主要有以下几个方面。

（1）需要频繁创建的一些类，使用单例可以降低系统的内存压力，减少 GC。

（2）某些类创建实例时占用资源较多，或实例化耗时较长，且经常使用。

（3）频繁访问数据库或文件的对象。

（4）对于一些控制硬件级别的操作，或者从系统上来讲应当是单一控制逻辑的操作，如果有多个实例，则系统会完全乱套。

8.1.3 单例模式的 UML 类图

单例模式的 UML 类图如下。

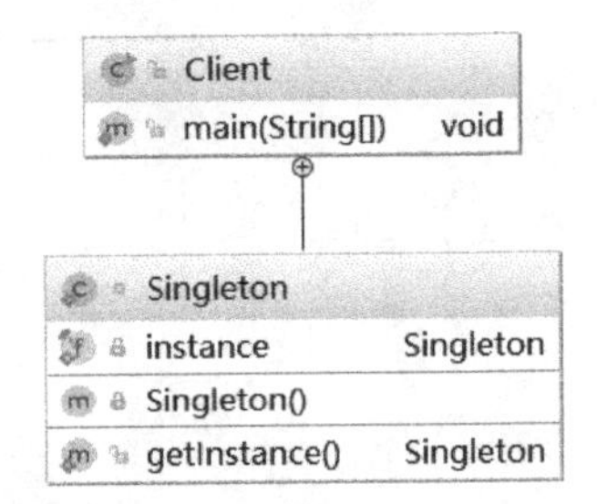

Singleton 类被称为单例类，通过隐藏构造方法，在内部初始化一次，并提供一个全局的访问点。

8.1.4 单例模式的通用写法

以下是单例模式的通用写法。

```
public class Client {
    public static void main(String[] args) {
        Singleton.getInstance();
    }

    static class Singleton {
        private static final Singleton instance = new Singleton();

        private Singleton(){}

        public static Singleton getInstance(){
            return  instance;
        }

    }
}
```

8.2　使用单例模式解决实际问题

8.2.1　饿汉式单例写法的弊端

其实我们前面看到的单例模式通用写法，就是饿汉式单例的标准写法。饿汉式单例写法在类加载的时候立即初始化，并且创建单例对象。它绝对线程安全，在线程还没出现之前就实例化了，不可能存在访问安全问题。饿汉式单例还有另外一种写法，代码如下。

```
//饿汉式静态块单例模式
public class HungryStaticSingleton {
    private static final HungryStaticSingleton hungrySingleton;
    static {
        hungrySingleton = new HungryStaticSingleton();
    }
    private HungryStaticSingleton(){}
    public static HungryStaticSingleton getInstance(){
        return  hungrySingleton;
    }
}
```

这种写法使用静态块的机制，非常简单也容易理解。饿汉式单例写法适用于单例对象较少的情况。这样写可以保证绝对线程安全，执行效率比较高。但是它的缺点也很明显，就是所有对象类在加载的时候就实例化。这样一来，如果系统中有大批量的单例对象存在，而且单例对象的数量也不确定，则系统初始化时会造成大量的内存浪费，从而导致系统内存不可控。也就是说，不管对象用或不用，都占着空间，浪费了内存，有可能占着内存又不使用。那有没有更优的写法呢？我们继续分析。

8.2.2　还原线程破坏单例的事故现场

为了解决饿汉式单例写法可能带来的内存浪费问题，于是出现了懒汉式单例的写法。懒汉式单例写法的特点是单例对象在被使用时才会初始化。懒汉式单例写法的简单实现 LazySimpleSingleton 如下。

```
//懒汉式单例在外部需要使用的时候才进行实例化
public class LazySimpleSingleton {
    private LazySimpleSingleton(){}
    //静态块，公共内存区域
    private static LazySimpleSingleton lazy = null;
    public static LazySimpleSingleton getInstance(){
        if(lazy == null){
```

```
            lazy = new LazySimpleSingleton();
        }
        return lazy;
    }
}
```

但这样写又带来了一个新的问题，如果在多线程环境下，则会出现线程安全问题。先来模拟一下，编写线程类 ExectorThread。

```
public class ExectorThread implements Runnable{
    @Override
    public void run() {
        LazySimpleSingleton singleton = LazySimpleSingleton.getInstance();
        System.out.println(Thread.currentThread().getName() + ":" + singleton);
    }
}
```

编写客户端测试代码如下。

```
public class LazySimpleSingletonTest {
    public static void main(String[] args) {
        Thread t1 = new Thread(new ExectorThread());
        Thread t2 = new Thread(new ExectorThread());
        t1.start();
        t2.start();
        System.out.println("End");
    }
}
```

我们反复多次运行程序上的代码，发现会有一定概率出现两种不同结果，有可能两个线程获取的对象是一致的，也有可能两个线程获取的对象是不一致的。下图是两个线程获取的对象不一致的运行结果。

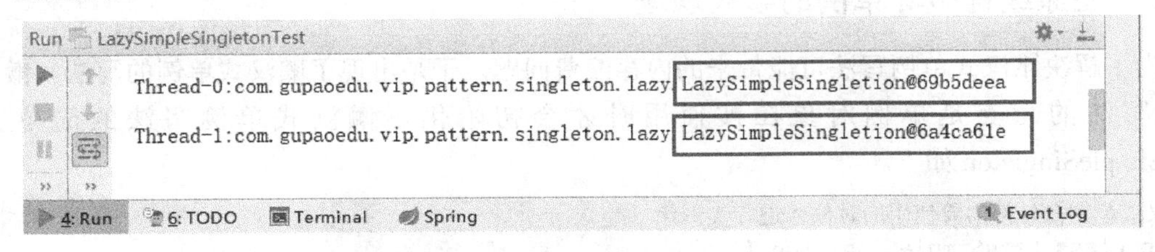

下图是两个线程获取的对象一致的结果。

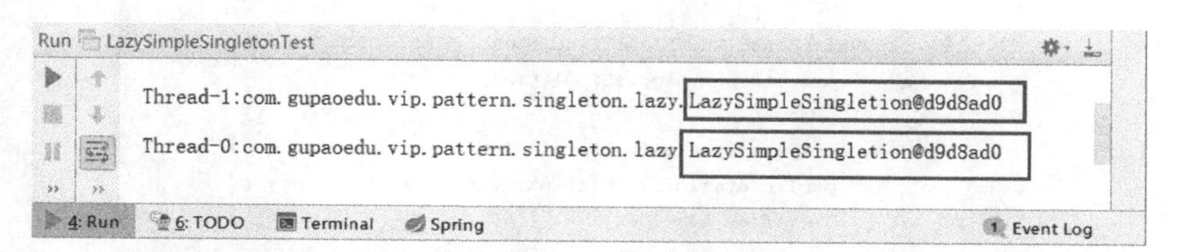

显然，这意味着上面的单例模式存在线程安全隐患。那么这个结果是怎么产生的呢？我们来分析一下，如下图所示，如果两个线程在同一时间同时进入 getInstance()方法，则会同时满足 if(null == instance)条件，创建两个对象。如果两个线程都继续往下执行后面的代码，则有可能后执行的线程的结果覆盖先执行的线程的结果。如果打印动作发生在覆盖之前，则最终得到的结果就是一致的；如果打印动作发生在覆盖之后，则得到两个不一样的结果。

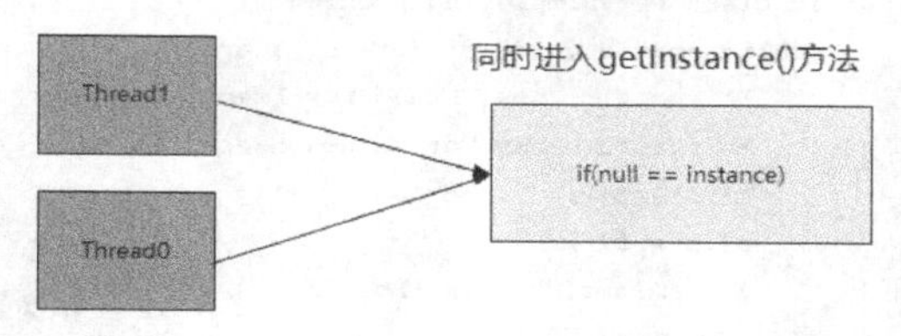

当然，也有可能没有发生并发，完全正常运行。下面通过调试方式来更深刻地理解一下。这里教大家一种新技能，用线程模式调试，手动控制线程的执行顺序来跟踪内存的变化。先把 ExectorThread 类打上断点，如下图所示。

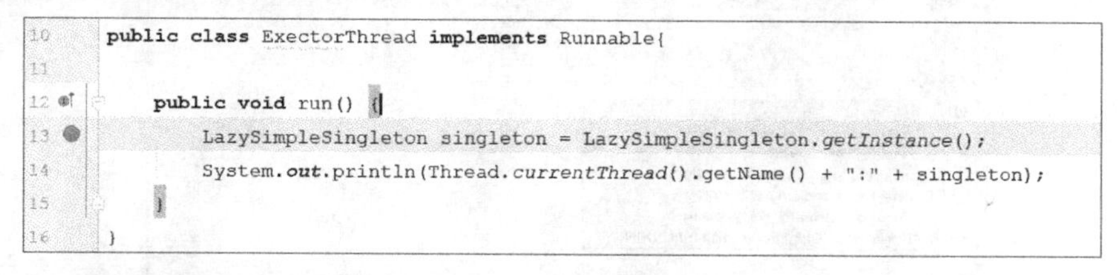

单击右键点击断点，切换为 Thread 模式，如下图所示。

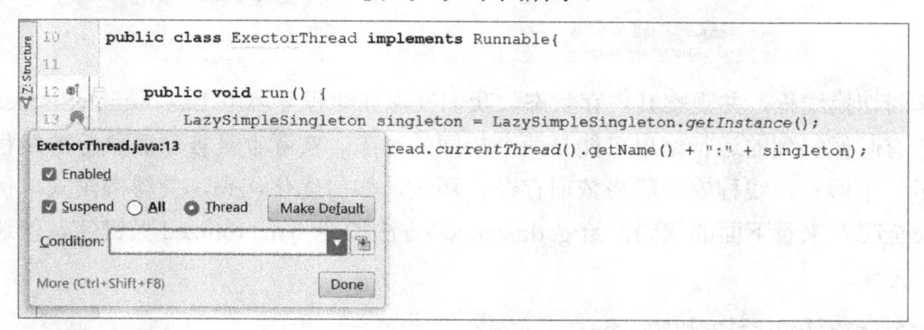

然后把 LazySimpleSingleton 类也打上断点，同样标记为 Thread 模式，如下图所示。

```java
public class LazySimpleSingleton {
    private LazySimpleSingleton(){}
    //静态块，公共内存区域
    private static LazySimpleSingleton lazy = null;
    public static LazySimpleSingleton getInstance(){
        if(null == lazy){
            lazy = new LazySimpleSingleton();
        }
        return lazy;
    }
}
```

切换回客户端测试代码，同样也打上断点，同时改为 Thread 模式，如下图所示。

```java
public class LazySimpleSingletonTest {
    public static void main(String[] args) {
        Thread t1 = new Thread(new ExectorThread());
        Thread t2 = new Thread(new ExectorThread());
        t1.start();
        t2.start();
        System.out.println("End");
    }
}
```

在开始 Debug 之后，我们会看到 Debug 控制台可以自由切换 Thread 的运行状态，如下图所示。

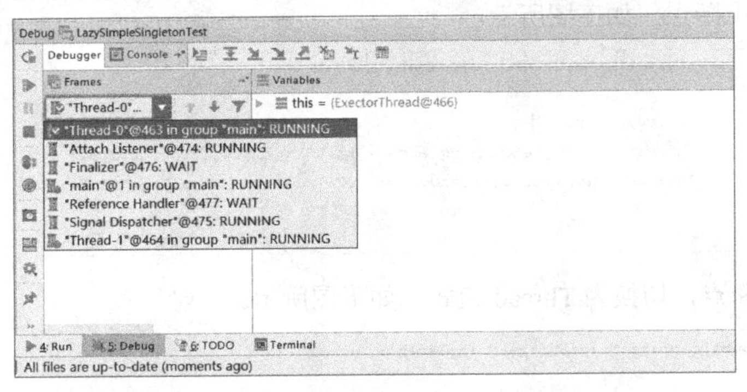

通过不断切换线程，并观察其内存状态，我们发现在线程环境下 LazySimpleSingleton 被实例化了两次。有时候得到的运行结果可能是两个相同的对象，实际上是被后面执行的线程覆盖了，我们看到了一个假象，线程安全隐患依旧存在。那么，如何优化代码，使得懒汉式单例写法在线程环境下安全呢？来看下面的代码，给 getInstance()方法加上 synchronized 关键字，使这个方法变成线程同步方法。

```java
public class LazySimpleSingleton {
    private LazySimpleSingleton(){}
```

```
    //静态块，公共内存区域
    private static LazySimpleSingleton lazy = null;
    public synchronized static LazySimpleSingleton getInstance(){
        if(lazy == null){
            lazy = new LazySimpleSingleton();
        }
        return lazy;
    }
}
```

我们再来调试。当执行其中一个线程并调用 getInstance()方法时，另一个线程在调用 getInstance()方法，线程的状态由 RUNNING 变成了 MONITOR，出现阻塞。直到第一个线程执行完，第二个线程才恢复到 RUNNING 状态继续调用 getInstance()方法，如下图所示。

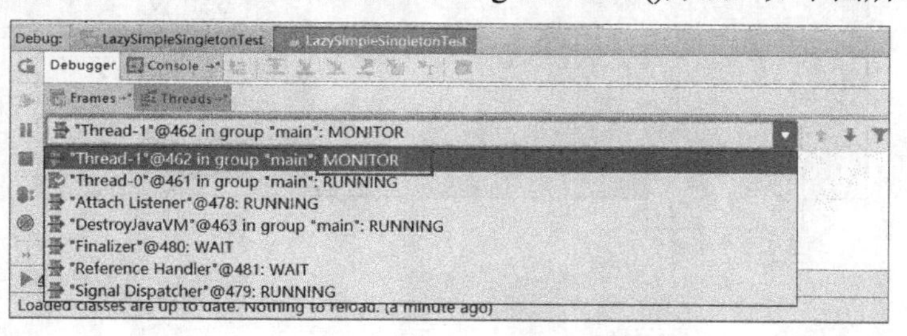

这样，通过使用 synchronized 就解决了线程安全问题。

8.2.3　双重检查锁单例写法闪亮登场

在上一节中，我们通过调试的方式完美地展现了 synchronized 监视锁的运行状态。但是，如果在线程数量剧增的情况下，用 synchronized 加锁，则会导致大批线程阻塞，从而导致程序性能大幅下降。就好比是地铁进站限流，在寒风刺骨的冬天，所有人都在站前广场转圈圈，用户体验很不好，如下图所示。

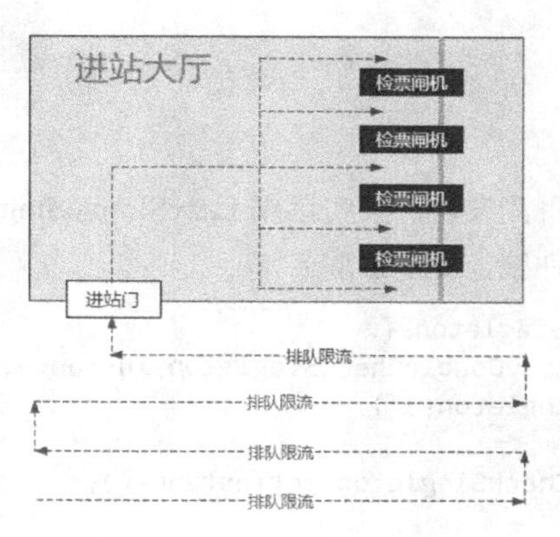

那有没有办法优化一下用户体验呢？其实可以让所有人先进入进站大厅，然后增设一些进站闸口，这样用户体验变好了，进站效率也提高了。当然，在现实生活中可能会受到很多硬性条件的限制，但是在虚拟世界中是完全可以实现的。其实这就叫作双重检查，在进站门安检一次，进入大厅后在闸口检票处再检查一次，如下图所示。

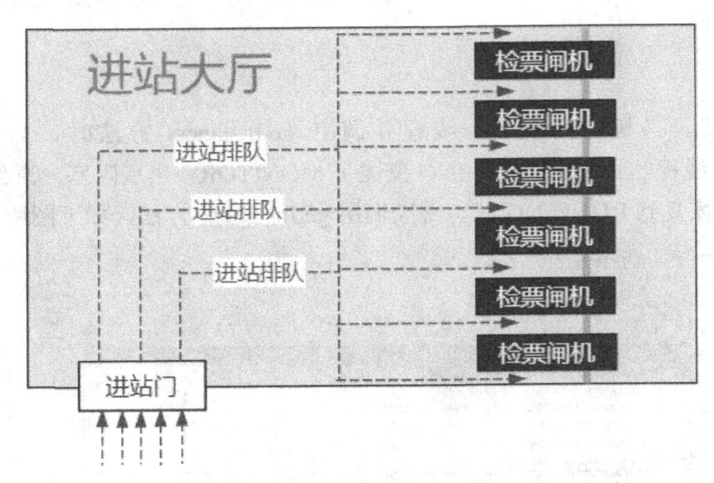

我们来改造一下代码，创建 LazyDoubleCheckSingleton 类。

```
public class LazyDoubleCheckSingleton {
    private volatile static LazyDoubleCheckSingleton instance;
    private LazyDoubleCheckSingleton(){}

    public static LazyDoubleCheckSingleton getInstance(){
        synchronized (LazyDoubleCheckSingleton.class) {
            if (instance == null) {
                instance = new LazyDoubleCheckSingleton();
            }
        }
        return instance;
    }
}
```

这样写就解决问题了吗？目测发现，其实这跟 LazySimpleSingletion 的写法并无差异，还是会大规模阻塞。那我们把判断条件往上提一级呢？

```
public class LazyDoubleCheckSingleton {
    private volatile static LazyDoubleCheckSingleton instance;
    private LazyDoubleCheckSingleton(){}

    public static LazyDoubleCheckSingleton getInstance(){
```

```
        if (instance == null) {
            synchronized (LazyDoubleCheckSingleton.class) {
                instance = new LazyDoubleCheckSingleton();
            }
        }
        return instance;
    }
}
```

在运行代码后，还是会存在线程安全问题。运行结果如下图所示。

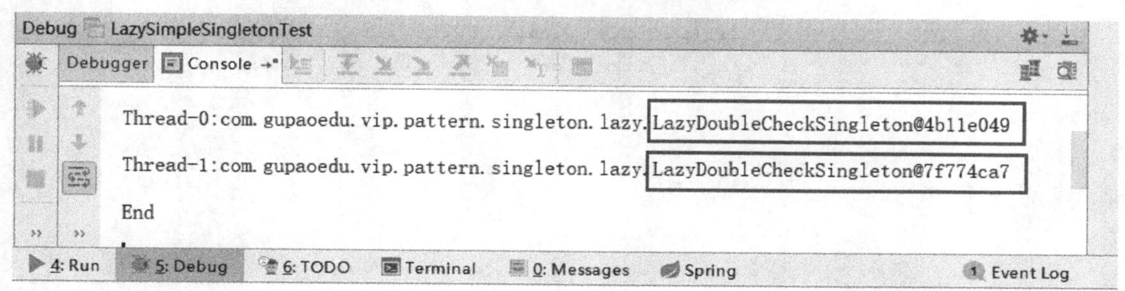

这是什么原因导致的呢？其实如果两个线程在同一时间都满足 if(instance == null)条件，则两个线程都会执行 synchronized 块中的代码，因此，还是会创建两次。再优化一下代码。

```
public class LazyDoubleCheckSingleton {
    private volatile static LazyDoubleCheckSingleton instance;
    private LazyDoubleCheckSingleton(){}

    public static LazyDoubleCheckSingleton getInstance(){
        //检查是否要阻塞
        if (instance == null) {
            synchronized (LazyDoubleCheckSingleton.class) {
                //检查是否要重新创建实例
                if (instance == null) {
                    instance = new LazyDoubleCheckSingleton();
                    //指令重排序的问题
                }
            }
        }
        return instance;
    }
}
```

我们进行断点调试，如下图所示。

```java
public class LazyDoubleCheckSingleton {
    private volatile static LazyDoubleCheckSingleton instance;
    private LazyDoubleCheckSingleton(){}

    public static LazyDoubleCheckSingleton getInstance(){
        //检查是否要阻塞
        if (instance == null) {
            synchronized (LazyDoubleCheckSingleton.class) {
                //检查是否要重新创建实例
                if (instance == null) {
                    instance = new LazyDoubleCheckSingleton();
                    //指令重排序的问题
                }
            }
        }
        return instance;
    }
}
```

当第一个线程调用 getInstance()方法时，第二个线程也可以调用。当第一个线程执行到 synchronized 时会上锁，第二个线程就会变成 MONITOR 状态，出现阻塞。此时，阻塞并不是基于整个 LazyDoubleCheckSingleton 类的阻塞，而是在 getInstance()方法内部的阻塞，只要逻辑不太复杂，对于调用者而言感觉不到。

8.2.4 看似完美的静态内部类单例写法

双重检查锁单例写法虽然解决了线程安全问题和性能问题，但是只要用到 synchronized 关键字就总是要上锁，对程序性能还是存在一定影响的。难道真的没有更好的方案吗？当然有。我们可以从类初始化的角度考虑，看下面的代码，采用静态内部类的方式。

```java
//这种形式兼顾饿汉式单例写法的内存浪费问题和 synchronized 的性能问题
//完美地屏蔽了这两个缺点
public class LazyStaticInnerClassSingleton {
    //使用 LazyInnerClassGeneral 的时候，默认会先初始化内部类
    //如果没使用，则内部类是不加载的
    private LazyStaticInnerClassSingleton(){

    }
    //每一个关键字都不是多余的，static 是为了使单例模式的空间共享，保证这个方法不会被重写、重载
    private static LazyStaticInnerClassSingleton getInstance(){
        //在返回结果之前，一定会先加载内部类
```

```
        return LazyHolder.INSTANCE;
    }

    //利用了 Java 本身的语法特点，默认不加载内部类
    private static class LazyHolder{
        private static final LazyStaticInnerClassSingleton INSTANCE = new LazyStaticInnerClassSingleton();
    }

}
```

这种方式兼顾了饿汉式单例写法的内存浪费问题和 synchronized 的性能问题。内部类一定要在方法调用之前被初始化，巧妙地避免了线程安全问题。由于这种方式比较简单，就不再一步步调试。但是，“金无足赤，人无完人”，单例模式亦如此。这种写法就真的完美了吗？

8.2.5　还原反射破坏单例模式的事故现场

我们来看一个事故现场。大家有没有发现，上面介绍的单例模式的构造方法除了加上 private 关键字，没有做任何处理。如果使用反射来调用其构造方法，再调用 getInstance()方法，应该有两个不同的实例。现在来看客户端测试代码，以 LazyStaticInnerClassSingleton 为例。

```
public static void main(String[] args) {
    try{
        //如果有人恶意用反射破坏
        Class<?> clazz = LazyStaticInnerClassSingleton.class;

        //通过反射获取私有的构造方法
        Constructor c = clazz.getDeclaredConstructor(null);
        //强制访问
        c.setAccessible(true);

        //暴力初始化
        Object o1 = c.newInstance();

        //调用了两次构造方法，相当于“new”了两次，犯了原则性错误
        Object o2 = c.newInstance();

        System.out.println(o1 == o2);
    }catch (Exception e){
        e.printStackTrace();
    }
}
```

运行结果如下图所示。

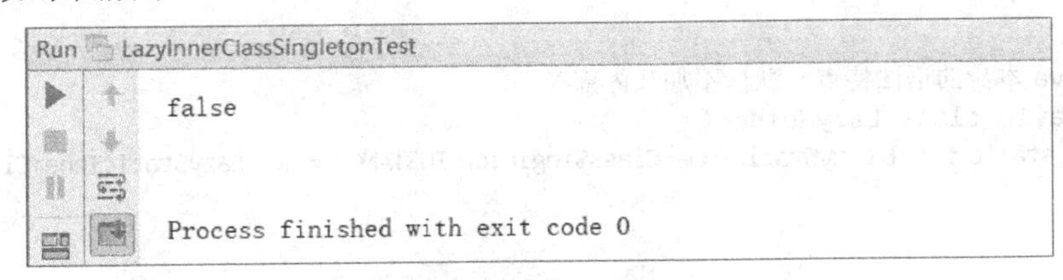

显然，内存中创建了两个不同的实例。那怎么办呢？我们来做一次优化。我们在其构造方法中做一些限制，一旦出现多次重复创建，则直接抛出异常。优化后的代码如下。

```
public class LazyStaticInnerClassSingleton {
    //使用 LazyInnerClassGeneral 的时候，默认会先初始化内部类
    //如果没使用，则内部类是不加载的
    private LazyStaticInnerClassSingleton(){
        if(LazyHolder.INSTANCE != null){
            throw new RuntimeException("不允许创建多个实例");
        }
    }
    //每一个关键字都不是多余的，static 是为了使单例模式的空间共享，保证这个方法不会被重写、重载
    private static LazyStaticInnerClassSingleton getInstance(){
        //在返回结果之前，一定会先加载内部类
        return LazyHolder.INSTANCE;
    }

    //利用了 Java 本身的语法特点，默认不加载内部类
    private static class LazyHolder{
        private static final LazyStaticInnerClassSingleton INSTANCE = new LazyStaticInnerClassSingleton();
    }

}
```

再运行客户端测试代码，结果如下图所示。

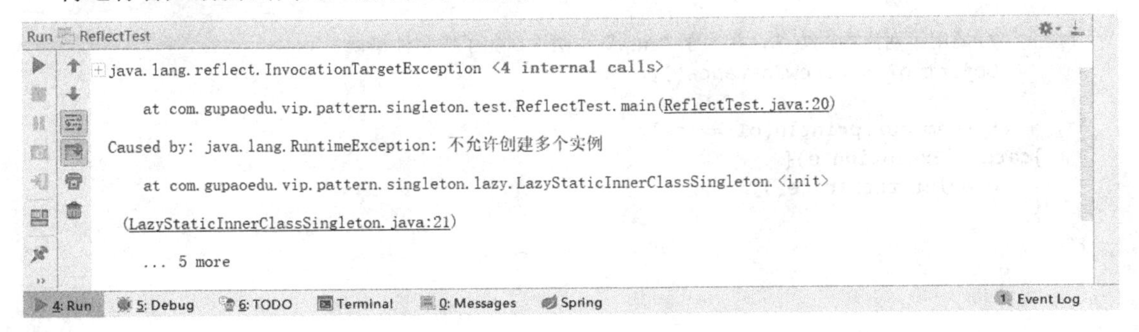

至此，自认为最优雅的单例模式写法便大功告成了。但是，上面看似完美的单例写法还是值得斟酌的。在构造方法中抛出异常，显然不够优雅。那么有没有比静态内部类更优雅的单例写法呢？

8.2.6 更加优雅的枚举式单例写法问世

枚举式单例写法可以解决上面的问题。首先来看枚举式单例的标准写法，创建 EnumSingleton 类。

```
public enum EnumSingleton {
    INSTANCE;
    private Object data;
    public Object getData() {
        return data;
    }
    public void setData(Object data) {
        this.data = data;
    }
    public static EnumSingleton getInstance(){
        return INSTANCE;
    }
}
```

然后看客户端测试代码。

```
public class EnumSingletonTest {
    public static void main(String[] args) {
        try {
            EnumSingleton instance1 = null;

            EnumSingleton instance2 = EnumSingleton.getInstance();
            instance2.setData(new Object());

            FileOutputStream fos = new FileOutputStream("EnumSingleton.obj");
            ObjectOutputStream oos = new ObjectOutputStream(fos);
            oos.writeObject(instance2);
            oos.flush();
            oos.close();

            FileInputStream fis = new FileInputStream("EnumSingleton.obj");
            ObjectInputStream ois = new ObjectInputStream(fis);
            instance1 = (EnumSingleton) ois.readObject();
            ois.close();

            System.out.println(instance1.getData());
```

```
            System.out.println(instance2.getData());
            System.out.println(instance1.getData() == instance2.getData());

        }catch (Exception e){
            e.printStackTrace();
        }
    }
}
```

最后得到运行结果，如下图所示。

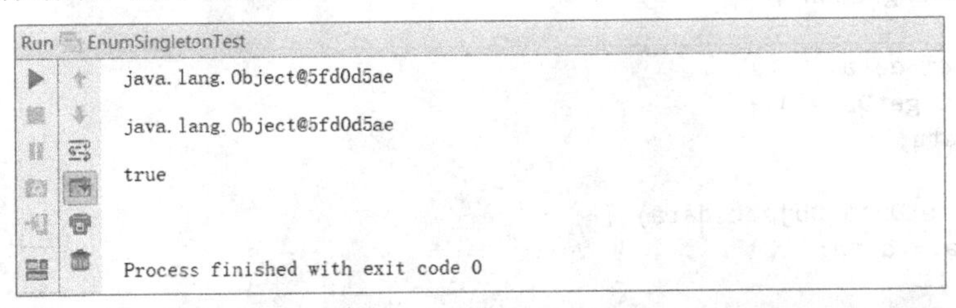

我们没有对代码逻辑做任何处理，但运行结果和预期一样。那么枚举式单例写法如此神奇，它的神秘之处体现在哪里呢？下面通过分析源码来揭开它的神秘面纱。

首先下载一个非常好用的 Java 反编译工具 Jad，在解压后配置好环境变量（这里不做详细介绍），就可以使用命令行调用了。找到工程所在的 Class 目录，复制 EnumSingleton.class 所在的路径，如下图所示。

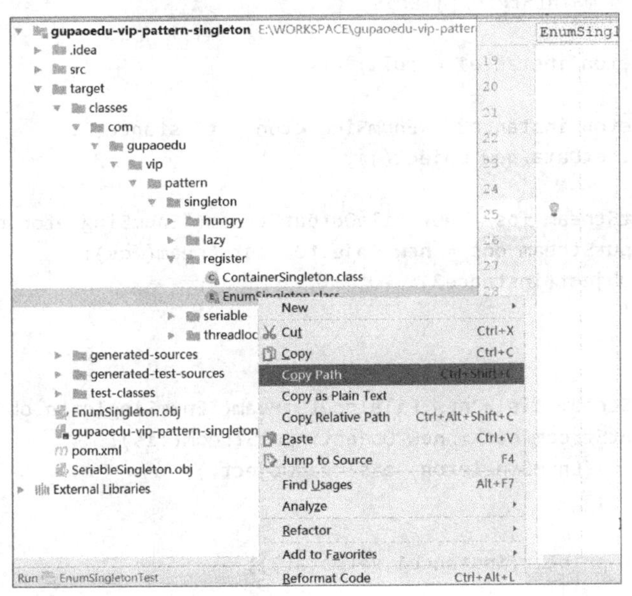

然后切换到命令行，切换到工程所在的 Class 目录，输入命令 jad 并输入复制好的路径，在 Class 目录下会多出一个 EnumSingleton.jad 文件。打开 EnumSingleton.jad 文件，我们惊奇地发现有如下代码。

```
static
{
    INSTANCE = new EnumSingleton("INSTANCE", 0);
    $VALUES = (new EnumSingleton[] {
        INSTANCE
    });
}
```

原来，枚举式单例写法在静态块中就对 INSTANCE 进行了赋值，是饿汉式单例写法的实现。至此，我们还可以试想，序列化能否破坏枚举式单例写法呢？不妨再来看一下 JDK 源码，还是回到 ObjectInputStream 的 readObject0()方法。

```
private Object readObject0(boolean unshared) throws IOException {
        ...

        case TC_ENUM:
            return checkResolve(readEnum(unshared));

        ...
}
```

我们看到，在 readObject0()中调用了 readEnum()方法，readEnum()方法的代码实现如下。

```
private Enum<?> readEnum(boolean unshared) throws IOException {
    if (bin.readByte() != TC_ENUM) {
        throw new InternalError();
    }

    ObjectStreamClass desc = readClassDesc(false);
    if (!desc.isEnum()) {
        throw new InvalidClassException("non-enum class: " + desc);
    }

    int enumHandle = handles.assign(unshared ? unsharedMarker : null);
    ClassNotFoundException resolveEx = desc.getResolveException();
    if (resolveEx != null) {
        handles.markException(enumHandle, resolveEx);
    }

    String name = readString(false);
    Enum<?> result = null;
```

```
    Class<?> cl = desc.forClass();
    if (cl != null) {
        try {
            @SuppressWarnings("unchecked")
            Enum<?> en = Enum.valueOf((Class)cl, name);
            result = en;
        } catch (IllegalArgumentException ex) {
            throw (IOException) new InvalidObjectException(
                "enum constant " + name + " does not exist in " +
                cl).initCause(ex);
        }
        if (!unshared) {
            handles.setObject(enumHandle, result);
        }
    }

    handles.finish(enumHandle);
    passHandle = enumHandle;
    return result;
}
```

由上可知，枚举类型其实通过类名和类对象找到一个唯一的枚举对象。因此，枚举对象不可能被类加载器加载多次。那么反射是否能破坏枚举式单例写法的单例对象呢？来看客户端测试代码。

```
public static void main(String[] args) {
    try {
        Class clazz = EnumSingleton.class;
        Constructor c = clazz.getDeclaredConstructor();
        c.newInstance();
    }catch (Exception e){
        e.printStackTrace();
    }
}
```

运行结果如下图所示。

```
Run  EnumSingletonTest
java.lang.NoSuchMethodException: com.gupaoedu.vip.pattern.singleton.register.EnumSingleton.<init>()
    at java.lang.Class.getConstructor0(Class.java:3082)
    at java.lang.Class.getDeclaredConstructor(Class.java:2178)
    at com.gupaoedu.vip.pattern.singleton.test.EnumSingletonTest.main(EnumSingletonTest.java:46)

Process finished with exit code 0
```

结果中报出的是 java.lang.NoSuchMethodException 异常，意思是没找到无参的构造方法。此时，打开 java.lang.Enum 的源码，查看它的构造方法，只有一个 protected 类型的构造方法，代码如下。

```
protected Enum(String name, int ordinal) {
    this.name = name;
    this.ordinal = ordinal;
}
```

再来做一个这样的测试。

```
public static void main(String[] args) {
    try {
        Class clazz = EnumSingleton.class;
        Constructor c = clazz.getDeclaredConstructor(String.class,int.class);
        c.setAccessible(true);
        EnumSingleton enumSingleton = (EnumSingleton)c.newInstance("Tom",666);

    }catch (Exception e){
        e.printStackTrace();
    }
}
```

运行结果如下图所示。

```
Run  EnumSingletonTest
java.lang.IllegalArgumentException: Cannot reflectively create enum objects <1 internal calls>
    at com.gupaoedu.vip.pattern.singleton.test.EnumSingletonTest.main(EnumSingletonTest.java:59)

Process finished with exit code 0
```

这时，错误已经非常明显了，“Cannot reflectively create enum objects”，即不能用反射来创建枚举类型。我们还是习惯性地想来看下 JDK 源码，进入 Constructor 的 newInstance()方法。

```
public T newInstance(Object ... initargs)
    throws InstantiationException, IllegalAccessException,
           IllegalArgumentException, InvocationTargetException
{
    if (!override) {
        if (!Reflection.quickCheckMemberAccess(clazz, modifiers)) {
            Class<?> caller = Reflection.getCallerClass();
            checkAccess(caller, clazz, null, modifiers);
        }
    }
    if ((clazz.getModifiers() & Modifier.ENUM) != 0)
        throw new IllegalArgumentException("Cannot reflectively create enum objects");
```

```
    ConstructorAccessor ca = constructorAccessor;
    if (ca == null) {
        ca = acquireConstructorAccessor();
    }
    @SuppressWarnings("unchecked")
    T inst = (T) ca.newInstance(initargs);
    return inst;
}
```

从上述代码可以看到，在 newInstance() 方法中做了强制性的判断，如果修饰符是 Modifier.ENUM 枚举类型，则直接抛出异常。这岂不是和静态内部类单例写法的处理方式有异曲同工之妙？对，但是我们在构造方法中写逻辑处理可能存在未知的风险，而 JDK 的处理是最官方、最权威、最稳定的。因此，枚举式单例写法也是 *Effective Java* 一书中推荐的一种单例模式写法。

到此为止，我们是不是已经非常清晰明了了呢？JDK 枚举的语法特殊性及反射也为枚举保驾护航，让枚举式单例写法成为一种更加优雅的实现。

8.2.7 还原反序列化破坏单例模式的事故现场

一个单例对象创建好后，有时候需要将对象序列化然后写入磁盘，当下次使用时再从磁盘中读取对象并进行反序列化，将其转化为内存对象。反序列化后的对象会重新分配内存，即重新创建。如果序列化的目标对象为单例对象，则违背了单例模式的初衷，相当于破坏了单例模式，来看一段代码。

```
//反序列化破坏了单例模式
public class SeriableSingleton implements Serializable {
    //序列化就是把内存中的状态通过转换成字节码的形式
    //从而转换为一个 I/O 流，写入其他地方（可以是磁盘、网络 I/O）
    //内存中的状态会被永久保存下来

    //反序列化就是将已经持久化的字节码内容转换为 I/O 流
    //通过 I/O 流的读取，进而将读取的内容转换为 Java 对象
    //在转换过程中会重新创建对象
    public  final static SeriableSingleton INSTANCE = new SeriableSingleton();
    private SeriableSingleton(){}

    public static SeriableSingleton getInstance(){
        return INSTANCE;
    }
}
```

编写客户端测试代码。

```
    public static void main(String[] args) {
```

```
        SeriableSingleton s1 = null;
        SeriableSingleton s2 = SeriableSingleton.getInstance();

        FileOutputStream fos = null;
        try {
            fos = new FileOutputStream("SeriableSingleton.obj");
            ObjectOutputStream oos = new ObjectOutputStream(fos);
            oos.writeObject(s2);
            oos.flush();
            oos.close();

            FileInputStream fis = new FileInputStream("SeriableSingleton.obj");
            ObjectInputStream ois = new ObjectInputStream(fis);
            s1 = (SeriableSingleton)ois.readObject();
            ois.close();

            System.out.println(s1);
            System.out.println(s2);
            System.out.println(s1 == s2);

        } catch (Exception e) {
            e.printStackTrace();
        }
    }
```

运行结果如下图所示。

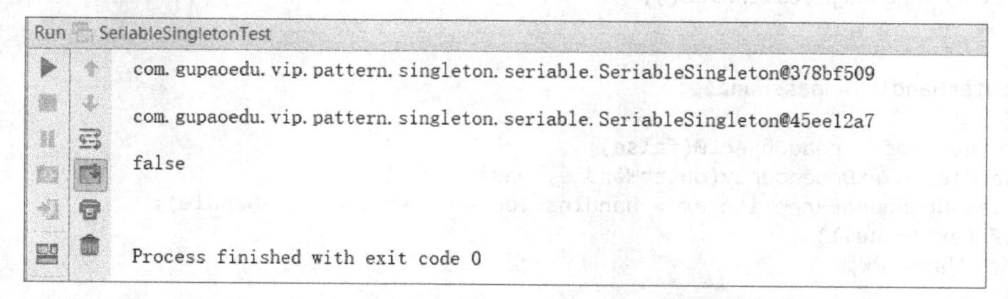

从运行结果可以看出，反序列化后的对象和手动创建的对象是不一致的，被实例化了两次，违背了单例模式的设计初衷。那么，如何保证在序列化的情况下也能够实现单例模式呢？其实很简单，只需要增加 readResolve()方法即可。优化后的代码如下。

```
public class SeriableSingleton implements Serializable {

    public  final static SeriableSingleton INSTANCE = new SeriableSingleton();
    private SeriableSingleton(){}

    public static SeriableSingleton getInstance(){
```

```
        return INSTANCE;
    }
    private  Object readResolve(){
        return  INSTANCE;
    }
}
```

再看运行结果，如下图所示。

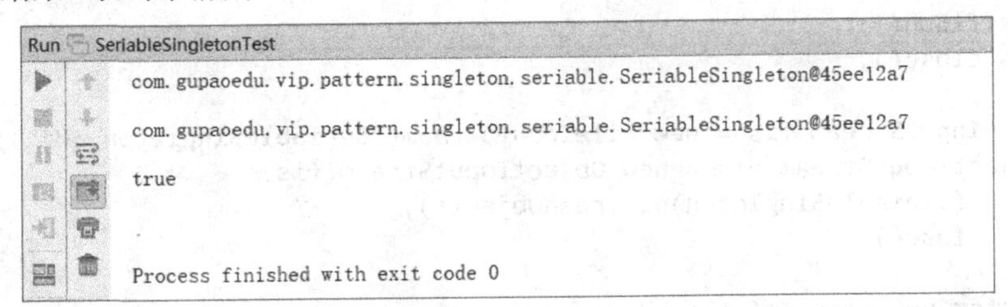

大家一定会想：这是什么原因呢？为什么要这样写？看上去很神奇的样子，也让人有些费解。不如一起来看 JDK 的源码实现以了解清楚。进入 ObjectInputStream 类的 readObject()方法，代码如下。

```
public final Object readObject()
    throws IOException, ClassNotFoundException
{
    if (enableOverride) {
        return readObjectOverride();
    }

    int outerHandle = passHandle;
    try {
        Object obj = readObject0(false);
        handles.markDependency(outerHandle, passHandle);
        ClassNotFoundException ex = handles.lookupException(passHandle);
        if (ex != null) {
            throw ex;
        }
        if (depth == 0) {
            vlist.doCallbacks();
        }
        return obj;
    } finally {
        passHandle = outerHandle;
        if (closed && depth == 0) {
            clear();
        }
    }
}
```

可以看到，在 readObject()方法中又调用了重写的 readObject0()方法。进入 readObject0()方法，源码如下。

```
private Object readObject0(boolean unshared) throws IOException {
        ...

        case TC_OBJECT:
            return checkResolve(readOrdinaryObject(unshared));

        ...
}
```

我们看到 TC_OBJECT 中调用了 ObjectInputStream 的 readOrdinaryObject()方法，源码如下。

```
private Object readOrdinaryObject(boolean unshared)
    throws IOException
{
    if (bin.readByte() != TC_OBJECT) {
        throw new InternalError();
    }

    ObjectStreamClass desc = readClassDesc(false);
    desc.checkDeserialize();

    Class<?> cl = desc.forClass();
    if (cl == String.class || cl == Class.class
            || cl == ObjectStreamClass.class) {
        throw new InvalidClassException("invalid class descriptor");
    }

    Object obj;
    try {
        obj = desc.isInstantiable() ? desc.newInstance() : null;
    } catch (Exception ex) {
        throw (IOException) new InvalidClassException(
            desc.forClass().getName(),
            "unable to create instance").initCause(ex);
    }

    ...

    return obj;
}
```

我们发现调用了 ObjectStreamClass 的 isInstantiable()方法，而 isInstantiable()方法的源码如下。

```
boolean isInstantiable() {
```

```
    requireInitialized();
    return (cons != null);
}
```

上述代码非常简单，就是判断一下构造方法是否为空。如果构造方法不为空，则返回 true。这意味着只要有无参构造方法就会实例化。

这时候其实还没有找到加上 readResolve()方法就可以避免单例模式被破坏的真正原因。再回到 ObjectInputStream 的 readOrdinaryObject()方法，继续往下看源码。

```
private Object readOrdinaryObject(boolean unshared)
    throws IOException
{
    if (bin.readByte() != TC_OBJECT) {
        throw new InternalError();
    }

    ObjectStreamClass desc = readClassDesc(false);
    desc.checkDeserialize();

    Class<?> cl = desc.forClass();
    if (cl == String.class || cl == Class.class
            || cl == ObjectStreamClass.class) {
        throw new InvalidClassException("invalid class descriptor");
    }

    Object obj;
    try {
        obj = desc.isInstantiable() ? desc.newInstance() : null;
    } catch (Exception ex) {
        throw (IOException) new InvalidClassException(
            desc.forClass().getName(),
            "unable to create instance").initCause(ex);
    }

    ...

    if (obj != null &&
        handles.lookupException(passHandle) == null &&
        desc.hasReadResolveMethod())
    {
        Object rep = desc.invokeReadResolve(obj);
        if (unshared && rep.getClass().isArray()) {
            rep = cloneArray(rep);
        }
        if (rep != obj) {
            if (rep != null) {
```

```
                if (rep.getClass().isArray()) {
                    filterCheck(rep.getClass(), Array.getLength(rep));
                } else {
                    filterCheck(rep.getClass(), -1);
                }
            }
            handles.setObject(passHandle, obj = rep);
        }
    }

    return obj;
}
```

在判断无参构造方法是否存在之后，又调用了 hasReadResolveMethod()方法，源码如下。

```
boolean hasReadResolveMethod() {
    requireInitialized();
    return (readResolveMethod != null);
}
```

上述代码的逻辑非常简单，就是判断 readResolveMethod 是否为空，如果不为空，则返回 true。那么 readResolveMethod 是在哪里被赋值的呢？通过全局查找知道，在私有方法 ObjectStreamClass()中对 readResolveMethod 进行了赋值，源码如下。

```
readResolveMethod = getInheritableMethod(
    cl, "readResolve", null, Object.class);
```

上面的逻辑其实就是通过反射找到一个无参的 readResolve()方法，并且保存下来。再回到 ObjectInputStream 的 readOrdinaryObject()方法，继续往下看，如果 readResolve()方法存在，则调用 invokeReadResolve()方法，代码如下。

```
Object invokeReadResolve(Object obj)
    throws IOException, UnsupportedOperationException
{
    requireInitialized();
    if (readResolveMethod != null) {
        try {
            return readResolveMethod.invoke(obj, (Object[]) null);
        } catch (InvocationTargetException ex) {
            Throwable th = ex.getTargetException();
            if (th instanceof ObjectStreamException) {
                throw (ObjectStreamException) th;
            } else {
                throwMiscException(th);
                throw new InternalError(th);
            }
```

```
        } catch (IllegalAccessException ex) {
            throw new InternalError(ex);
        }
    } else {
        throw new UnsupportedOperationException();
    }
}
```

可以看到，在 invokeReadResolve()方法中用反射调用了 readResolveMethod 方法。

通过 JDK 源码分析可以看出，虽然增加 readResolve()方法返回实例解决了单例模式被破坏的问题，但是实际上单例对象被实例化了两次，只不过新创建的对象没有被返回而已。如果创建对象的动作发生频率加快，则意味着内存分配开销也会随之增大，难道真的就没办法从根本上解决问题吗？其实，枚举式单例写法也是能够避免这个问题发生的，因为它在类加载的时候就已经创建好了所有的对象。

8.2.8 使用容器式单例写法解决大规模生产单例的问题

虽然枚举式单例写法更加优雅，但是也会存在一些问题。因为它在类加载时将所有的对象初始化都放在类内存中，这其实和饿汉式单例写法并无差异，不适合大量创建单例对象的场景。接下来看注册式单例模式的另一种写法，即容器式单例写法，创建 ContainerSingleton 类。

```
public class ContainerSingleton {
    private ContainerSingleton(){}
    private static Map<String,Object> ioc = new ConcurrentHashMap<String,Object>();
    public static Object getBean(String className){
        synchronized (ioc) {
            if (!ioc.containsKey(className)) {
                Object obj = null;
                try {
                    obj = Class.forName(className).newInstance();
                    ioc.put(className, obj);
                } catch (Exception e) {
                    e.printStackTrace();
                }
                return obj;
            } else {
                return ioc.get(className);
            }
        }
    }
}
```

容器式单例写法适用于需要大量创建单例对象的场景，便于管理，但它是非线程安全的。到此，注册式单例写法介绍完毕。再来看 Spring 中的容器式单例写法的源码。

```
public abstract class AbstractAutowireCapableBeanFactory extends AbstractBeanFactory
      implements AutowireCapableBeanFactory {
   /** Cache of unfinished FactoryBean instances: FactoryBean name --> BeanWrapper */
   private final Map<String, BeanWrapper> factoryBeanInstanceCache = new ConcurrentHashMap<>(16);
...
}
```

从上面代码来看，存储单例对象的容器其实就是一个 Map。

8.2.9 ThreadLocal 单例详解

最后赠送大家一个彩蛋，线程单例实现 ThreadLocal。ThreadLocal 不能保证其创建的对象是全局唯一的，但能保证在单个线程中是唯一的，是线程安全的。下面来看代码。

```
public class ThreadLocalSingleton {
   private static final ThreadLocal<ThreadLocalSingleton> threadLocalInstance =
         new ThreadLocal<ThreadLocalSingleton>(){
            @Override
            protected ThreadLocalSingleton initialValue() {
               return new ThreadLocalSingleton();
            }
         };
   private ThreadLocalSingleton(){}

   public static ThreadLocalSingleton getInstance(){
      return threadLocalInstance.get();
   }
}
```

客户端测试代码如下。

```
public static void main(String[] args) {

   System.out.println(ThreadLocalSingleton.getInstance());
   System.out.println(ThreadLocalSingleton.getInstance());
   System.out.println(ThreadLocalSingleton.getInstance());
   System.out.println(ThreadLocalSingleton.getInstance());
   System.out.println(ThreadLocalSingleton.getInstance());

   Thread t1 = new Thread(new ExectorThread());
   Thread t2 = new Thread(new ExectorThread());
   t1.start();
   t2.start();
   System.out.println("End");
}
```

运行结果如下图所示。

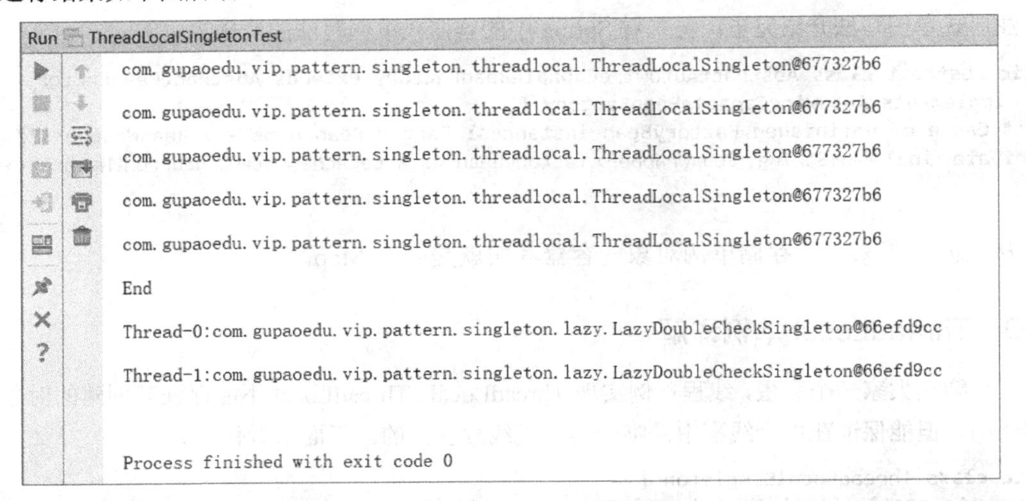

由上图可知，在主线程中无论调用多少次，获取的实例都是同一个，都在两个子线程中分别获取了不同的实例。那么，ThreadLocal 是如何实现这样的效果的呢？我们知道，单例模式为了达到线程安全的目的，会给方法上锁，以时间换空间。ThreadLocal 将所有对象全部放在 ThreadLocalMap 中，为每个线程都提供一个对象，实际上是以空间换时间来实现线程隔离的。

8.3 单例模式在框架源码中的应用

8.3.1 单例模式在 JDK 源码中的应用

首先来看 JDK 的一个经典应用 Runtime 类，源码如下。

```
public class Runtime {
    private static Runtime currentRuntime = new Runtime();

    public static Runtime getRuntime() {
        return currentRuntime;
    }

    private Runtime() {}
    ...
}
```

8.3.2　单例模式在 Spring 源码中的应用

接下来看单例模式在 Spring 中的应用，Spring 中加载单例的过程都是在 BeanFactory 的 getBean()方法中被定义的，其默认的功能实现在 AbstractBeanFactory 中，主要包含两个功能。

（1）从缓存中获取单例 Bean。

（2）从 Bean 的实例中获取对象。

getBean()方法最终会调用 AbstractBeanFactory 的 doGetBean()方法，源码如下。

```
protected <T> T doGetBean(final String name, @Nullable final Class<T> requiredType,
        @Nullable final Object[] args, boolean typeCheckOnly) throws BeansException {

    //对传入的 beanName 稍作修改，防止有一些非法字段，然后提取 Bean 的 Name
    final String beanName = transformedBeanName(name);
    Object bean;

    //直接从缓存中获取单例工厂中的 objectFactory 单例
    Object sharedInstance = getSingleton(beanName);
    if (sharedInstance != null && args == null) {
        if (logger.isDebugEnabled()) {
            if (isSingletonCurrentlyInCreation(beanName)) {
                logger.debug("Returning eagerly cached instance of singleton bean '" +
                        beanName +"' that is not fully initialized yet - a consequence of a
                        circular reference");
            }
            else {
                logger.debug("Returning cached instance of singleton bean '" + beanName + "'");
            }
        }
        //返回对应的实例，从 Bean 实例中获取对象
        //有时候存在诸如 BeanFactory 的情况，并不是直接返回实例本身，而是返回指定方法返回的实例
        bean = getObjectForBeanInstance(sharedInstance, name, beanName, null);
    }

    else {

  ...
    }
    ...
}
```

getBean()方法不仅处理单例对象的逻辑，还处理原型对象的逻辑。继续看 getSingleton()方法的代码实现。

```
/**
    * 单例对象的缓存
    */
private final Map<String, Object> singletonObjects = new ConcurrentHashMap<>(256);

protected Object getSingleton(String beanName, boolean allowEarlyReference) {
        //首先通过名字查找这个 Bean 是否存在
      Object singletonObject = this.singletonObjects.get(beanName);
      if (singletonObject == null && isSingletonCurrentlyInCreation(beanName)) {
          synchronized (this.singletonObjects) {
              //查看缓存中是否存在这个 Bean
              singletonObject = this.earlySingletonObjects.get(beanName);
              //如果这个时候的 Bean 实例还为空并且允许懒加载
              if (singletonObject == null && allowEarlyReference) {
                  ObjectFactory<?> singletonFactory = this.singletonFactories.get(beanName);
                  if (singletonFactory != null) {
                      singletonObject = singletonFactory.getObject();
                      this.earlySingletonObjects.put(beanName, singletonObject);
                      this.singletonFactories.remove(beanName);
                  }
              }
          }
      }
      return singletonObject;
  }
```

在上面代码片段中，synchronized (this.singletonObjects)是关键，但是前提条件isSingletonCurrentlyInCreation的返回值也是true，也就是这个Bean正在被创建。因此，第一次调用doGetBean()的时候，getSingleton()基本上都是返回null，所以会继续执行doGetBean()方法中后面的逻辑。

```
protected <T> T doGetBean(final String name, @Nullable final Class<T> requiredType,
        @Nullable final Object[] args, boolean typeCheckOnly) throws BeansException {
        ...
        //前面省略部分代码
        //获取 beanDefinition
            final RootBeanDefinition mbd = getMergedLocalBeanDefinition(beanName);
            checkMergedBeanDefinition(mbd, beanName, args);


            String[] dependsOn = mbd.getDependsOn();
            if (dependsOn != null) {
                for (String dep : dependsOn) {
                    if (isDependent(beanName, dep)) {
                        throw new BeanCreationException(mbd.getResourceDescription(), beanName,
                                "Circular depends-on relationship between '" + beanName + "'
```

```
                    and '" + dep + "'");
                }
                registerDependentBean(dep, beanName);
                try {
                    getBean(dep);
                }
                catch (NoSuchBeanDefinitionException ex) {
                    throw new BeanCreationException(mbd.getResourceDescription(), beanName,
                            "'" + beanName + "' depends on missing bean '" + dep + "'", ex);
                }
            }
        }
        if (mbd.isSingleton()) {
            sharedInstance = getSingleton(beanName, () -> {
                try {
                    return createBean(beanName, mbd, args);
                }
                catch (BeansException ex) {

                    destroySingleton(beanName);
                    throw ex;
                }
            });
            bean = getObjectForBeanInstance(sharedInstance, name, beanName, mbd);
        }
    }
```

可以看到，在 BeanFactory 中，从 XML 中解析出来的相关配置信息被放在 BeanDefinitionMap 中，通过这个 Map 获取 RootBeanDefinition，然后执行判断语句 if (mbd.isSingleton())。如果是单例的，则接着调用 getSingleton()的重载方法，传入 mbd 参数。当从缓存中加载单例对象时，会把当前的单例对象在 singletonObjects 中存放一份，这样可以保证在调用 getBean()方法的时候，singletonObjects 中永远只有一个实例，在获取对象时才会给它分配内存，既保证了内存高效利用，又是线程安全的。

```
public Object getSingleton(String beanName, ObjectFactory<?> singletonFactory) {
    Assert.notNull(beanName, "Bean name must not be null");
    synchronized (this.singletonObjects) {
//直接从缓存中获取单例 Bean
        Object singletonObject = this.singletonObjects.get(beanName);
        if (singletonObject == null) {
            if (this.singletonsCurrentlyInDestruction) {
                throw new BeanCreationNotAllowedException(beanName,
                        "Singleton bean creation not allowed while singletons of this factory
                        are in destruction " +
                        "(Do not request a bean from a BeanFactory in a destroy method
```

```
                    implementation!)");
            }
            if (logger.isDebugEnabled()) {
                logger.debug("Creating shared instance of singleton bean '" + beanName + "'");
            }
            beforeSingletonCreation(beanName);
            boolean newSingleton = false;
            boolean recordSuppressedExceptions = (this.suppressedExceptions == null);
            if (recordSuppressedExceptions) {
                this.suppressedExceptions = new LinkedHashSet<>();
            }
            try {
                singletonObject = singletonFactory.getObject();
                newSingleton = true;
            }
            catch (IllegalStateException ex) {

                singletonObject = this.singletonObjects.get(beanName);
                if (singletonObject == null) {
                    throw ex;
                }
            }
            catch (BeanCreationException ex) {
                if (recordSuppressedExceptions) {
                    for (Exception suppressedException : this.suppressedExceptions) {
                        ex.addRelatedCause(suppressedException);
                    }
                }
                throw ex;
            }
            finally {
                if (recordSuppressedExceptions) {
                    this.suppressedExceptions = null;
                }
                afterSingletonCreation(beanName);
            }
            if (newSingleton) {
                //在 singletonObject 中添加要加载的单例
                addSingleton(beanName, singletonObject);
            }
        }
        return singletonObject;
    }
}
```

如此一来，当下次需要这个单例 Bean 时，可以直接从缓存中获取。在 Spring 中创建单例的过程虽然有点绕，但是逻辑非常清楚，就是将需要的对象放在 Map 中，下次需要的时候直接从 Map 中获取即可。

8.4　单例模式扩展

8.4.1　单例模式的优点

（1）单例模式可以保证内存里只有一个实例，减少了内存的开销。

（2）可以避免对资源的多重占用。

（3）单例模式设置全局访问点，可以优化和共享资源的访问。

8.4.2　单例模式的缺点

（1）单例模式一般没有接口，扩展困难。如果要扩展，则除了修改原来的代码，没有第二种途径，违背开闭原则。

（2）在并发测试中，单例模式不利于代码调试。在调试过程中，如果单例中的代码没有执行完，也不能模拟生成一个新的对象。

（3）单例模式的功能代码通常写在一个类中，如果功能设计不合理，则很容易违背单一职责原则。

单例模式看起来非常简单，实现起来也非常简单。但是，单例模式在面试中也是一个高频面试题。希望通过本章的学习，能够让大家颠覆以往的认知，对单例模式有非常深刻的认识。掌握单例模式，能够让大家在面试中彰显技术深度，提升核心竞争力，给面试加分，顺利拿到 Offer。

第9章 原型模式

9.1 原型模式概述

9.1.1 原型模式的定义

原型模式（Prototype Pattern）指原型实例指定创建对象的种类，并且通过复制这些原型创建新的对象，属于创建型设计模式。

原文：Specify the kinds of objects to create using a prototypical instance, and create new objects by copying this prototype.

原型模式的核心在于复制原型对象。以系统中已存在的一个对象为原型，直接基于内存二进制流进行复制，不需要再经历耗时的对象初始化过程（不调用构造函数），性能提升许多。当对象的构建过程比较耗时时，可以把当前系统中已存在的对象作为原型，对其进行复制（一般是基于二进制流的复制），躲避初始化过程，使得新对象的创建时间大大缩短。

9.1.2 原型模式的应用场景

你一定遇到过大篇幅 Getter、Setter 赋值的场景，例如下面的代码。

```
import lombok.Data;

@Data
public class ExamFromVo{

   //省略属性设计
   ...

   public ExamFromVo copy(){

      ExamFromVo vo = new ExamFromVo();
      //剩余时间
      vo.setLeavTime(this.getLeavTime());
      //单位主键
      vo.setOrganizationId(this.getOrganizationId());
      //考试主键
      vo.setId(this.getId());
      //用户主键
      vo.setUserId(this.getUserId());
      //专业代码
      vo.setSpecialtyCode(this.getSpecialtyCode());
      //报考岗位
      vo.setPostionCode(this.getPostionCode());
      //报考等级
      vo.setGradeCode(this.getGradeCode());
      //考试开始时间
      vo.setExamStartTime(this.getExamStartTime());
      //考试结束时间
      vo.setExamEndTime(this.getExamEndTime());

      ...

      return vo;
   }

}
```

上面的代码看上去非常工整，命名非常规范，注释也写得很全面。但是，大家觉得这样的代码优雅吗？笔者认为，这样的代码属于纯体力劳动。那原型模式能够帮助我们大量地避免这种纯体力劳动，提高开发效率。

原型模式主要适用于以下应用场景。

（1） 创建对象成本较大（例如，初始化时间长，占用 CPU 太多，或者占用网络资源太多等），需要优化资源。

（2）创建一个对象需要烦琐的数据准备或访问权限等，需要提高性能或者提高安全性。

（3）系统中大量使用该类对象，且各个调用者都需要给它的属性重新赋值。

在 Spring 中，原型模式应用得非常广泛，例如 scope="prototype"、JSON.parseObject()，都是原型模式的具体应用。

9.1.3 原型模式的 UML 类图

原型模式的 UML 类图如下。

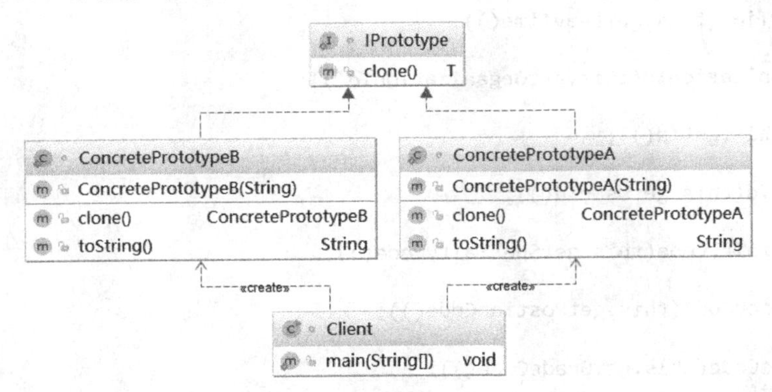

由上图可以看到，原型模式主要包含 3 个角色。

（1）客户（Client）：客户类提出创建对象的请求。

（2）抽象原型（IPrototype）：规定复制接口。

（3）具体原型（ConcretePrototype）：被复制的对象。

> 注：不是通过 new 关键字而是通过对象复制来实现创建对象的模式被称作原型模式。

9.1.4 原型模式的通用写法

按照原型模式的 UML 类图，实现原型模式的通用代码如下。

```
class Client {
```

```
//客户类
public static void main(String[] args) {
    //创建原型对象
    ConcretePrototypeA prototypeA = new ConcretePrototypeA("originalA");
    System.out.println(prototypeA);
    //复制原型对象
    ConcretePrototypeA cloneTypeA = prototypeA.clone();
    cloneTypeA.desc = "clone";
    System.out.println(cloneTypeA);

    //创建原型对象
    ConcretePrototypeB prototypeB = new ConcretePrototypeB("originalB");

    //复制原型对象
    ConcretePrototypeB cloneTypeB = prototypeB.clone();
    System.out.println(cloneTypeB);

}

//抽象原型
interface IPrototype<T> {
    T clone();
}

//具体原型
static class ConcretePrototypeB implements IPrototype<ConcretePrototypeB> {
    private String desc;

    public ConcretePrototypeB(String desc) {
        this.desc = desc;
    }

    @Override
    public ConcretePrototypeB clone() {
        //进行复制
        return new ConcretePrototypeB(this.desc);
    }

    @Override
    public String toString() {
        return "ConcretePrototypeB{" +
                "desc='" + desc + '\'' +
                '}';
    }
}

//具体原型
```

```
    static class ConcretePrototypeA implements IPrototype<ConcretePrototypeA> {
        private String desc;

        public ConcretePrototypeA(String desc) {
            this.desc = desc;
        }

        @Override
        public ConcretePrototypeA clone() {
            //进行复制
            return new ConcretePrototypeA(this.desc);
        }

        @Override
        public String toString() {
            return "ConcretePrototypeA{" +
                    "desc='" + desc + '\'' +
                    '}';
        }
    }
}
```

9.2 使用原型模式解决实际问题

9.2.1 分析 JDK 浅克隆 API 带来的问题

在 Java 提供的 API 中，不需要手动创建抽象原型接口，因为 Java 已经内置了 Cloneable 抽象原型接口，自定义的类型只需实现该接口并重写 Object.clone()方法即可完成本类的复制。

通过查看 JDK 的源码可以发现，其实 Cloneable 是一个空接口。Java 之所以提供 Cloneable 接口，只是为了在运行时通知 Java 虚拟机可以安全地在该类上使用 clone()方法。而如果该类没有实现 Cloneable 接口，则调用 clone()方法会抛出 CloneNotSupportedException 异常。

一般情况下，如果使用 clone()方法，则需满足以下条件。

（1）对任何对象 o，都有 o.clone() != o。换言之，克隆对象与原型对象不是同一个对象。

（2）对任何对象 o，都有 o.clone().getClass() == o.getClass()。换言之，克隆对象与原型对象的类型一样。

（3）如果对象 o 的 equals()方法定义恰当，则 o.clone().equals(o)应当成立。

我们在设计自定义类的 clone()方法时，应当遵守这 3 个条件。一般来说，这 3 个条件中的前 2 个是必需的，第 3 个是可选的。

下面使用 Java 提供的 API 应用来实现原型模式，代码如下。

```
class Client {
    public static void main(String[] args) {
        //创建原型对象
        ConcretePrototype type = new ConcretePrototype("original");
        System.out.println(type);
        //复制原型对象
        ConcretePrototype cloneType = type.clone();
        cloneType.desc = "clone";
        System.out.println(cloneType);

    }
    static class ConcretePrototype implements Cloneable {
        private String desc;

        public ConcretePrototype(String desc) {
            this.desc = desc;
        }

        @Override
        protected ConcretePrototype clone() {
            ConcretePrototype cloneType = null;
            try {
                cloneType = (ConcretePrototype) super.clone();
            } catch (CloneNotSupportedException e) {
                e.printStackTrace();
            }
            return cloneType;
        }

        @Override
        public String toString() {
            return "ConcretePrototype{" +
                    "desc='" + desc + '\'' +
                    '}';
        }
    }
}
```

super.clone()方法直接从堆内存中以二进制流的方式进行复制，重新分配一个内存块，因此其效率很高。由于 super.clone()方法基于内存复制，因此不会调用对象的构造函数，也就是不需要经历初始化过程。

在日常开发中，使用 super.clone()方法并不能满足所有需求。如果类中存在引用对象属性，则原型对象与克隆对象的该属性会指向同一对象的引用。

```
@Data
public class ConcretePrototype implements Cloneable {

    private int age;
    private String name;
    private List<String> hobbies;

    @Override
    public ConcretePrototype clone() {
        try {
            return (ConcretePrototype)super.clone();
        } catch (CloneNotSupportedException e) {
            e.printStackTrace();
            return null;
        }
    }

    @Override
    public String toString() {
        return "ConcretePrototype{" +
                "age=" + age +
                ", name='" + name + '\'' +
                ", hobbies=" + hobbies +
                '}';
    }
}
```

修改客户端测试代码。

```
public static void main(String[] args) {
        //创建原型对象
        ConcretePrototype prototype = new ConcretePrototype();
        prototype.setAge(18);
        prototype.setName("Tom");
        List<String> hobbies = new ArrayList<String>();
        hobbies.add("书法");
        hobbies.add("美术");
        prototype.setHobbies(hobbies);
        System.out.println(prototype);
        //复制原型对象
        ConcretePrototype cloneType = prototype.clone();
        cloneType.getHobbies().add("技术控");

        System.out.println("原型对象：" + prototype);
```

```
        System.out.println("克隆对象：" + cloneType);
    }
```

我们给克隆对象新增一个属性 hobbies（爱好）之后，发现原型对象也发生了变化，这显然不符合预期。因为我们希望克隆对象和原型对象是两个独立的对象，不再有联系。从测试结果来看，应该是 hobbies 共用了一个内存地址，意味着复制的不是值，而是引用的地址。这样的话，如果我们修改任意一个对象中的属性值，protoType 和 cloneType 的 hobbies 值都会改变。这就是我们常说的浅克隆，只是完整复制了值类型数据，没有赋值引用对象。换言之，所有的引用对象仍然指向原来的对象，显然不是我们想要的结果。那如何解决这个问题呢？

Java 自带的 clone()方法进行的就是浅克隆。而如果我们想进行深克隆，可以直接在 super.clone()后，手动给克隆对象的相关属性分配另一块内存，不过如果当原型对象维护很多引用属性的时候，手动分配会比较烦琐。因此，在 Java 中，如果想完成原型对象的深克隆，则通常使用序列化（Serializable）的方式。

9.2.2　使用序列化实现深克隆

在上节的基础上继续改造，增加一个 deepClone()方法。

```
@Data
public class ConcretePrototype implements Cloneable,Serializable {

    private int age;
    private String name;
    private List<String> hobbies;

    @Override
    public ConcretePrototype clone() {
        try {
            return (ConcretePrototype)super.clone();
        } catch (CloneNotSupportedException e) {
            e.printStackTrace();
            return null;
        }
    }

    public ConcretePrototype deepClone(){
        try {
            ByteArrayOutputStream bos = new ByteArrayOutputStream();
            ObjectOutputStream oos = new ObjectOutputStream(bos);
            oos.writeObject(this);
```

```
            ByteArrayInputStream bis = new ByteArrayInputStream(bos.toByteArray());
            ObjectInputStream ois = new ObjectInputStream(bis);

            return (ConcretePrototype)ois.readObject();
        }catch (Exception e){
            e.printStackTrace();
            return null;
        }

    }

    @Override
    public String toString() {
        return "ConcretePrototype{" +
                "age=" + age +
                ", name='" + name + '\'' +
                ", hobbies=" + hobbies +
                '}';
    }
}
```

客户端调用代码如下。

```
public static void main(String[] args) {
        //创建原型对象
        ConcretePrototype prototype = new ConcretePrototype();
        prototype.setAge(18);
        prototype.setName("Tom");
        List<String> hobbies = new ArrayList<String>();
        hobbies.add("书法");
        hobbies.add("美术");
        prototype.setHobbies(hobbies);

        //复制原型对象
        ConcretePrototype cloneType = prototype.deepCloneHobbies();
        cloneType.getHobbies().add("技术控");

        System.out.println("原型对象：" + prototype);
        System.out.println("克隆对象：" + cloneType);
        System.out.println(prototype == cloneType);

        System.out.println("原型对象的爱好：" + prototype.getHobbies());
        System.out.println("克隆对象的爱好：" + cloneType.getHobbies());
        System.out.println(prototype.getHobbies() == cloneType.getHobbies());

    }
```

运行程序，得到如下图所示的结果，与期望的结果一致。

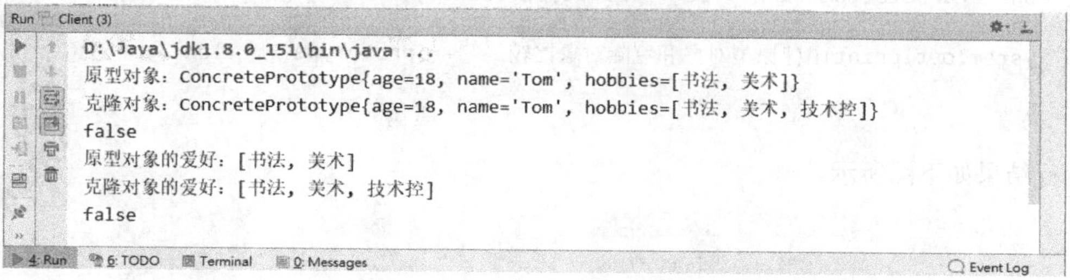

```
Run  Client (3)
D:\Java\jdk1.8.0_151\bin\java ...
原型对象：ConcretePrototype{age=18, name='Tom', hobbies=[书法, 美术]}
克隆对象：ConcretePrototype{age=18, name='Tom', hobbies=[书法, 美术, 技术控]}
false
原型对象的爱好：[书法, 美术]
克隆对象的爱好：[书法, 美术, 技术控]
false
4: Run   6: TODO   Terminal   0: Messages   Event Log
```

从运行结果来看，我们的确完成了深克隆。

9.2.3　还原克隆破坏单例的事故现场

假设有这样一个场景，如果复制的目标对象恰好是单例对象，那会不会使单例对象被破坏呢？当然，我们在已知的情况下肯定不会这么干，但如果发生了意外怎么办？不妨来修改一下代码。

```
@Data
public class ConcretePrototype implements Cloneable {

    private static  ConcretePrototype instance = new ConcretePrototype();

    private ConcretePrototype(){}

    public static ConcretePrototype getInstance(){
        return instance;
    }

    @Override
    public ConcretePrototype clone() {
        try {
            return (ConcretePrototype)super.clone();
        } catch (CloneNotSupportedException e) {
            e.printStackTrace();
            return null;
        }
    }

}
```

我们把构造方法私有化，并且提供 getInstance()方法。编写客户端测试代码如下。

```
public static void main(String[] args) {
    //创建原型对象
    ConcretePrototype prototype = ConcretePrototype.getInstance();
```

```
        //复制原型对象
        ConcretePrototype cloneType = prototype.clone();

        System.out.println("原型对象和克隆对象比较：" + (prototype == cloneType));

    }
```

运行结果如下图所示。

从运行结果来看，确实创建了两个不同的对象。实际上防止复制破坏单例对象的解决思路非常简单，禁止复制便可。要么我们的单例类不实现 Cloneable 接口，要么我们重写 clone()方法，在 clone()方法中返回单例对象即可，具体代码如下。

```
    @Override
    public ConcretePrototype clone() {
        return instance;
    }
```

9.3 原型模式在框架源码中的应用

9.3.1 原型模式在 JDK 源码中的应用

首先定义 JDK 中的 Cloneable 接口。

```
public interface Cloneable {
}
```

定义接口还是很简单的，我们找源码其实只需要看哪些接口实现了 Cloneable 即可。来看 ArrayList 类的实现。

```
public Object clone() {
    try {
        ArrayList<?> v = (ArrayList<?>) super.clone();
        v.elementData = Arrays.copyOf(elementData, size);
        v.modCount = 0;
        return v;
    } catch (CloneNotSupportedException e) {
        throw new InternalError(e);
    }
}
```

我们发现，clone()方法只是将 List 中的元素循环遍历了一遍。此时，再思考一下，是不是这种形式就是深克隆呢？其实用代码验证一下就知道了，继续修改 ConcretePrototype 类，增加一个 deepCloneHobbies()方法。

```
public class ConcretePrototype implements Cloneable,Serializable {

  ...
   public ConcretePrototype deepCloneHobbies(){
       try {
           ConcretePrototype result = (ConcretePrototype)super.clone();
           result.hobbies = (List)((ArrayList)result.hobbies).clone();
           return result;
       } catch (CloneNotSupportedException e) {
           e.printStackTrace();
           return null;
       }
   }
...
}
```

客户端代码修改如下。

```
public static void main(String[] args) {
     ...

     //复制原型对象
     ConcretePrototype cloneType = prototype.deepCloneHobbies();
     ...

   }
```

运行代码也能得到期望的结果。但是这样的代码其实是硬编码。如果在对象中声明了各种集合类型，则每种情况都需要单独处理。因此，深克隆的写法一般会直接用序列化来操作。

9.3.2　原型模式在 Spring 源码中的应用

在 Spring 中，如果用户将创建对象的方式设置为原型，则每次调用 getBean()的时候都要重新创建一个新的对象返回。和单例模式一样，还是继续来看 doGetBean()方法的代码片段。

```
protected Object doCreateBean(final String beanName, final RootBeanDefinition mbd, final
Object[] args) {
   BeanWrapper instanceWrapper = null;
      if (mbd.isSingleton()) {
         instanceWrapper = this.factoryBeanInstanceCache.remove(beanName);
```

```
        }
        if (instanceWrapper == null) {
            instanceWrapper = createBeanInstance(beanName, mbd, args);//每次都创建一个新的对象
        }
    ...
    }
```

Spring 中创建对象的方式默认采用单例模式，可以通过设置@Scope(“prototype”)注解将其改为原型模式。但是，采用单例模式创建或者采用原型模式创建，只能二选一。

9.4 原型模式扩展

9.4.1 原型模式的优点

（1）Java 自带的原型模式基于内存二进制流的复制，在性能上比直接 new 一个对象更加优良。

（2）可以使用深克隆方式保存对象的状态，使用原型模式将对象复制一份，并将其状态保存起来，简化了创建对象的过程，以便在需要的时候使用（例如恢复到历史某一状态），可辅助实现撤销操作。

9.4.2 原型模式的缺点

（1）需要为每一个类都配置一个 clone 方法。

（2）clone 方法位于类的内部，当对已有类进行改造的时候，需要修改代码，违背了开闭原则。

（3）当实现深克隆时，需要编写较为复杂的代码，而且当对象之间存在多重嵌套引用时，为了实现深克隆，每一层对象对应的类都必须支持深克隆，实现起来会比较麻烦。因此，深克隆、浅克隆需要运用得当。

10

第 10 章 建造者模式

10.1 建造者模式概述

10.1.1 建造者模式的定义

建造者模式（Builder Pattern）将一个复杂对象的构建过程与它的表示分离，使得同样的构建过程可以创建不同的表示，属于创建型设计模式。

> **原文**：Separate the construction of a complex object from its representation so that the same construction process can create different representations.

对于用户而言，使用建造者模式只需指定需要创建的类型就可以获得对象，创建过程及细节不需要了解。根据建造者模式的定义，可以简单地理解为两层含义。

（1）构建与表示分离：构建代表对象创建，表示代表对象行为、方法，也就是将对象的创建与行为进行分离（对应到 Java 代码，其实就是使用接口规定行为，然后由具体的实现类进行构建）。

（2）创建不同的表示：也就是具备同样的行为，但是却由于构建的行为顺序不同或其他原因可以创建出不同的表示。

10.1.2 建造者模式的应用场景

建造者模式在日常生活中是很常见的。比如我们在买车的时候会有低配、标配和高配，也有可能会增配、减配。这个选择的过程，就是在运用建造者模式。

从定义来看，建造者模式和工厂模式是非常相似的，和工厂模式一样，具备创建与表示分离的特性。建造者模式唯一区别于工厂模式的是针对复杂对象的创建。也就是说，如果创建简单对象，通常都是使用工厂模式进行创建；而如果创建复杂对象，就可以考虑使用建造者模式。

当需要创建的产品具备复杂创建过程时，可以抽取出共性创建过程，然后交由具体实现类自定义创建流程，使得同样的创建行为可以生产出不同的产品，分离了创建与表示，使创建产品的灵活性大大增加。建造者模式主要适用于以下应用场景。

（1）相同的方法，不同的执行顺序，产生不同的结果。

（2）多个部件或零件，都可以装配到一个对象中，但是产生的结果又不相同。

（3）产品类非常复杂，或者产品类中不同的调用顺序产生不同的作用。

（4）初始化一个对象特别复杂，参数多，而且很多参数都具有默认值。

10.1.3 建造者模式的 UML 类图

建造者模式的 UML 类图如下。

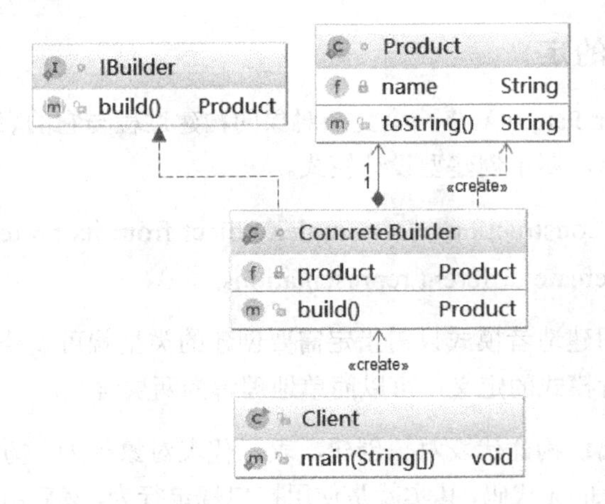

由上图可以看到，建造者模式主要包含 4 个角色。

（1）产品（Product）：要创建的产品类对象。

（2）抽象建造者（IBuilder）：建造者的抽象类，规范产品对象的各个组成部分的创建，一般由子类实现具体的创建过程。

（3）建造者（Concrete Builder）：具体的 Builder 类，根据不同的业务逻辑，具体化对象的各个组成部分的创建。

（4）调用者（Director）：调用具体的建造者，来创建对象的各个部分，在指导者中不涉及具体产品的信息，只负责保证对象各部分完整创建或者按某种顺序创建。在类图中，Client 相当于调用者的角色。

10.1.4　建造者模式的通用写法

以下是建造者模式的通用写法。

```
public class Client {
    public static void main(String[] args) {
        IBuilder builder = new ConcreteBuilder();

        System.out.println(builder.build());

    }

    @Data
    static class Product{

        private String name;

        @Override
        public String toString() {
            return "Product{" +
                    "name='" + name + '\'' +
                    '}';
        }

    }

    interface IBuilder {
        Product build();
    }

    static class ConcreteBuilder implements IBuilder {

        private Product product = new Product();
```

```
    public Product build() {
        return product;
    }
  }
}
```

10.2 使用建造者模式解决实际问题

10.2.1 建造者模式的链式写法

还是以课程为例，一个完整的课程由 PPT 课件、回放视频、课堂笔记、课后作业组成，但是这些内容的设置顺序可以随意调整，我们用建造者模式来代入理解一下。首先创建一个产品类 Course。

```
@Data
public class Course {

    private String name;
    private String ppt;
    private String video;
    private String note;

    private String homework;

    @Override
    public String toString() {
        return "CourseBuilder{" +
                "name='" + name + '\'' +
                ", ppt='" + ppt + '\'' +
                ", video='" + video + '\'' +
                ", note='" + note + '\'' +
                ", homework='" + homework + '\'' +
                '}';
    }
}
```

然后创建建造者类 CourseBuilder，将复杂的创建过程封装起来，创建步骤由用户决定。

```
public class CourseBuilder {

    private Course course = new Course();
```

```
    public CourseBuilder addName(String name){
        course.setName(name);
        return this;
    }

    public CourseBuilder addPpt(String ppt){
        course.setPpt(ppt);
        return this;
    }

    public CourseBuilder addVideo(String video){
        course.setVideo(video);
        return this;
    }

    public CourseBuilder addNote(String note){
        course.setNote(note);
        return this;
    }

    public CourseBuilder addHomework(String homework){
        course.setHomework(homework);
        return this;
    }

    public Course builder(){
        return course;
    }

}
```

最后编写客户端测试代码。

```
    public static void main(String[] args) {
        CourseBuilder builder = new CourseBuilder()
                    .addName("设计模式")
                    .addPPT("【PPT 课件】")
                    .addVideo("【回放视频】")
                    .addNote("【课堂笔记】")
                    .addHomework("【课后作业】");

        System.out.println(builder.build());
    }
```

这样的写法是不是很眼熟？后面分析建造者模式在框架源码中的应用时大家就会明白。再来看一下类图的变化，如下图所示。

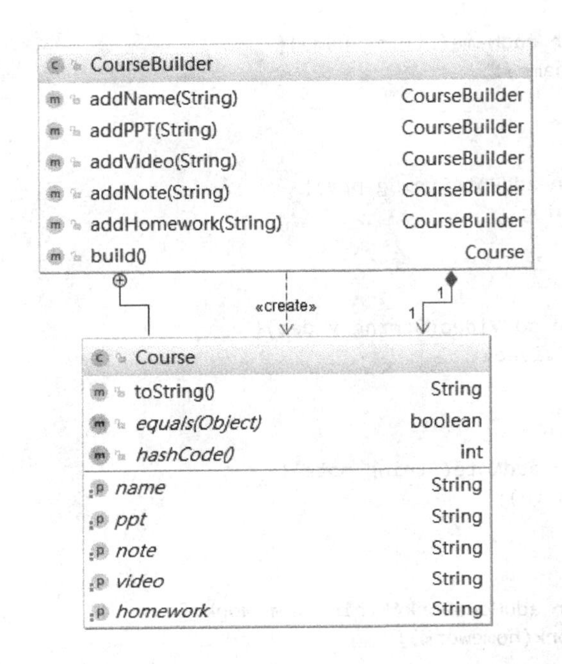

10.2.2 使用静态内部类实现建造者模式

事实上，在平常的编码中，我们通常都会忽略对象的复杂性，优先考虑使用工厂模式创建对象，而不是建造者模式。因为工厂模式和建造者模式的作用都是创建一个产品对象，而工厂模式的结构更加简洁直接（没有 Builder 和 Director），因此更常使用。

一般情况下，我们更习惯使用静态内部类的方式实现建造者模式，即一个产品类内部自动带有一个具体建造者，由它负责该产品的组装创建，不再需要 Builder 和 Director，这样，产品表示与创建之间的联系更加紧密，结构更加紧凑，同时使得建造者模式的形式更加简洁。

如果采用静态内部类形式实现建造者模式，则前面的案例可以改写如下。

```
@Data
public class Course {
    private String name;
    private String ppt;
    private String video;
    private String note;

    private String homework;

    @Override
```

```
    public String toString() {
        return "Course{" +
                "name='" + name + '\'' +
                ", ppt='" + ppt + '\'' +
                ", video='" + video + '\'' +
                ", note='" + note + '\'' +
                ", homework='" + homework + '\'' +
                '}';
    }

    public static class Builder {

        private Course course = new Course();

        public Builder addName(String name){
            course.setName(name);
            return this;
        }

        public Builder addPpt(String ppt){
            course.setPpt(ppt);
            return this;
        }

        public Builder addVideo(String video){
            course.setVideo(video);
            return this;
        }

        public Builder addNote(String note){
            course.setNote(note);
            return this;
        }

        public Builder addHomework(String homework){
            course.setHomework(homework);
            return this;
        }

        public Course builder(){
            return course;
        }

    }
}
```

客户端测试代码如下。

```
public static void main(String[] args) {
    Course course = new Course.Builder()
            .addName("设计模式")
            .addPpt("【PPT 课件】")
            .addVideo("【录播视频】")
             .builder();

    System.out.println(course);
}
```

这样，代码也会看上去更加简洁，不会让人感觉到多了一个类。

10.2.3 使用建造者模式动态构建 SQL 语句

下面来看一个实战案例，这个案例参考了开源框架 JPA 的 SQL 构造模式。我们在构造 SQL 查询条件的时候，需要根据不同的条件来拼接 SQL 字符串。如果查询条件复杂，则 SQL 拼接的过程也会变得非常复杂，从而给代码维护带来非常大的困难。因此，我们用建造者类 QueryRuleSqlBuilder 将复杂的 SQL 构造过程进行封装，用 QueryRule 对象专门保存 SQL 查询时的条件，最后根据查询条件，自动生成 SQL 语句。首先创建 QueryRule 类，代码如下。

```
import java.io.Serializable;
import java.util.ArrayList;
import java.util.List;

/**
 * QueryRule，主要功能用于构造查询条件
 */
public final class QueryRule implements Serializable
{
    private static final long serialVersionUID = 1L;
    public static final int ASC_ORDER = 101;
    public static final int DESC_ORDER = 102;
    public static final int LIKE = 1;
    public static final int IN = 2;
    public static final int NOTIN = 3;
    public static final int BETWEEN = 4;
    public static final int EQ = 5;
    public static final int NOTEQ = 6;
    public static final int GT = 7;
    public static final int GE = 8;
    public static final int LT = 9;
    public static final int LE = 10;
    public static final int ISNULL = 11;
```

```
public static final int ISNOTNULL = 12;
public static final int ISEMPTY = 13;
public static final int ISNOTEMPTY = 14;
public static final int AND = 201;
public static final int OR = 202;
private List<Rule> ruleList = new ArrayList<Rule>();
private List<QueryRule> queryRuleList = new ArrayList<QueryRule>();
private String propertyName;

private QueryRule() {}

private QueryRule(String propertyName) {
    this.propertyName = propertyName;
}

public static QueryRule getInstance() {
    return new QueryRule();
}

/**
 * 添加升序规则
 * @param propertyName
 * @return
 */
public QueryRule addAscOrder(String propertyName) {
    this.ruleList.add(new Rule(ASC_ORDER, propertyName));
    return this;
}

/**
 * 添加降序规则
 * @param propertyName
 * @return
 */
public QueryRule addDescOrder(String propertyName) {
    this.ruleList.add(new Rule(DESC_ORDER, propertyName));
    return this;
}

public QueryRule andIsNull(String propertyName) {
    this.ruleList.add(new Rule(ISNULL, propertyName).setAndOr(AND));
    return this;
}

public QueryRule andIsNotNull(String propertyName) {
    this.ruleList.add(new Rule(ISNOTNULL, propertyName).setAndOr(AND));
    return this;
```

```
}

public QueryRule andIsEmpty(String propertyName) {
    this.ruleList.add(new Rule(ISEMPTY, propertyName).setAndOr(AND));
    return this;
}

public QueryRule andIsNotEmpty(String propertyName) {
    this.ruleList.add(new Rule(ISNOTEMPTY, propertyName).setAndOr(AND));
    return this;
}

public QueryRule andLike(String propertyName, Object value) {
    this.ruleList.add(new Rule(LIKE, propertyName, new Object[] { value }).setAndOr(AND));
    return this;
}

public QueryRule andEqual(String propertyName, Object value) {
    this.ruleList.add(new Rule(EQ, propertyName, new Object[] { value }).setAndOr(AND));
    return this;
}

public QueryRule andBetween(String propertyName, Object... values) {
    this.ruleList.add(new Rule(BETWEEN, propertyName, values).setAndOr(AND));
    return this;
}

public QueryRule andIn(String propertyName, List<Object> values) {
    this.ruleList.add(new Rule(IN, propertyName, new Object[] { values }).setAndOr(AND));
    return this;
}

public QueryRule andIn(String propertyName, Object... values) {
    this.ruleList.add(new Rule(IN, propertyName, values).setAndOr(AND));
    return this;
}

public QueryRule andNotIn(String propertyName, List<Object> values) {
    this.ruleList.add(new Rule(NOTIN,
                        propertyName,
                        new Object[] { values }).setAndOr(AND));
    return this;
}

//此处省略部分代码，可以在随书源码中下载查看
```

```
public List<Rule> getRuleList() {
    return this.ruleList;
}

public List<QueryRule> getQueryRuleList() {
    return this.queryRuleList;
}

public String getPropertyName() {
    return this.propertyName;
}

protected class Rule implements Serializable {
    private static final long serialVersionUID = 1L;
    private int type;  //规则的类型
    private String property_name;
    private Object[] values;
    private int andOr = AND;

    public Rule(int paramInt, String paramString) {
        this.property_name = paramString;
        this.type = paramInt;
    }

    public Rule(int paramInt, String paramString,
            Object[] paramArrayOfObject) {
        this.property_name = paramString;
        this.values = paramArrayOfObject;
        this.type = paramInt;
    }

    public Rule setAndOr(int andOr){
        this.andOr = andOr;
        return this;
    }

    public int getAndOr(){
        return this.andOr;
    }

    public Object[] getValues() {
        return this.values;
    }

    public int getType() {
        return this.type;
    }
```

```
        public String getPropertyName() {
            return this.property_name;
        }
    }
}
```

然后创建 QueryRuleSqlBuilder 类。

```
package com.gupaoedu.vip.pattern.builder.sql;


/**
 * 根据 QueryRule 自动构建 SQL 语句
 */
public class QueryRuleSqlBuilder {
    private int CURR_INDEX = 0; //记录参数所在的位置
    private List<String> properties; //保存列名列表
    private List<Object> values; //保存参数值列表
    private List<Order> orders; //保存排序规则列表

    private String whereSql = "";
    private String orderSql = "";
    private Object [] valueArr = new Object[]{};
    private Map<Object,Object> valueMap = new HashMap<Object,Object>();

    /**
     * 获得查询条件
     * @return
     */
    private String getWhereSql(){
        return this.whereSql;
    }

    /**
     * 获得排序条件
     * @return
     */
    private String getOrderSql(){
        return this.orderSql;
    }

    /**
     * 获得参数值列表
     * @return
     */
    public Object [] getValues(){
```

```
        return this.valueArr;
    }

    /**
     * 获得参数列表
     * @return
     */
    private Map<Object,Object> getValueMap(){
        return this.valueMap;
    }

    /**
     * 创建 SQL 构造器
     * @param queryRule
     */
    public QueryRuleSqlBuilder(QueryRule queryRule) {
        CURR_INDEX = 0;
        properties = new ArrayList<String>();
        values = new ArrayList<Object>();
        orders = new ArrayList<Order>();
        for (QueryRule.Rule rule : queryRule.getRuleList()) {
            switch (rule.getType()) {
            case QueryRule.BETWEEN:
                processBetween(rule);
                break;
            case QueryRule.EQ:
                processEqual(rule);
                break;
            case QueryRule.LIKE:
                processLike(rule);
                break;
            case QueryRule.NOTEQ:
                processNotEqual(rule);
                break;
            case QueryRule.GT:
                processGreaterThen(rule);
                break;
            case QueryRule.GE:
                processGreaterEqual(rule);
                break;
            case QueryRule.LT:
                processLessThen(rule);
                break;
            case QueryRule.LE:
                processLessEqual(rule);
                break;
            case QueryRule.IN:
```

```
                processIN(rule);
                break;
            case QueryRule.NOTIN:
                processNotIN(rule);
                break;
            case QueryRule.ISNULL:
                processIsNull(rule);
                break;
            case QueryRule.ISNOTNULL:
                processIsNotNull(rule);
                break;
            case QueryRule.ISEMPTY:
                processIsEmpty(rule);
                break;
            case QueryRule.ISNOTEMPTY:
                processIsNotEmpty(rule);
                break;
            case QueryRule.ASC_ORDER:
                processOrder(rule);
                break;
            case QueryRule.DESC_ORDER:
                processOrder(rule);
                break;
            default:
                throw new IllegalArgumentException("type"+rule.getType()+"not supported.");
            }
        }
        //拼装where语句
        appendWhereSql();
        //拼装排序语句
        appendOrderSql();
        //拼装参数值
        appendValues();
    }

    /**
     * 去掉order
     *
     * @param sql
     * @return
     */
    private String removeOrders(String sql) {
        Pattern p = Pattern.compile("order\\s*by[\\w|\\W|\\s|\\S]*", Pattern.CASE_INSENSITIVE);
        Matcher m = p.matcher(sql);
        StringBuffer sb = new StringBuffer();
        while (m.find()) {
            m.appendReplacement(sb, "");
```

```
        }
        m.appendTail(sb);
        return sb.toString();
    }

    /**
     * 去掉 select
     *
     * @param sql
     * @return
     */
    private String removeSelect(String sql) {
        if(sql.toLowerCase().matches("from\\s+")){
            int beginPos = sql.toLowerCase().indexOf("from");
            return sql.substring(beginPos);
        }else{
            return sql;
        }
    }

    /**
     * 处理 like
     * @param rule
     */
    private  void processLike(QueryRule.Rule rule) {
        if (ArrayUtils.isEmpty(rule.getValues())) {
            return;
        }
        Object obj = rule.getValues()[0];

        if (obj != null) {
            String value = obj.toString();
            if (!StringUtils.isEmpty(value)) {
                value = value.replace('*', '%');
                obj = value;
            }
        }
        add(rule.getAndOr(),rule.getPropertyName(),"like","%"+rule.getValues()[0]+"%");
    }

    /**
     * 处理 between
     * @param rule
     */
    private  void processBetween(QueryRule.Rule rule) {
        if ((ArrayUtils.isEmpty(rule.getValues()))
                || (rule.getValues().length < 2)) {
```

```
            return;
        }

    add(rule.getAndOr(),rule.getPropertyName(),"","between",rule.getValues()[0],"and");
        add(0,"","","",rule.getValues()[1],"");
    }

//此处省略部分代码，可以在随书源码中下载查看

    /**
     * 加入SQL查询规则队列
     * @param andOr and或者or
     * @param key 列名
     * @param split 列名与值之间的间隔
     * @param value 值
     */
    private  void add(int andOr,String key,String split ,Object value){
        add(andOr,key,split,"",value,"");
    }

    /**
     * 加入SQL查询规则队列
     * @param andOr and或者or
     * @param key 列名
     * @param split 列名与值之间的间隔
     * @param prefix 值前缀
     * @param value 值
     * @param suffix 值后缀
     */
    private  void add(int andOr,String key,String split ,String prefix,Object value,String
    suffix){
        String andOrStr = (0 == andOr ? "" :(QueryRule.AND == andOr ? " and " : " or "));
        properties.add(CURR_INDEX,
         andOrStr + key + " " + split + prefix + (null != value ? " ? " : " ") + suffix);
        if(null != value){
            values.add(CURR_INDEX,value);
            CURR_INDEX ++;
        }
    }


    /**
     * 拼装where语句
     */
    private void appendWhereSql(){
```

```java
        StringBuffer whereSql = new StringBuffer();
        for (String p : properties) {
            whereSql.append(p);
        }
        this.whereSql = removeSelect(removeOrders(whereSql.toString()));
    }

    /**
     * 拼装排序语句
     */
    private void appendOrderSql(){
        StringBuffer orderSql = new StringBuffer();
        for (int i = 0 ; i < orders.size(); i ++) {
            if(i > 0 && i < orders.size()){
                orderSql.append(",");
            }
            orderSql.append(orders.get(i).toString());
        }
        this.orderSql = removeSelect(removeOrders(orderSql.toString()));
    }

    /**
     * 拼装参数值
     */
    private void appendValues(){
        Object [] val = new Object[values.size()];
        for (int i = 0; i < values.size(); i ++) {
            val[i] = values.get(i);
            valueMap.put(i, values.get(i));
        }
        this.valueArr = val;
    }

    public String builder(String tableName){
        String ws = removeFirstAnd(this.getWhereSql());
        String whereSql = ("".equals(ws) ? ws : (" where " + ws));
        String sql = "select * from " + tableName + whereSql;
        Object [] values = this.getValues();
        String orderSql = this.getOrderSql();
        orderSql = (StringUtils.isEmpty(orderSql) ? " " : (" order by " + orderSql));
        sql += orderSql;
        return sql;
    }


    private String removeFirstAnd(String sql){
        if(StringUtils.isEmpty(sql)){return sql;}
```

```
        return sql.trim().toLowerCase().replaceAll("^\\s*and", "") + " ";
    }

}
```

接着创建 Order 类。

```
/**
 * SQL 排序组件
 */
public class Order {
    private boolean ascending; //升序还是降序
    private String propertyName; //哪个字段升序，哪个字段降序

    public String toString() {
        return propertyName + ' ' + (ascending ? "asc" : "desc");
    }

    /**
     * Constructor for Order.
     */
    protected Order(String propertyName, boolean ascending) {
        this.propertyName = propertyName;
        this.ascending = ascending;
    }

    /**
     * Ascending order
     *
     * @param propertyName
     * @return Order
     */
    public static Order asc(String propertyName) {
        return new Order(propertyName, true);
    }

    /**
     * Descending order
     *
     * @param propertyName
     * @return Order
     */
    public static Order desc(String propertyName) {
        return new Order(propertyName, false);
    }

}
```

最后编写客户端测试代码。

```
public static void main(String[] args) {
      QueryRule queryRule = QueryRule.getInstance();
      queryRule.addAscOrder("age");
      queryRule.andEqual("addr","Changsha");
      queryRule.andLike("name","Tom");
      QueryRuleSqlBuilder builder = new QueryRuleSqlBuilder(queryRule);

      System.out.println(builder.builder("t_member"));

      System.out.println("Params: " + Arrays.toString(builder.getValues()));


   }
```

这样一来，客户端代码就非常清楚，运行结果如下图所示。

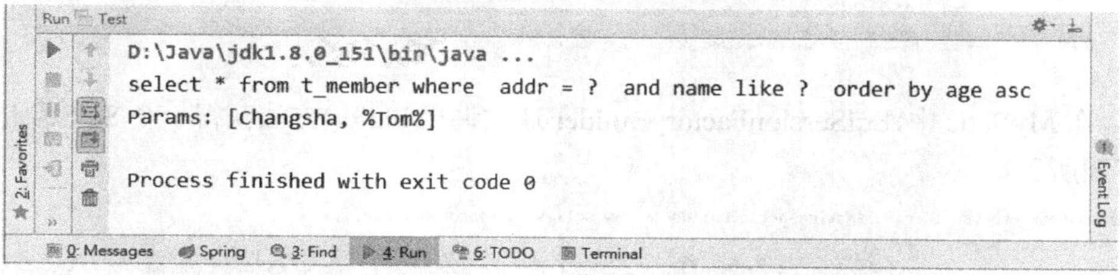
Run
Test
D:\Java\jdk1.8.0_151\bin\java ...
select * from t_member where addr = ? and name like ? order by age asc
Params: [Changsha, %Tom%]
Process finished with exit code 0
2: Favorites
Event Log
0: Messages
Spring
3: Find
4: Run
6: TODO
Terminal

10.3　建造者模式在框架源码中的应用

10.3.1　建造者模式在 JDK 源码中的应用

首先来看 JDK 的 StringBuilder，它提供 append()方法，开放构造步骤，最后调用 toString()方法就可以获得一个构造好的完整字符串，源码如下。

```
   public final class StringBuilder
   extends AbstractStringBuilder
   implements java.io.Serializable, CharSequence{

   ...

   public StringBuilder append(String str) {
      super.append(str);
      return this;
   }

   ...
}
```

10.3.2 建造者模式在 MyBatis 源码中的应用

来看 MyBatis 中的 CacheBuilder 类，如下图所示。

```
CacheBuilder cacheBuilder = new CacheBuilder( id: "");
cacheBuilder.blocking(false);
cacheBuilder.
    blocking(boolean blocking)                              CacheBuilder
    addDecorator(Class<? extends Cache> decorator)          CacheBuilder
    build()                                                 Cache
    clearInterval(Long clearInterval)                       CacheBuilder
    implementation(Class<? extends Cache> implementation)   CacheBuilder
    properties(Properties properties)                       CacheBuilder
    readWrite(boolean readWrite)                            CacheBuilder
    size(Integer size)                                      CacheBuilder
    equals(Object obj)                                      boolean
    hashCode()                                              int
Press Ctrl+句点 to choose the selected (or first) suggestion and insert a dot afterwards >>
```

同样，在 MyBatis 中，SqlSessionFactoryBuilder 通过调用 build()方法获得一个 SqlSessionFactory 类，如下图所示。

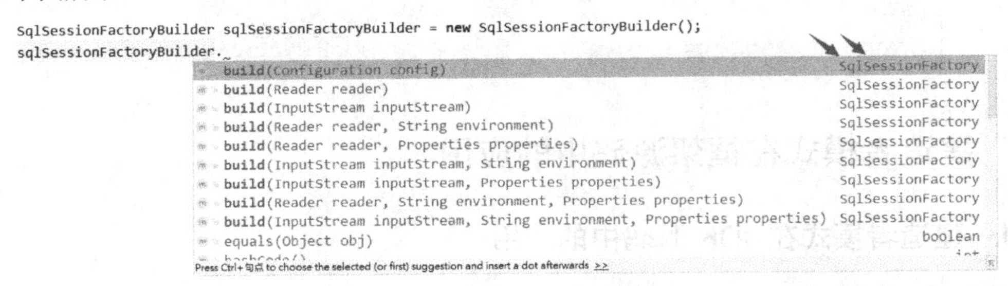

10.3.3 建造者模式在 Spring 源码中的应用

当然，在 Spring 中自然也少不了建造者模式。比如 BeanDefinitionBuilder 通过调用 getBeanDefinition()方法获得一个 BeanDefinition 对象，如下图所示。

```
BeanDefinitionBuilder beanDefinitionBuilder = BeanDefinitionBuilder.genericBeanDefinition();
beanDefinitionBuilder.
    getBeanDefinition()                                 AbstractBeanDefinition
    addConstructorArgReference(String beanName)         BeanDefinitionBuilder
    addConstructorArg(Object value)                     BeanDefinitionBuilder
    addConstructorArgValue(Object value)                BeanDefinitionBuilder
    addDependsOn(String beanName)                       BeanDefinitionBuilder
    addPropertyReference(String name, String beanName)  BeanDefinitionBuilder
    addPropertyValue(String name, Object value)         BeanDefinitionBuilder
    getRawBeanDefinition()                              AbstractBeanDefinition
    setAbstract(boolean flag)                           BeanDefinitionBuilder
    setAutowireMode(int autowireMode)                   BeanDefinitionBuilder
Press Ctrl+句点 to choose the selected (or first) suggestion and insert a dot afterwards >>
```

10.4　建造者模式扩展

10.4.1　建造者模式与工厂模式的区别

通过前面的学习，我们已经了解建造者模式，那么它和工厂模式有什么区别呢？

（1）建造者模式更加注重方法的调用顺序，工厂模式注重创建对象。

（2）创建对象的力度不同，建造者模式创建复杂的对象，由各种复杂的部件组成，工厂模式创建出来的对象都一样。

（3）关注重点不一样，工厂模式只需要把对象创建出来就可以了，而建造者模式不仅要创建出对象，还要知道对象由哪些部件组成。

（4）建造者模式根据建造过程中的顺序不一样，最终的对象部件组成也不一样。

10.4.2　建造者模式的优点

（1）封装性好，构建和表示分离。

（2）扩展性好，建造类之间独立，在一定程度上解耦。

（3）便于控制细节，建造者可以对创建过程逐步细化，而不对其他模块产生任何影响。

10.4.3　建造者模式的缺点

（1）需要多创建一个 IBuilder 对象。

（2）如果产品内部发生变化，则建造者也要同步修改，后期维护成本较大。

3

第3篇 结构型设计模式

第 11 章 代理模式

11.1 代理模式概述

11.1.1 代理模式的定义

代理模式（Proxy Pattern）指为其他对象提供一种代理，以控制对这个对象的访问，属于结构型设计模式。

原文： Provide a surrogate or placeholder for another object to control access to it.

在某些情况下，一个对象不适合或者不能直接引用另一个对象，而代理对象可以在客户端与目标对象之间起到中介的作用。

11.1.2 代理模式的应用场景

生活中的租房中介、婚姻介绍、经纪人、快递、事务代理、日志监听等，都是代理模式的实际体现。

当无法或不想直接引用某个对象或访问某个对象存在困难时，可以通过代理对象来间接访问。使用代理模式主要有两个目的：一是保护目标对象，二是增强目标对象。

11.1.3　代理模式的 UML 类图

代理模式的 UML 类图如下。

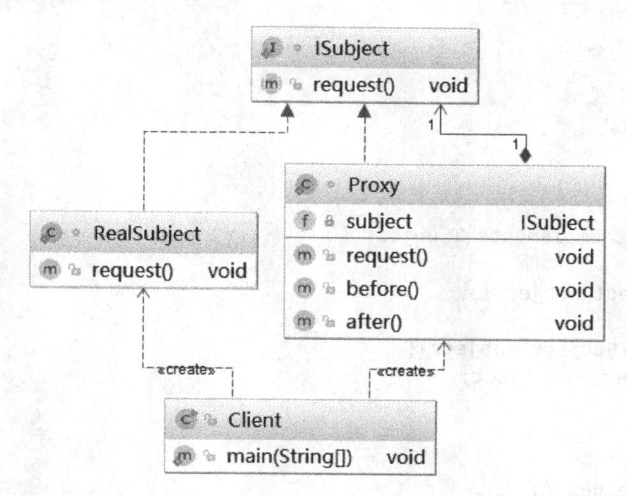

由上图可以看到，代理模式一般包含 3 个角色。

（1）抽象主题角色（ISubject）：抽象主题类的主要职责是声明真实主题与代理的共同接口方法，该类可以是接口，也可以是抽象类。

（2）真实主题角色（RealSubject）：该类也被称为被代理类，该类定义了代理所表示的真实对象，是负责执行系统的真正的逻辑业务对象。

（3）代理主题角色（Proxy）：也被称为代理类，其内部持有 RealSubject 的引用，因此具备完全的对 RealSubject 的代理权。客户端调用代理对象的方法，也调用被代理对象的方法，但是会在代理对象前后增加一些处理代码。

在代码中，一般代理会被理解为代码增强，实际上就是在原代码逻辑前后增加一些代码逻辑，而使调用者无感知。代理模式分为静态代理和动态代理。

11.1.4　代理模式的通用写法

下面是代理模式的通用代码展示，首先创建抽象主题角色 ISubject 类。

```
public class Client {

    public static void main(String[] args) {
```

```
        Proxy proxy = new Proxy(new RealSubject());
        proxy.request();

    }

    //抽象主题角色
    interface ISubject {
        void request();
    }

    //代理主题角色
    static class Proxy implements ISubject {

        private ISubject subject;

        public Proxy(ISubject subject){
            this.subject = subject;
        }


        public void request() {
            before();
            subject.request();
            after();
        }

        public void before(){
            System.out.println("called before request().");
        }

        public void after(){
            System.out.println("called after request().");
        }
    }

    //真实主题角色
    static class RealSubject implements ISubject {

        public void request() {
            System.out.println("real service is called.");
        }

    }

}
```

11.2 使用代理模式解决实际问题

11.2.1 从静态代理到动态代理

举个例子，有些人到了适婚年龄，会被父母催婚。而现在在各种压力之下，很多人都选择晚婚晚育。于是着急的父母就开始到处为子女相亲，比子女自己还着急。下面来看代码实现。创建顶层接口 IPerson 的代码如下。

```
public interface IPerson {

    void findLove();

}
```

儿子张三要找对象，实现 ZhangSan 类。

```
public class ZhangSan implements IPerson {

    public void findLove() {
        System.out.println("儿子张三提出要求");
    }

}
```

父亲张老三要帮儿子张三相亲，实现 ZhangLaosan 类。

```
public class ZhangLaosan implements IPerson {

    private ZhangSan zhangsan;

    public ZhangLaosan(ZhangSan zhangsan) {
        this.zhangsan = zhangsan;
    }

    public void findLove() {
        System.out.println("张老三开始物色");
        zhangsan.findLove();
        System.out.println("开始交往");
    }

}
```

来看客户端测试代码。

```
public class Test {
    public static void main(String[] args) {
        ZhangLaosan zhangLaosan = new ZhangLaosan(new ZhangSan());
        zhangLaosan.findLove();
    }
}
```

运行结果如下图所示。

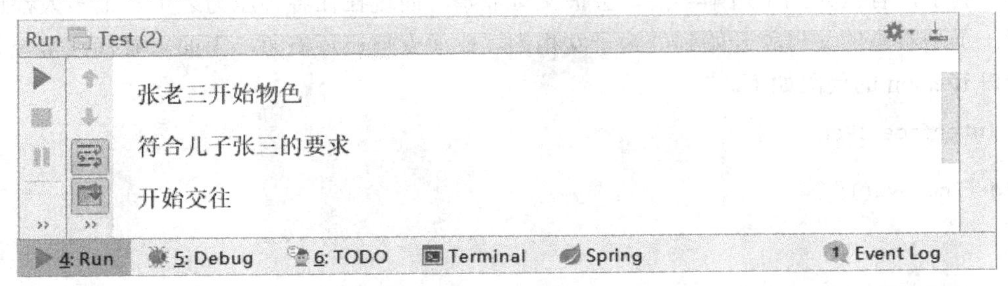

但是，上面的场景有个弊端，就是自己的父亲只会帮自己的子女去物色对象，别人家的孩子是不会管的。但社会上这项业务发展成了一个产业，出现了媒婆、婚介所等，还有各种各样的定制套餐。如果还使用静态代理成本就太高了，需要一个更加通用的解决方案，满足任何单身人士找对象的需求。这就由静态代理升级到了动态代理。采用动态代理基本上只要是人（IPerson）就可以提供相亲服务。动态代理的底层实现一般不用我们亲自去实现，已经有很多现成的 API。在 Java 生态中，目前普遍使用的是 JDK 自带的代理和 CGLib 提供的类库。首先基于 JDK 动态代理支持来升级一下代码。

首先创建媒婆（婚介所）类 JdkMeipo。

```
public class JdkMeipo implements InvocationHandler {
    private IPerson target;
    public IPerson getInstance(IPerson target){
        this.target = target;
        Class<?> clazz =  target.getClass();
        return (IPerson) Proxy.newProxyInstance(clazz.getClassLoader(),clazz.getInterfaces(),this);
    }

    public Object invoke(Object proxy, Method method, Object[] args) throws Throwable {
        before();
        Object result = method.invoke(this.target,args);
        after();
        return result;
    }

    private void after() {
        System.out.println("双方同意，开始交往");
```

```
    }

    private void before() {
        System.out.println("我是媒婆，已经收集到你的需求，开始物色");
    }
}
```

然后创建一个类 ZhaoLiu。

```
public class ZhaoLiu implements IPerson {

    public void findLove() {
        System.out.println("符合赵六的要求");
    }

    public void buyInsure() {

    }

}
```

最后客户端测试代码如下。

```
    public static void main(String[] args) {
        JdkMeipo jdkMeipo = new JdkMeipo();

        IPerson zhaoliu = jdkMeipo.getInstance(new ZhaoLiu());
        zhaoliu.findLove();

    }
```

运行结果如下图所示。

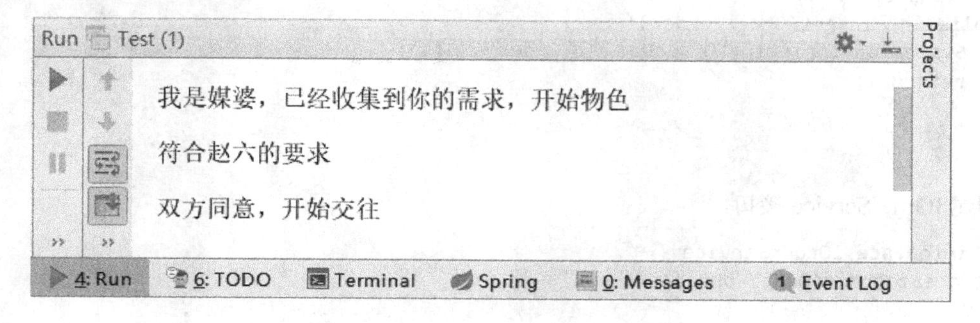

11.2.2　三层架构中的静态代理

小伙伴们可能会觉得还是不知道如何将代理模式应用到业务场景中，我们来看一个实际的业务场景。在分布式业务场景中，通常会对数据库进行分库分表，分库分表之后使用 Java 操作时就可能需要配置多个数据源，我们通过设置数据源路由来动态切换数据源。首先创建 Order 订单类。

```
public class Order {
    private Object orderInfo;
    private Long createTime;
    private String id;

    public Object getOrderInfo() {
        return orderInfo;
    }
    public void setOrderInfo(Object orderInfo) {
        this.orderInfo = orderInfo;
    }
    public Long getCreateTime() {
        return createTime;
    }
    public void setCreateTime(Long createTime) {
        this.createTime = createTime;
    }
    public String getId() {
        return id;
    }
    public void setId(String id) {
        this.id = id;
    }
}
```

创建 OrderDao 持久层操作类。

```
public class OrderDao {
    public int insert(Order order){
        System.out.println("OrderDao 创建 Order 成功!");
        return 1;
    }
}
```

创建 IOrderService 接口。

```
public interface IOrderService {
    int createOrder(Order order);
}
```

创建 OrderService 实现类。

```
public class OrderService implements IOrderService {
    private OrderDao orderDao;
```

```
    public OrderService(){
        //如果使用 Spring，则应该是自动注入的
        //为了使用方便，我们在构造方法中直接将 orderDao 初始化
        orderDao = new OrderDao();
    }

    @Override
    public int createOrder(Order order) {
        System.out.println("OrderService 调用 orderDao 创建订单");
        return orderDao.insert(order);
    }
}
```

然后使用静态代理，主要完成的功能是：根据订单创建时间自动按年进行分库。根据开闭原则，我们修改原来写好的代码逻辑，通过代理对象来完成。创建数据源路由对象，使用 ThreadLocal 的单例实现 DynamicDataSourceEntry 类。

```
//动态切换数据源
public class DynamicDataSourceEntry {

    //默认数据源
    public final static String DEFAULT_SOURCE = null;

    private final static ThreadLocal<String> local = new ThreadLocal<String>();

    private DynamicDataSourceEntry(){}

    //清空数据源
    public static void clear() {
        local.remove();
    }

    //获取当前正在使用的数据源名字
    public static String get() {
        return local.get();
    }

    //还原当前切换的数据源
    public static void restore() {
        local.set(DEFAULT_SOURCE);
    }

    //设置已知名字的数据源
    public static void set(String source) {
        local.set(source);
    }
```

```
    //根据年份动态设置数据源
    public static void set(int year) {
        local.set("DB_" + year);
    }
}
```

创建切换数据源的代理类OrderServiceSaticProxy。

```
public class OrderServiceStaticProxy implements IOrderService {

    private SimpleDateFormat yearFormat = new SimpleDateFormat("yyyy");

    private IOrderService orderService;
    public OrderServiceStaticProxy(IOrderService orderService){
        this.orderService = orderService;
    }

    public int createOrder(Order order) {
        before();
        Long time = order.getCreateTime();
        Integer dbRouter = Integer.valueOf(yearFormat.format(new Date(time)));
        System.out.println("静态代理类自动分配到【DB_" + dbRouter + "】数据源处理数据");
        DynamicDataSourceEntry.set(dbRouter);
        orderService.createOrder(order);
        after();
        return 0;
    }

    private void before(){
        System.out.println("Proxy before method.");
    }

    private void after(){
        System.out.println("Proxy after method.");
    }

}
```

来看客户端测试代码。

```
    public static void main(String[] args) {

        try {

            Order order = new Order();
            SimpleDateFormat sdf = new SimpleDateFormat("yyyy/MM/dd");
```

```
            Date date = sdf.parse("2017/02/01");
            order.setCreateTime(date.getTime());

            IOrderService orderService = new OrderServiceStaticProxy(new OrderService());
            orderService.createOrder(order);
        }catch (Exception e){
            e.printStackTrace();;
        }

}
```

运行结果如下图所示。

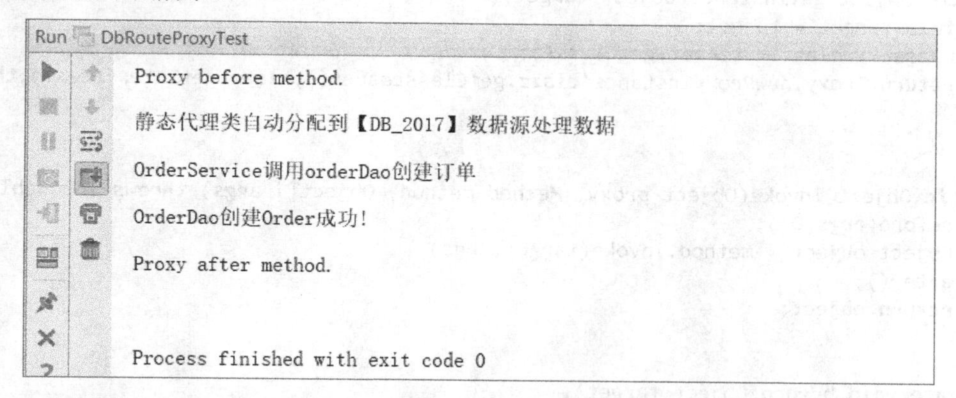

Run DbRouteProxyTest
Proxy before method.
静态代理类自动分配到【DB_2017】数据源处理数据
OrderService调用orderDao创建订单
OrderDao创建Order成功！
Proxy after method.
Process finished with exit code 0

由上图可知，结果符合预期。再来回顾一下类图，看是否与我们最先画的一致，如下图所示。

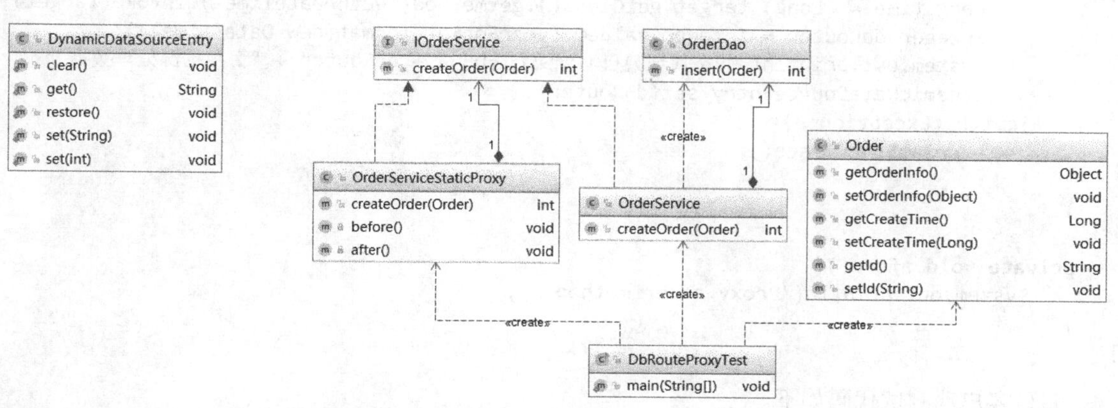

动态代理和静态代理的基本思路是一致的，只不过动态代理的功能更强大，随着业务的扩展，适应性更强。

11.2.3 使用动态代理实现无感知切换数据源

在理解了上面的案例后，再来看数据源动态路由业务，帮助小伙伴们加深对动态代理的印象。创建动态代理的类 OrderServiceDynamicProxy，代码如下。

```
public class OrderServiceDynamicProxy implements InvocationHandler {

    private SimpleDateFormat yearFormat = new SimpleDateFormat("yyyy");
    private Object target;

    public Object getInstance(Object target){
        this.target = target;
        Class<?> clazz = target.getClass();
        return Proxy.newProxyInstance(clazz.getClassLoader(),clazz.getInterfaces(),this);
    }


    public Object invoke(Object proxy, Method method, Object[] args) throws Throwable {
        before(args[0]);
        Object object = method.invoke(target,args);
        after();
        return object;
    }

    private void before(Object target){
        try {
            System.out.println("Proxy before method.");
            Long time = (Long) target.getClass().getMethod("getCreateTime").invoke(target);
            Integer dbRouter = Integer.valueOf(yearFormat.format(new Date(time)));
            System.out.println("静态代理类自动分配到【DB_" + dbRouter + "】数据源处理数据");
            DynamicDataSourceEntry.set(dbRouter);
        }catch (Exception e){
            e.printStackTrace();
        }
    }

    private void after(){
        System.out.println("Proxy after method.");
    }
}
```

编写客户端测试代码如下。

```
public static void main(String[] args) {

    try {
```

```
        Order order = new Order();

        SimpleDateFormat sdf = new SimpleDateFormat("yyyy/MM/dd");
        Date date = sdf.parse("2018/02/01");
        order.setCreateTime(date.getTime());

        IOrderService orderService = (IOrderService)new OrderServiceDynamicProxy().
        getInstance(new OrderService());
        orderService.createOrder(order);
    }catch (Exception e){
        e.printStackTrace();
    }

}
```

由上面代码可以看出，依然能够达到相同的运行效果。但是，使用动态代理实现之后，不仅能实现 Order 的数据源动态路由，还可以实现其他任何类的数据源路由。当然，有一个比较重要的约定，必须实现 getCreateTime()方法，因为路由规则是根据时间来运算的。可以通过接口规范达到约束的目的，在此不再举例。

11.2.4 手写 JDK 动态代理核心原理

不仅知其然，还得知其所以然。既然 JDK 动态代理的功能如此强大，那么它是如何实现的呢？现在来探究一下原理，并模仿 JDK 动态代理手写一个属于自己的动态代理。

我们都知道 JDK 动态代理采用字节重组，重新生成对象来替代原始对象，以达到动态代理的目的。JDK 动态代理生成对象的步骤如下。

（1）获取被代理对象的引用，并且获取它的所有接口，反射获取。

（2）JDK 动态代理类重新生成一个新的类，同时新的类要实现被代理类实现的所有接口。

（3）动态生成 Java 代码，新加的业务逻辑方法由一定的逻辑代码调用（在代码中体现）。

（4）编译新生成的 Java 代码.class 文件。

（5）重新加载到 JVM 中运行。

以上过程就叫作字节码重组。JDK 中有一个规范，在 ClassPath 下只要是$开头的.class 文件，一般都是自动生成的。那么有没有办法看到代替后的对象的“真容”呢？做一个这样的测试，将内存中的对象字节码通过文件流输出到一个新的.class 文件，然后使用反编译工具查看源码。

```
public static void main(String[] args) {
```

```
        try {
            IPerson obj = (IPerson)new JdkMeipo().getInstance(new Zhangsan());
            obj.findLove();

            //通过反编译工具查看源代码
            byte [] bytes = ProxyGenerator.generateProxyClass("$Proxy0",new Class[]{IPerson.class});
            FileOutputStream os = new FileOutputStream("E://$Proxy0.class");
            os.write(bytes);
            os.close();
        } catch (Exception e) {
            e.printStackTrace();
        }
    }
```

运行以上代码，可以在 E 盘找到一个$Proxy0.class 文件。使用 Jad 反编译，得到$Proxy0.jad 文件，打开文件看到如下内容。

```
import com.gupaoedu.vip.pattern.proxy.dynamicproxy.jdkproxy.IPerson;
import java.lang.reflect.InvocationHandler;
import java.lang.reflect.Method;
import java.lang.reflect.Proxy;
import java.lang.reflect.UndeclaredThrowableException;

public final class $Proxy0 extends Proxy implements IPerson {
    private static Method m1;
    private static Method m3;
    private static Method m2;
    private static Method m4;
    private static Method m0;

    public $Proxy0(InvocationHandler var1) throws  {
        super(var1);
    }

    public final boolean equals(Object var1) throws  {
        try {
            return ((Boolean)super.h.invoke(this, m1, new Object[]{var1})).booleanValue();
        } catch (RuntimeException | Error var3) {
            throw var3;
        } catch (Throwable var4) {
            throw new UndeclaredThrowableException(var4);
        }
    }

    public final void findLove() throws  {
        try {
            super.h.invoke(this, m3, (Object[])null);
```

```
    } catch (RuntimeException | Error var2) {
        throw var2;
    } catch (Throwable var3) {
        throw new UndeclaredThrowableException(var3);
    }
}

public final String toString() throws  {
    try {
        return (String)super.h.invoke(this, m2, (Object[])null);
    } catch (RuntimeException | Error var2) {
        throw var2;
    } catch (Throwable var3) {
        throw new UndeclaredThrowableException(var3);
    }
}

public final void buyInsure() throws  {
    try {
        super.h.invoke(this, m4, (Object[])null);
    } catch (RuntimeException | Error var2) {
        throw var2;
    } catch (Throwable var3) {
        throw new UndeclaredThrowableException(var3);
    }
}

public final int hashCode() throws  {
    try {
        return ((Integer)super.h.invoke(this, m0, (Object[])null)).intValue();
    } catch (RuntimeException | Error var2) {
        throw var2;
    } catch (Throwable var3) {
        throw new UndeclaredThrowableException(var3);
    }
}

static {
    try {
        m1 = Class.forName("java.lang.Object").getMethod("equals",
                new Class[]{Class.forName("java.lang.Object")});
        m3 = Class.forName("com.gupaoedu.vip.pattern.proxy.dynamicproxy.jdkproxy.IPerson")
                .getMethod("findLove", new Class[0]);
        m2 = Class.forName("java.lang.Object").getMethod("toString", new Class[0]);
        m4 = Class.forName("com.gupaoedu.vip.pattern.proxy.dynamicproxy.jdkproxy.IPerson")
                .getMethod("buyInsure", new Class[0]);
        m0 = Class.forName("java.lang.Object").getMethod("hashCode", new Class[0]);
```

```
        } catch (NoSuchMethodException var2) {
            throw new NoSuchMethodError(var2.getMessage());
        } catch (ClassNotFoundException var3) {
            throw new NoClassDefFoundError(var3.getMessage());
        }
    }
}
```

我们发现，$Proxy0 继承了 Proxy 类，同时实现了 Person 接口，而且重写了 findLove()等方法。在静态块中用反射查找到了目标对象的所有方法，而且保存了所有方法的引用，重写的方法用反射调用目标对象的方法。小伙伴们此时一定会好奇：这些代码是从哪里来的？其实是 JDK 自动生成的。现在我们不依赖 JDK，自己来动态生成源码、动态完成编译，然后替代目标对象并执行。

创建 GPInvocationHandler 接口。

```
public interface GPInvocationHandler {
    public Object invoke(Object proxy, Method method, Object[] args)
            throws Throwable;
}
```

创建 GPProxy 类。

```
/**
 * 用来生成源码的工具类
 * Created by Tom.
 */
public class GPProxy {

    public static final String ln = "\r\n";

    public static Object newProxyInstance(GPClassLoader classLoader, Class<?> [] interfaces,
    GPInvocationHandler h){
       try {
            //1.动态生成源码.java 文件
            String src = generateSrc(interfaces);
            //2.Java 文件输出磁盘
            String filePath = GPProxy.class.getResource("").getPath();

            File f = new File(filePath + "$Proxy0.java");
            FileWriter fw = new FileWriter(f);
            fw.write(src);
            fw.flush();
            fw.close();

            //3.把生成的.java 文件编译成.class 文件
            JavaCompiler compiler = ToolProvider.getSystemJavaCompiler();
```

```
            StandardJavaFileManager manage = compiler.getStandardFileManager(null,null,null);
            Iterable iterable = manage.getJavaFileObjects(f);

           JavaCompiler.CompilationTask task = compiler.getTask(null,manage,null,null,null,iterable);
           task.call();
           manage.close();

            //4.编译生成的.class 文件加载到 JVM 中
           Class proxyClass =  classLoader.findClass("$Proxy0");
           Constructor c = proxyClass.getConstructor(GPInvocationHandler.class);
           f.delete();

            //5.返回字节码重组以后的新的代理对象
            return c.newInstance(h);
        }catch (Exception e){
            e.printStackTrace();
        }
         return null;
    }

    private static String generateSrc(Class<?>[] interfaces){
            StringBuffer sb = new StringBuffer();
            sb.append(GPProxy.class.getPackage() + ";" + ln);
            sb.append("import " + interfaces[0].getName() + ";" + ln);
            sb.append("import java.lang.reflect.*;" + ln);
            sb.append("public class $Proxy0 implements " + interfaces[0].getName() + "{" + ln);
                sb.append("GPInvocationHandler h;" + ln);
                sb.append("public $Proxy0(GPInvocationHandler h) { " + ln);
                    sb.append("this.h = h;");
                sb.append("}" + ln);
                for (Method m : interfaces[0].getMethods()){
                    Class<?>[] params = m.getParameterTypes();

                    StringBuffer paramNames = new StringBuffer();
                    StringBuffer paramValues = new StringBuffer();
                    StringBuffer paramClasses = new StringBuffer();

                    for (int i = 0; i < params.length; i++) {
                        Class clazz = params[i];
                        String type = clazz.getName();
                        String paramName = toLowerFirstCase(clazz.getSimpleName());
                        paramNames.append(type + " " +  paramName);
                        paramValues.append(paramName);
                        paramClasses.append(clazz.getName() + ".class");
                        if(i > 0 && i < params.length-1){
                            paramNames.append(",");
                            paramClasses.append(",");
```

```
                    paramValues.append(",");
                }
            }

            sb.append("public " + m.getReturnType().getName() + " " + m.getName() + "("
                    + paramNames.toString() + ") {" + ln);
                sb.append("try{" + ln);
                    sb.append("Method  m  =  "  +  interfaces[0].getName()  +  ".class.
                      getMethod(\""   +    m.getName()    +    "\",new    Class[]{"    +
                      paramClasses.toString() + "});" + ln);
                    sb.append((hasReturnValue(m.getReturnType()) ? "return " : "") +
                      getCaseCode("this.h.invoke(this,m,new Object[]{" + paramValues +
                      "})",m.getReturnType()) + ";" + ln);
                sb.append("}catch(Error _ex) { }");
                sb.append("catch(Throwable e){" + ln);
                sb.append("throw new UndeclaredThrowableException(e);" + ln);
                sb.append("}");
                sb.append(getReturnEmptyCode(m.getReturnType()));
            sb.append("}");
        }
        sb.append("}" + ln);
        return sb.toString();
    }

    private static Map<Class,Class> mappings = new HashMap<Class, Class>();
    static {
        mappings.put(int.class,Integer.class);
    }

    private static String getReturnEmptyCode(Class<?> returnClass){
        if(mappings.containsKey(returnClass)){
            return "return 0;";
        }else if(returnClass == void.class){
            return "";
        }else {
            return "return null;";
        }
    }

    private static String getCaseCode(String code,Class<?> returnClass){
        if(mappings.containsKey(returnClass)){
            return "((" + mappings.get(returnClass).getName() +  ")" + code + ")." +
                    returnClass.getSimpleName() + "Value()";
        }
        return code;
    }
```

```
    private static boolean hasReturnValue(Class<?> clazz){
        return clazz != void.class;
    }

    private static String toLowerFirstCase(String src){
        char [] chars = src.toCharArray();
        chars[0] += 32;
        return String.valueOf(chars);
    }

}
```

创建 GPClassLoader 类。

```
public class GPClassLoader extends ClassLoader {

    private File classPathFile;
    public GPClassLoader(){
        String classPath = GPClassLoader.class.getResource("").getPath();
        this.classPathFile = new File(classPath);
    }

    @Override
    protected Class<?> findClass(String name) throws ClassNotFoundException {

        String className = GPClassLoader.class.getPackage().getName() + "." + name;
        if(classPathFile  != null){
            File classFile = new File(classPathFile,name.replaceAll("\\.","/") + ".class");
            if(classFile.exists()){
                FileInputStream in = null;
                ByteArrayOutputStream out = null;
                try{
                    in = new FileInputStream(classFile);
                    out = new ByteArrayOutputStream();
                    byte [] buff = new byte[1024];
                    int len;
                    while ((len = in.read(buff)) != -1){
                        out.write(buff,0,len);
                    }
                    return defineClass(className,out.toByteArray(),0,out.size());
                }catch (Exception e){
                    e.printStackTrace();
                }
            }
        }
        return null;
    }
}
```

创建 GPMeipo 类。

```
public class GpMeipo implements GPInvocationHandler {
    private IPerson target;
    public IPerson getInstance(IPerson target){
        this.target = target;
        Class<?> clazz =  target.getClass();
        return (IPerson) GPProxy.newProxyInstance(new GPClassLoader(),clazz.getInterfaces(),this);
    }

    public Object invoke(Object proxy, Method method, Object[] args) throws Throwable {
        before();
        Object result = method.invoke(this.target,args);
        after();
        return result;
    }

    private void after() {
        System.out.println("双方同意，开始交往");
    }

    private void before() {
        System.out.println("我是媒婆，已经收集到你的需求，开始物色");
    }
}
```

客户端测试代码如下。

```
    public static void main(String[] args) {
        GpMeipo gpMeipo = new GpMeipo();
        IPerson zhangsan = gpMeipo.getInstance(new Zhangsan());
        zhangsan.findLove();


    }
```

至此，手写 JDK 动态代理就完成了。小伙伴们是不是又多了一个面试用的“撒手锏”呢？

11.2.5 CGLib 动态代理 API 原理分析

简单看一下 CGLib 动态代理的使用，还是以媒婆为例，创建 CglibMeipo 类。

```
public class CGlibMeipo implements MethodInterceptor {

    public Object getInstance(Class<?> clazz) throws Exception{
        //相当于 JDK 中的 Proxy 类，是完成代理的工具类
        Enhancer enhancer = new Enhancer();
        enhancer.setSuperclass(clazz);
```

```
        enhancer.setCallback(this);
        return enhancer.create();
    }

    public Object intercept(Object o, Method method, Object[] objects, MethodProxy methodProxy)
                            throws Throwable {
        before();
        Object obj = methodProxy.invokeSuper(o,objects);
        after();
        return obj;
    }

    private void before(){
        System.out.println("我是媒婆，我要给你找对象，现在已经确认你的需求");
        System.out.println("开始物色");
    }

    private void after(){
        System.out.println("双方同意，准备办婚事");
    }
}
```

创建单身客户类 Customer。

```
public class Customer {

    public void findLove(){
        System.out.println("符合要求");
    }
}
```

这里有一个小细节，CGLib 动态代理的目标对象不需要实现任何接口，它是通过动态继承目标对象实现动态代理的，客户端测试代码如下。

```
public static void main(String[] args) {

    try {

        //JDK 采用读取接口的信息
        //CGLib 覆盖父类方法
        //目的都是生成一个新的类，去实现增强代码逻辑的功能

        //JDK Proxy 对于用户而言，必须要有一个接口实现，目标类相对来说复杂
        //CGLib 可以代理任意一个普通的类，没有任何要求

        //CGLib 生成代理的逻辑更复杂，调用效率更高，生成一个包含了所有逻辑的 FastClass，不再需
          要反射调用
```

```
            //JDK Proxy 生成代理的逻辑简单，执行效率相对要低，每次都要反射动态调用

            //CGLib 有一个缺点，CGLib 不能代理 final 的方法

            Customer obj = (Customer) new CGlibMeipo().getInstance(Customer.class);
            System.out.println(obj);
            obj.findLove();
        } catch (Exception e) {
            e.printStackTrace();
        }

    }
```

CGLib 动态代理的实现原理又是怎样的呢？我们可以在客户端测试代码中加上一句代码，将CGLib 动态代理后的.class 文件写入磁盘，然后反编译来一探究竟，代码如下。

```
public static void main(String[] args) {
    try {

        //使用 CGLib 的代理类可以将内存中的.class 文件写入本地磁盘
        System.setProperty(DebuggingClassWriter.DEBUG_LOCATION_PROPERTY, "E://cglib_proxy_class/");

        Customer obj = (Customer)new CglibMeipo().getInstance(Customer.class);
        obj.findLove();

    } catch (Exception e) {
        e.printStackTrace();
    }
}
```

重新执行代码，我们会发现在 E://cglib_proxy_class 目录下多了三个.class 文件，如下图所示。

cglib_proxy_class › com › gupaoedu › vip › pattern › proxy › dynamicproxy › cglibproxy

名称

Customer$$EnhancerByCGLIB$$3feeb52a$$FastClassByCGLIB$$6aad62f1.class
Customer$$EnhancerByCGLIB$$3feeb52a.class
Customer$$FastClassByCGLIB$$2669574a.class

通过调试跟踪发现，Customer$$EnhancerByCGLIB$$3feeb52a.class 就是 CGLib 动态代理生成的代理类，继承了 Customer 类。

```
package com.gupaoedu.vip.pattern.proxy.dynamicproxy.cglibproxy;

import java.lang.reflect.Method;
import net.sf.cglib.core.ReflectUtils;
```

```
import net.sf.cglib.core.Signature;
import net.sf.cglib.proxy.*;

public class Customer$$EnhancerByCGLIB$$3feeb52a extends Customer
    implements Factory
{

    ...

    final void CGLIB$findLove$0()
    {
        super.findLove();
    }

    public final void findLove()
    {
        CGLIB$CALLBACK_0;
        if(CGLIB$CALLBACK_0 != null) goto _L2; else goto _L1
_L1:
        JVM INSTR pop ;
        CGLIB$BIND_CALLBACKS(this);
        CGLIB$CALLBACK_0;
_L2:
        JVM INSTR dup ;
        JVM INSTR ifnull 37;
           goto _L3 _L4
_L3:
        break MISSING_BLOCK_LABEL_21;
_L4:
        break MISSING_BLOCK_LABEL_37;
        this;
        CGLIB$findLove$0$Method;
        CGLIB$emptyArgs;
        CGLIB$findLove$0$Proxy;
        intercept();
        return;
        super.findLove();
        return;
    }

    ...

}
```

我们重写了 Customer 类的所有方法，通过代理类的源码可以看到，代理类会获得所有从父类继承来的方法，并且会有 MethodProxy 与之对应，比如 Method CGLIB$findLove$0$Method、

MethodProxy CGLIB$findLove$0$Proxy 等方法在代理类的 findLove()方法中都有调用。

```
//代理方法（methodProxy.invokeSuper()方法会调用）
    final void CGLIB$findLove$0()
    {
        super.findLove();
    }

     //被代理方法（methodProxy.invoke()方法会调用）
     //这就是为什么在拦截器中调用 methodProxy.invoke 会发生死循环，一直在调用拦截器
    public final void findLove()
    {
        ...
        //调用拦截器
        intercept();
        return;
        super.findLove();
        return;
    }
```

调用过程为：代理对象调用 this.findLove()方法→调用拦截器→methodProxy.invokeSuper()→CGLIB$findLove$0→被代理对象 findLove()方法。

此时，我们发现 MethodInterceptor 拦截器就是由 MethodProxy 的 invokeSuper()方法调用代理方法的，因此，MethodProxy 类中的代码非常关键，我们分析它具体做了什么。

```
package net.sf.cglib.proxy;

import java.lang.reflect.InvocationTargetException;
import java.lang.reflect.Method;
import net.sf.cglib.core.AbstractClassGenerator;
import net.sf.cglib.core.CodeGenerationException;
import net.sf.cglib.core.GeneratorStrategy;
import net.sf.cglib.core.NamingPolicy;
import net.sf.cglib.core.Signature;
import net.sf.cglib.reflect.FastClass;
import net.sf.cglib.reflect.FastClass.Generator;

public class MethodProxy {
    private Signature sig1;
    private Signature sig2;
    private MethodProxy.CreateInfo createInfo;
    private final Object initLock = new Object();
    private volatile MethodProxy.FastClassInfo fastClassInfo;

    public static MethodProxy create(Class c1, Class c2, String desc, String name1, String name2) {
```

```
        MethodProxy proxy = new MethodProxy();
        proxy.sig1 = new Signature(name1, desc);
        proxy.sig2 = new Signature(name2, desc);
        proxy.createInfo = new MethodProxy.CreateInfo(c1, c2);
        return proxy;
    }

    ...

    private static class CreateInfo {
        Class c1;
        Class c2;
        NamingPolicy namingPolicy;
        GeneratorStrategy strategy;
        boolean attemptLoad;

        public CreateInfo(Class c1, Class c2) {
            this.c1 = c1;
            this.c2 = c2;
            AbstractClassGenerator fromEnhancer = AbstractClassGenerator.getCurrent();
            if(fromEnhancer != null) {
                this.namingPolicy = fromEnhancer.getNamingPolicy();
                this.strategy = fromEnhancer.getStrategy();
                this.attemptLoad = fromEnhancer.getAttemptLoad();
            }

        }
    }
    ...

}
```

继续看 invokeSuper()方法。

```
public Object invokeSuper(Object obj, Object[] args) throws Throwable {
    try {
        this.init();
        MethodProxy.FastClassInfo fci = this.fastClassInfo;
        return fci.f2.invoke(fci.i2, obj, args);
    } catch (InvocationTargetException var4) {
        throw var4.getTargetException();
    }
}

...

private static class FastClassInfo {
```

```
    FastClass f1;
    FastClass f2;
    int i1;
    int i2;

    private FastClassInfo() {
    }
}
```

上面代码调用获取代理类对应的FastClass，并执行代理方法。还记得之前生成的三个.class文件吗？Customer$$EnhancerByCGLIB$$3feeb52a$$FastClassByCGLIB$$6aad62f1.class 就是代理类的FastClass，Customer$$FastClassByCGLIB$$2669574a.class就是被代理类的FastClass。

CGLib动态代理执行代理方法的效率之所以比JDK高，是因为CGlib采用了FastClass机制，它的原理简单来说就是：为代理类和被代理类各生成一个类，这个类会为代理类或被代理类的方法分配一个index（int类型）；这个index被当作一个入参，FastClass可以直接定位要调用的方法并直接进行调用，省去了反射调用，因此调用效率比JDK动态代理通过反射调用高。下面我们来反编译一个FastClass。

```
public int getIndex(Signature signature)
    {
        String s = signature.toString();
        s;
        s.hashCode();
        JVM INSTR lookupswitch 11: default 223
        …
        JVM INSTR pop ;
        return -1;
    }

//部分代码省略

    //根据index直接定位执行方法
    public Object invoke(int i, Object obj, Object aobj[])
        throws InvocationTargetException
    {
        (Customer)obj;
        i;
        JVM INSTR tableswitch 0 10: default 161
           goto _L1 _L2 _L3 _L4 _L5 _L6 _L7 _L8 _L9 _L10 _L11 _L12
_L2:
        eat();
        return null;
```

```
_L3:
      findLove();
      return null;
      …
      throw new IllegalArgumentException("Cannot find matching method/constructor");
   }
```

FastClass 并不是跟代理类一起生成的，而是在第一次执行 MethodProxy 的 invoke()或 invokeSuper()方法时生成的，并被放在了缓存中。

```
//MethodProxy 的 invoke()或 invokeSuper()方法都调用了 init()方法
private void init() {
    if(this.fastClassInfo == null) {
        Object var1 = this.initLock;
        synchronized(this.initLock) {
            if(this.fastClassInfo == null) {
                MethodProxy.CreateInfo ci = this.createInfo;
                MethodProxy.FastClassInfo fci = new MethodProxy.FastClassInfo();
                //如果在缓存中，则取出；如果没在缓存中，则生成新的 FastClass
                fci.f1 = helper(ci, ci.c1);
                fci.f2 = helper(ci, ci.c2);
                fci.i1 = fci.f1.getIndex(this.sig1);//获取方法的 index
                fci.i2 = fci.f2.getIndex(this.sig2);
                this.fastClassInfo = fci;
            }
        }
    }
}
```

至此，我们基本清楚了 CGLib 动态代理的原理，对代码细节感兴趣的小伙伴们可以自行深入研究。

11.2.6　CGLib 和 JDK 动态代理对比分析

（1）JDK 动态代理实现了被代理对象的接口，CGLib 动态代理继承了被代理对象。

（2）JDK 动态代理和 CGLib 动态代理都在运行期生成字节码，JDK 动态代理直接写 Class 字节码，CGLib 动态代理使用 ASM 框架写 Class 字节码。CGLib 动态代理实现更复杂，生成代理类比 JDK 动态代理效率低。

（3）JDK 动态代理调用代理方法是通过反射机制调用的，CGLib 动态代理是通过 FastClass 机制直接调用方法的，CGLib 动态代理的执行效率更高。

11.3 代理模式在框架源码中的应用

11.3.1 代理模式在 Spring 源码中的应用

先看 ProxyFactoryBean 的核心方法 getObject()，源码如下。

```
public Object getObject() throws BeansException {
  initializeAdvisorChain();
  if (isSingleton()) {
    return getSingletonInstance();
  }
  else {
    if (this.targetName == null) {
      logger.warn("Using non-singleton proxies with singleton targets is often undesirable. " +
            "Enable prototype proxies by setting the 'targetName' property.");
    }
    return newPrototypeInstance();
  }
}
```

在 getObject()方法中，主要调用 getSingletonInstance()和 newPrototypeInstance()。在 Spring 的配置中，如果不做任何设置，则 Spring 代理生成的 Bean 都是单例对象。如果修改 scope，则每次都创建一个新的原型对象。newPrototypeInstance()里的逻辑比较复杂，我们后面再做深入研究，这里先简单了解。

Spring 使用动态代理实现 AOP 时有两个非常重要的类：JdkDynamicAopProxy 类和 CglibAopProxy 类，其类图如下。

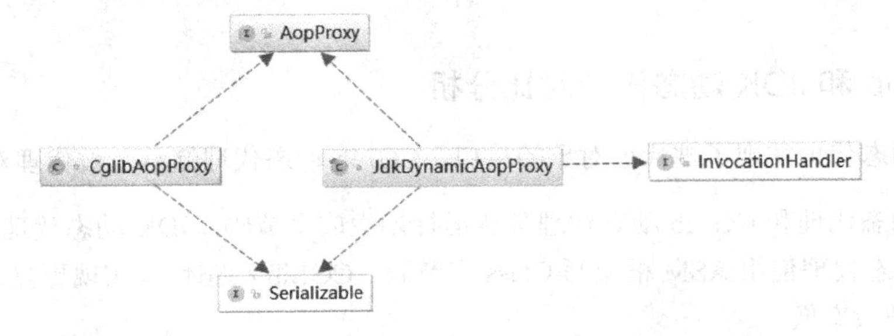

Spring 中的代理选择原则如下。

（1）当 Bean 有实现接口时，Spring 会用 JDK 动态代理。

（2）当 Bean 没有实现接口时，Spring 会选择 CGLib 动态代理。

（3） Spring 可以通过配置强制使用 CGLib 动态代理，只需在 Spring 的配置文件中加入如下代码即可。

```
<aop:aspectj-autoproxy proxy-target-class="true"/>
```

11.3.2　代理模式在 MyBatis 源码中的应用

MyBatis 是一个应用非常广泛的持久层框架，它避免了几乎所有 JDBC 代码和手动设置参数及获取结果集。MyBatis 可以使用简单的 XML 文件配置或注解配置来映射类、接口和 POJO（Plain Old Java Objects，普通老式 Java 对象）与数据库记录的对应关系。

如果使用过 MyBatis，则会发现 MyBatis 的使用非常简单。首先定义一个 Dao 接口，然后编写一个与 Dao 接口对应的配置文件，Java 对象与数据库字段的映射关系和 Dao 接口对应的 SQL 语句都是以配置的形式写在配置文件中的，非常简单清晰。但是我们在使用过程中曾有这样的疑问，Dao 接口是怎么和 Mapper 文件映射的呢？只有一个 Dao 接口，又是怎么以对象的形式来实现数据库的读写操作的呢？当然，在了解代理模式之后，应该很容易猜到，可以通过动态代理来创建 Dao 接口的代理对象，并通过这个代理对象实现数据库的操作。首先来看 MapperProxyFactory 类。

```
public class MapperProxyFactory<T> {

  private final Class<T> mapperInterface;
  private Map<Method, MapperMethod> methodCache = new ConcurrentHashMap<Method, MapperMethod>();

  public MapperProxyFactory(Class<T> mapperInterface) {
    this.mapperInterface = mapperInterface;
  }

  public Class<T> getMapperInterface() {
    return mapperInterface;
  }

  public Map<Method, MapperMethod> getMethodCache() {
    return methodCache;
  }

  @SuppressWarnings("unchecked")
  protected T newInstance(MapperProxy<T> mapperProxy) {
    return   (T)   Proxy.newProxyInstance(mapperInterface.getClassLoader(),   new   Class[]
{ mapperInterface }, mapperProxy);
  }

  public T newInstance(SqlSession sqlSession) {
```

```
    final MapperProxy<T> mapperProxy = new MapperProxy<T>(sqlSession, mapperInterface,
methodCache);
    return newInstance(mapperProxy);
  }

}
```

MapperProxyFactory类的代码不多，主要看它的构造方法和newInstance()方法。构造方法传入了一个Class，通过参数名mapperInterface可以很容易猜到这个类就是Dao接口。newInstance()方法首先创建了一个MapperProxy类，然后通过Proxy.newProxyInstance()方法创建了一个对象并返回，也是通过同样的方法返回了mapperInterface接口的代理对象，而上面提到的MapperProxy类显然是InvocationHandler接口的实现。因此，MapperProxyFactory类就是一个创建代理对象的工厂类，它通过构造函数传入自定义的Dao接口，并通过newInstance方法返回Dao接口的代理对象。

看到这里，有了一种豁然开朗的感觉，但新的疑问又来了，我们定义的Dao接口的方法并没有被实现，那这个代理对象又是如何实现增删改查的呢？带着这个疑问，我们来看一下MapperProxy类，它是怎么来改造增强接口方法的呢？

```
public class MapperProxy<T> implements InvocationHandler, Serializable {

  private static final long serialVersionUID = -6424540398559729838L;
  private final SqlSession sqlSession;
  private final Class<T> mapperInterface;
  private final Map<Method, MapperMethod> methodCache;

//此处省略构造函数
  public Object invoke(Object proxy, Method method, Object[] args) throws Throwable {
    if (Object.class.equals(method.getDeclaringClass())) {
      try {
        return method.invoke(this, args);
      } catch (Throwable t) {
        throw ExceptionUtil.unwrapThrowable(t);
      }
    }
    final MapperMethod mapperMethod = cachedMapperMethod(method);
    return mapperMethod.execute(sqlSession, args);
  }

  private MapperMethod cachedMapperMethod(Method method) {
    MapperMethod mapperMethod = methodCache.get(method);
    if (mapperMethod == null) {
      mapperMethod = new MapperMethod(mapperInterface, method, sqlSession.getConfiguration());
      methodCache.put(method, mapperMethod);
```

```
    }
    return mapperMethod;
  }

}
```

这个类主要重写了 invoke()方法来实现接口方法的增强，在 invoke()方法中，只要看最后两行就可以。前面的 Object.class.equals(method.getDeclaringClass())主要是为了让代理对象可以实现一些 Object 类的公共方法，所有自定义的接口方法都只执行 invoke()方法的最后两行。

MapperProxy 类的 3 个成员变量分别如下。

（1）SqlSession，通过名字可以知道这个变量是一个定义了执行 SQL 的接口，简单看一下它的接口定义。

```
public interface SqlSession extends Closeable {

  <T> T selectOne(String statement);

  <T> T selectOne(String statement, Object parameter);

  //下面省略
}
```

这个接口方法的入参是 statement 和参数，返回值是数据对象，这里的 statement 有些人可能会误解为 SQL 语句，但其实这里的 statement 指 Dao 接口方法的名称，自定义的 SQL 语句都被缓存在 Configuration 对象中。在 SqlSession 中，可以通过 Dao 接口的方法名称找到对应的 SQL 语句。因此，可以想到代理对象本质上就是将要执行的方法名称和参数传入 SqlSession 的对应方法中，根据方法名称找到对应的 SQL 语句并替换参数，最后得到返回的结果。

（2）mapperInterface，它的作用要结合第 3 个成员变量来说明。

（3）methodCache，它其实就是一个 Map 键值对结构，键是 Method，值是 MapperMethod。

再回到 invoke()方法的最后两行，它首先通过 cachedMapperMethod()方法找到与要执行的 Dao 接口方法对应的 MapperMethod，然后调用 MapperMethod 的 execute()方法来实现数据库的操作，这里显然是将 SqlSession 传入 MapperMethod 内部，并在 MapperMethod 内部将要执行的方法名称和参数再传入 SqlSession 对应的方法中去执行。

接着来看 MapperMethod 类的内部，看它是怎么完成 SQL 的执行的。

```
public class MapperMethod {
```

```
  private final SqlCommand command;
  private final MethodSignature method;

}
```

这个类有两个成员变量，分别是 SqlCommand 和 MethodSignature。虽然这两个类的代码看起来很多，但实际上这两个内部类非常简单，SqlCommand 主要解析了接口的方法名称和方法类型，定义了诸如 INSERT、SELECT、DELETE 等数据库操作的枚举类型。MethodSignature 则解析了接口方法的签名，即接口方法的参数名称和参数值的映射关系，即通过 MethodSignature 类可以将入参的值转换成参数名称和参数值的映射，这里不再详细分析 SqlCommand 和 MethodSignature 的具体实现。

最后来看 MapperMethod 类中最重要的 execute()方法。

```
public Object execute(SqlSession sqlSession, Object[] args) {
    Object result;
    if (SqlCommandType.INSERT == command.getType()) {
      Object param = method.convertArgsToSqlCommandParam(args);
      result = rowCountResult(sqlSession.insert(command.getName(), param));
    } else if (SqlCommandType.UPDATE == command.getType()) {
      Object param = method.convertArgsToSqlCommandParam(args);
      result = rowCountResult(sqlSession.update(command.getName(), param));
    } else if (SqlCommandType.DELETE == command.getType()) {
      Object param = method.convertArgsToSqlCommandParam(args);
      result = rowCountResult(sqlSession.delete(command.getName(), param));
    } else if (SqlCommandType.SELECT == command.getType()) {
      if (method.returnsVoid() && method.hasResultHandler()) {
        executeWithResultHandler(sqlSession, args);
        result = null;
      } else if (method.returnsMany()) {
        result = executeForMany(sqlSession, args);
      } else if (method.returnsMap()) {
        result = executeForMap(sqlSession, args);
      } else {
        Object param = method.convertArgsToSqlCommandParam(args);
        result = sqlSession.selectOne(command.getName(), param);
      }
    } else {
      throw new BindingException("Unknown execution method for: " + command.getName());
    }
    if (result == null && method.getReturnType().isPrimitive() && !method.returnsVoid()) {
      throw new BindingException("Mapper method '" + command.getName()
          + " attempted to return null from a method with a primitive return type (" +
method.getReturnType() + ").");
    }
```

```
    return result;
  }
```

通过上面对 SqlCommand 和 MethodSignature 的简单分析，我们很容易理解这段代码。首先，根据 SqlCommand 中解析出来的方法类型选择对应 SqlSession 中的方法，即如果是 INSERT 类型，则选择 SqlSession.insert()方法来执行数据库操作。其次，通过 MethodSignature 将参数值转换为 Map<Key,Value>的映射，Key 是方法的参数名称，Value 是参数值。最后，将方法名称和参数传入对应的 SqlSession 的方法中执行。至于在配置文件中定义的 SQL 语句，则被缓存在 SqlSession 的成员变量 Configuration 中。Configuration 中有非常多参数，其中一个是 mappedStatements，它保存了我们在配置文件中定义的所有方法。还有一个是 mappedStatement，它保存了我们在配置文件中定义的各种参数，包括 SQL 语句。

到这里，我们应该对 MyBatis 中如何通过将配置与 Dao 接口映射起来、如何通过代理模式生成代理对象来执行数据库读写操作有了较为宏观的认识，至于 SqlSession 中如果将参数与 SQL 语句结合组装成完整的 SQL 语句，以及如何将数据库字段与 Java 对象映射，感兴趣的小伙伴可以自行分析相关的源码。

11.4　代理模式扩展

11.4.1　静态代理和动态代理的区别

（1） 静态代理只能通过手动完成代理操作，如果被代理类增加了新的方法，则代理类需要同步增加，违背开闭原则。

（2）动态代理采用在运行时动态生成代码的方式，取消了对被代理类的扩展限制，遵循开闭原则。

（3）若动态代理要对目标类的增强逻辑进行扩展，结合策略模式，只需要新增策略类便可完成，无须修改代理类的代码。

11.4.2　代理模式的优点

（1）代理模式能将代理对象与真实被调用目标对象分离。

（2）在一定程度上降低了系统的耦合性，扩展性好。

（3）可以起到保护目标对象的作用。

（4）可以增强目标对象的功能。

11.4.3 代理模式的缺点

（1）代理模式会造成系统设计中类的数量增加。

（2）在客户端和目标对象中增加一个代理对象，会导致处理请求的速度变慢。

（3）增加了系统的复杂度。

12

第 12 章 门面模式

12.1 门面模式概述

12.1.1 门面模式的定义

门面模式（Facade Pattern）又叫作外观模式，提供了一个统一的接口，用来访问子系统中的一群接口。其主要特征是定义了一个高层接口，让子系统更容易使用，属于结构型设计模式。

原文：Provide a unified interface to a set of interfaces in a subsystem. Facade defines a higher-level interface that makes the subsystem easier to use.

其实，在日常编码工作中，我们都在有意无意地大量使用门面模式。但凡只要高层模块需要调度多个子系统（2 个以上类对象），我们都会自觉地创建一个新类封装这些子系统，提供精简的接口，让高层模块可以更加容易地间接调用这些子系统的功能。尤其是现阶段各种第三方 SDK、开源类库，很大概率都会使用门面模式。大家觉得调用方便的，一般门面模式使用得更多。

12.1.2 门面模式的应用场景

在日常生活中，门面模式也是很常见的。比如，我们去医院就诊，很多医院都设置了导诊台，这个导诊台就好比一个门面。有了这个导诊台，我们全程就诊都不需要到处乱转，就诊路线变得非常清楚。再比如，现在中国就要全面进入小康社会，很多农村家家户户都建起了小别墅。那么，建别墅也是一项很复杂的工程。在以前，都是相互帮忙把房子建起来，但是建别墅一般要找一个承建方，负责设计、施工等。我们通常说的包工头其实就是一个门面，在施工过程中有任何需要协调对接的找包工头就可以了。

在软件系统中，门面模式适用于以下应用场景。

（1）为一个复杂的模块或子系统提供一个简洁的供外界访问的接口。

（2）希望提高子系统的独立性时。

（3）当子系统由于不可避免的暂时原因导致可能存在 Bug 或性能相关问题时，可以通过门面模式提供一个高层接口，隔离客户端与子系统的直接交互，预防代码污染。

12.1.3 门面模式的 UML 类图

门面模式的 UML 类图如下。

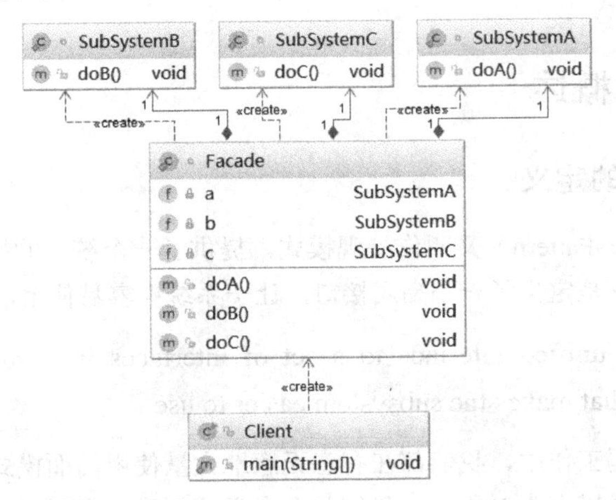

由上图可以看到，门面模式主要包含 2 个角色。

（1）外观角色（Facade）：也叫作门面角色，是系统对外的统一接口。

（2）子系统角色（SubSystem）：可以同时有一个或多个 SubSystem。每个 SubSytem 都不是一个单独的类，而是一个类的集合。SubSystem 并不知道 Facade 的存在，对于 SubSystem 而言，Facade 只是另一个客户端而已（即 Facade 对 SubSystem 透明）。

12.1.4　门面模式的通用写法

下面是门面模式的通用写法，代码很简单。

```
public class Client {
    //客户
    public static void main(String[] args) {
        Facade facade = new Facade();
        facade.doA();
        facade.doB();
        facade.doC();
    }

    //子系统 A
    static class SubSystemA {
        public void doA() {
            System.out.println("doing A stuff");
        }
    }

    //子系统 B
    static class SubSystemB {
        public void doB() {
            System.out.println("doing B stuff");
        }
    }

    //子系统 C
    static class SubSystemC {
        public void doC() {
            System.out.println("doing C stuff");
        }
    }

    //外观角色
    static class Facade {
        private SubSystemA a = new SubSystemA();
        private SubSystemB b = new SubSystemB();
        private SubSystemC c = new SubSystemC();

        //对外接口
        public void doA() {
            this.a.doA();
        }

        //对外接口
        public void doB() {
```

```
            this.b.doB();
        }

        //对外接口
        public void doC() {
            this.c.doC();
        }
    }
}
```

12.2 使用门面模式整合已知 API 的功能

GPer 社区上线了一个积分兑换礼品的商城，这个商城中的大部分功能都不是重新开发的，而是要去对接已有的各个子系统，如下图所示。

这些子系统可能涉及积分系统、支付系统、物流系统的接口调用。如果所有的接口调用全部由前端发送网络请求去调用现有接口，一则会增加前端开发人员的难度，二则会增加一些网络请求，影响页面性能。此时就可以发挥门面模式的优势了。将所有现成的接口全部整合到一个类中，由后端提供统一的接口供前端调用，这样前端开发人员就不需要关心各接口的业务关系，只需要把精力集中在页面交互上。我们用代码来模拟一下这个场景。

首先创建礼品的实体类 GiftInfo。

```
public class GiftInfo {
    private String name;

    public GiftInfo(String name) {
        this.name = name;
```

```
    }

    public String getName() {
        return name;
    }
}
```

然后编写各个子系统的业务逻辑代码，创建积分系统 QualifyService 类。

```
public class QualifyService {
    public boolean isAvailable(GiftInfo giftInfo){
        System.out.println("校验" + giftInfo.getName() + " 积分资格通过,库存通过");
        return true;
    }
}
```

创建支付系统 PaymentService 类。

```
public class PaymentService {
    public boolean pay(GiftInfo pointsGift){
        //扣减积分
        System.out.println("支付" + pointsGift.getName() + " 积分成功");
        return true;
    }
}
```

创建物流系统 ShippingService 类。

```
public class ShippingService {

    //发货
    public String delivery(GiftInfo giftInfo){
        //物流系统的对接逻辑
        System.out.println(giftInfo.getName() + "进入物流系统");
        String shippingOrderNo = "666";
        return shippingOrderNo;
    }
}
```

接着创建外观角色 GiftFacadeService 类，对外只开放一个兑换礼物的 exchange()方法，在 exchange()方法内部整合 3 个子系统的所有功能。

```
public class GiftFacadeService {
    private QualifyService qualifyService = new QualifyService();
    private PaymentService pointsPaymentService = new PaymentService();
    private ShippingService shippingService = new ShippingService();

    //兑换
```

```
    public void exchange(GiftInfo giftInfo){
        if(qualifyService.isAvailable(giftInfo)){
            //资格校验通过
            if(pointsPaymentService.pay(giftInfo)){
                //如果支付积分成功
                String shippingOrderNo = shippingService.delivery(giftInfo);
                System.out.println("物流系统下单成功,订单号是:"+shippingOrderNo);
            }
        }
    }
}
```

最后来看客户端代码。

```
    public static void main(String[] args) {
        GiftInfo giftInfo = new GiftInfo("《Spring 5 核心原理》");
        GiftFacadeService giftFacadeService = new GiftFacadeService();
        giftFacadeService.exchange(giftInfo);
    }
```

运行结果如下图所示。

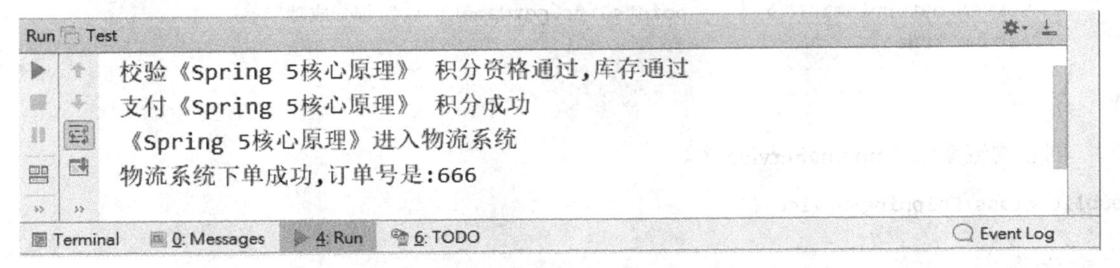

通过这样一个案例对比，相信大家对门面模式的印象就非常深刻了。

12.3 门面模式在框架源码中的应用

12.3.1 门面模式在 Spring 源码中的应用

先来看 Spring JDBC 模块下的 JdbcUtils 类，它封装了与 JDBC 相关的所有操作，代码片段如下。

```
public abstract class JdbcUtils {
    public static final int TYPE_UNKNOWN = -2147483648;
    private static final Log logger = LogFactory.getLog(JdbcUtils.class);
```

```
    public JdbcUtils() {
    }

    public static void closeConnection(Connection con) {
        if(con != null) {
            try {
                con.close();
            } catch (SQLException var2) {
                logger.debug("Could not close JDBC Connection", var2);
            } catch (Throwable var3) {
                logger.debug("Unexpected exception on closing JDBC Connection", var3);
            }
        }

    }

    public static void closeStatement(Statement stmt) {
        if(stmt != null) {
            try {
                stmt.close();
            } catch (SQLException var2) {
                logger.trace("Could not close JDBC Statement", var2);
            } catch (Throwable var3) {
                logger.trace("Unexpected exception on closing JDBC Statement", var3);
            }
        }

    }

    public static void closeResultSet(ResultSet rs) {
        if(rs != null) {
            try {
                rs.close();
            } catch (SQLException var2) {
                logger.trace("Could not close JDBC ResultSet", var2);
            } catch (Throwable var3) {
                logger.trace("Unexpected exception on closing JDBC ResultSet", var3);
            }
        }

    }
    ...
}
```

更多其他操作，看它的结构就非常清楚了，如下图所示。

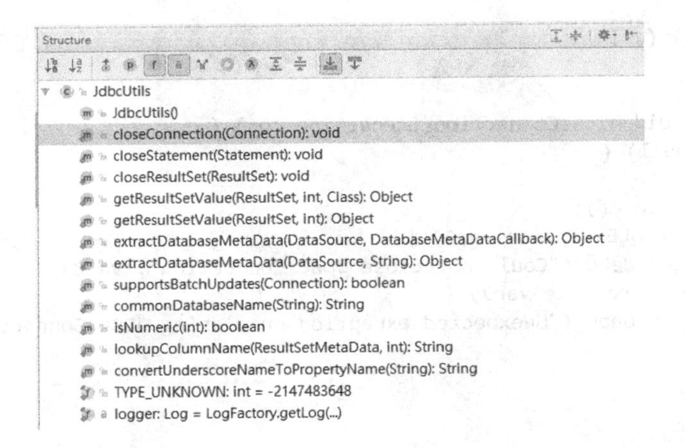

12.3.2 门面模式在 MyBatis 源码中的应用

再来看一个 MyBatis 中的 Configuration 类，其中有很多 new 开头的方法，源码如下。

```
public MetaObject newMetaObject(Object object) {
      return MetaObject.forObject(object, this.objectFactory, this.objectWrapperFactory,
this.reflectorFactory);
   }

public  ParameterHandler  newParameterHandler(MappedStatement  mappedStatement,  Object
parameterObject,
BoundSql boundSql) {
      ParameterHandler parameterHandler =
             mappedStatement.getLang().createParameterHandler(mappedStatement,
             parameterObject, boundSql);
      parameterHandler                                                              =
(ParameterHandler)this.interceptorChain.pluginAll(parameterHandler);
      return parameterHandler;
   }

   public ResultSetHandler newResultSetHandler(Executor executor, MappedStatement mappedStatement,
      RowBounds rowBounds, ParameterHandler parameterHandler, ResultHandler resultHandler,
BoundSql boundSql) {
      ResultSetHandler resultSetHandler =
      new DefaultResultSetHandler(executor, mappedStatement, parameterHandler, resultHandler,
boundSql, rowBounds);
      ResultSetHandler  resultSetHandler  =  (ResultSetHandler)this.interceptorChain.pluginAll
(resultSetHandler);
      return resultSetHandler;
   }
```

```
    public StatementHandler newStatementHandler(Executor executor, MappedStatement mappedStatement,
                Object parameterObject, RowBounds rowBounds, ResultHandler resultHandler,
                    BoundSql boundSql) {
        StatementHandler statementHandler =
                    new RoutingStatementHandler(executor, mappedStatement, parameterObject,
                    rowBounds, resultHandler, boundSql);
        StatementHandler statementHandler = (StatementHandler)this.interceptorChain.pluginAll
                    (statementHandler);
        return statementHandler;
    }

    public Executor newExecutor(Transaction transaction) {
        return this.newExecutor(transaction, this.defaultExecutorType);
}
```

上面这些方法都是对 JDBC 中关键组件操作的封装。

12.3.3　门面模式在 Tomcat 源码中的应用

另外，门面模式在 Tomcat 的源码中也有体现，也非常有意思。以 RequestFacade 类为例，来看其源码。

```
public class RequestFacade implements HttpServletRequest {
...
@Override
    public String getContentType() {

        if (request == null) {
            throw new IllegalStateException(
                            sm.getString("requestFacade.nullRequest"));
        }

        return request.getContentType();
    }


    @Override
    public ServletInputStream getInputStream() throws IOException {

        if (request == null) {
            throw new IllegalStateException(
                            sm.getString("requestFacade.nullRequest"));
        }

        return request.getInputStream();
    }
```

```
    @Override
    public String getParameter(String name) {

        if (request == null) {
            throw new IllegalStateException(
                        sm.getString("requestFacade.nullRequest"));
        }

        if (Globals.IS_SECURITY_ENABLED){
            return AccessController.doPrivileged(
                new GetParameterPrivilegedAction(name));
        } else {
            return request.getParameter(name);
        }
    }
...
}
```

从名字就知道它用了门面模式。它封装了非常多的 request 操作，也整合了很多 servlet-api 以外的内容，给用户使用提供了很大便捷。同样，Tomcat 针对 Response 和 Session 也封装了对应的 ResponseFacade 类和 StandardSessionFacade 类，感兴趣的小伙伴可以深入了解一下。

12.4 门面模式扩展

12.4.1 门面模式的优点

（1）简化了调用过程，不用深入了解子系统，以防给子系统带来风险。

（2）减少系统依赖，松散耦合。

（3）更好地划分访问层次，提高了安全性。

（4）遵循迪米特法则。

12.4.2 门面模式的缺点

（1）当增加子系统和扩展子系统行为时，可能容易带来未知风险。

（2）不符合开闭原则。

（3）某些情况下，可能违背单一职责原则。

13

第 13 章 装饰器模式

13.1 装饰器模式概述

13.1.1 装饰器模式的定义

装饰器模式（Decorator Pattern）也叫作包装器模式（Wrapper Pattern），指在不改变原有对象的基础上，动态地给一个对象添加一些额外的职责。就增加功能来说，装饰器模式相比生成子类更为灵活，属于结构型设计模式。

原文： Attach additional responsibilities to an object dynamically keeping the same interface. Decorators provide a flexible alternative to subclassing for extending functionality.

装饰器模式提供了比继承更有弹性的替代方案（扩展原有对象的功能）将功能附加到对象上。因此，装饰器模式的核心是功能扩展。使用装饰器模式可以透明且动态地扩展类的功能。

13.1.2 装饰器模式的应用场景

来看这样一个场景，上班族大多有睡懒觉的习惯，每天早上上班时间都很紧张，于是很多人为了多睡一会儿，就用更方便的方式解决早餐问题，有些人早餐可能会吃煎饼。煎饼中可以加鸡蛋，也可以加香肠，但是不管怎么“加码”，都还是一个煎饼。再比如，给蛋糕加上一些水果，给房子装修，都是装饰器模式。

装饰器模式在代码程序中适用于以下应用场景。

（1）用于扩展一个类的功能，或者给一个类添加附加职责。

（2）动态地给一个对象添加功能，这些功能可以再动态地被撤销。

（3）需要为一批平行的兄弟类进行改装或加装功能。

13.1.3 装饰器模式的 UML 类图

装饰器模式的 UML 类图如下。

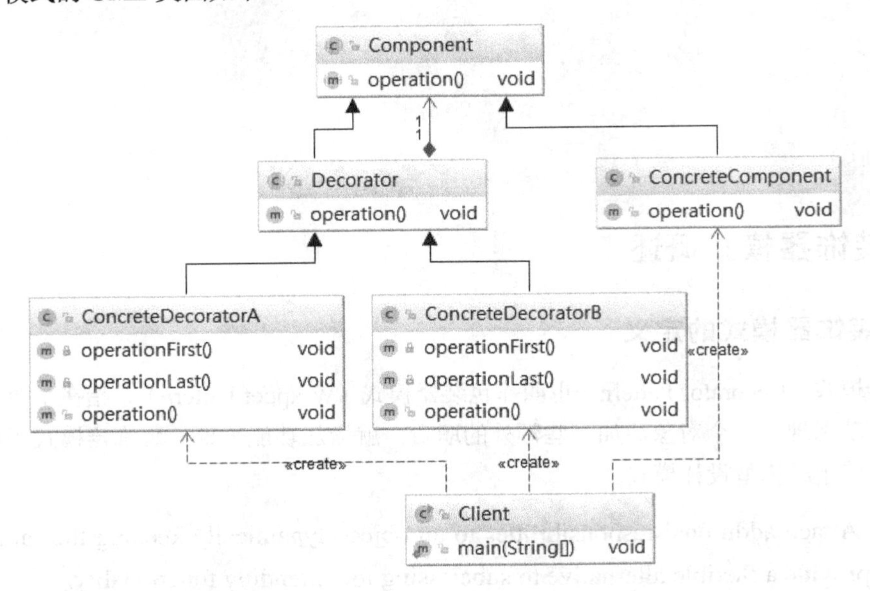

由上图可以看到，装饰器模式主要包含 4 个角色。

（1）抽象组件（Component）：可以是一个接口或者抽象类，充当被装饰类的原始对象，规定了被装饰对象的行为。

（2）具体组件（ConcreteComponent）：实现/继承 Component 的一个具体对象，即被装饰对象。

（3）抽象装饰器（Decorator）：通用的装饰 ConcreteComponent 的装饰器，其内部必然有一个属性指向 Component；其实现一般是一个抽象类，主要为了让其子类按照其构造形式传入一个

Component，这是强制的通用行为。如果系统中装饰逻辑单一，则并不需要实现许多装饰器，可以直接省略该类，而直接实现一个具体装饰器即可。

（4）具体装饰器（ConcreteDecorator）：Decorator 的具体实现类，理论上，每个 ConcreteDecorator 都扩展了 Component 对象的一种功能。

总结：装饰器模式角色分配符合设计模式的里氏替换原则、依赖倒置原则，从而使得其具备很强的扩展性，最终满足开闭原则。

装饰器模式的实现原理是，让装饰器实现与被装饰类（例如 ConcreteComponent）相同的接口（例如 Component），使得装饰器与被扩展类类型一致，并在构造函数中传入该接口对象，然后在实现这个接口的被包装类对象的现有功能上添加新功能。由于装饰器与被包装类属于同一类型（均为 Component），且构造函数的参数为其实现接口类（Component），因此装饰器模式具备嵌套扩展功能，这样就能使用装饰器模式一层一层地对底层被包装类进行功能扩展了。

13.1.4　装饰器模式的通用写法

下面是装饰器模式的通用写法。

```
public class Client{
    public static void main(String[] args){
        Component c1 = new ConcreteComponent (); //首先创建需要被装饰的原始对象（即要被装饰的对象）
        Decorator decoratorA = new ConcreteDecoratorA(c1); //给对象透明地增加功能 A 并调用
        decoratorA .operation();
        Decorator decoratorB = new ConcreteDecoratorB(c1); //给对象透明地增加功能 B 并调用
        decoratorB .operation();
        //装饰器也可以装饰具体的装饰对象，此时相当于给对象在增加功能 A 的基础上再增加功能 B
        Decorator decoratorBandA = new ConcreteDecoratorB(decoratorA);
        decoratorBandA.operation();
    }

    //抽象组件
    static abstract class Component {
        /**
         * 示例方法
         */
        public abstract void operation();
    }

    //具体组件
    static class ConcreteComponent extends Component {
        public void operation() {
            //相应的功能处理
```

```
        System.out.println("处理业务逻辑");
    }
}

static abstract class Decorator extends Component {
    /**
     * 持有组件对象
     */
    protected Component component;

    /**
     * 构造方法，传入组件对象
     * @param component 组件对象
     */
    public Decorator(Component component) {
        this.component = component;
    }

    public void operation() {
        //转发请求给组件对象，可以在转发前后执行一些附加动作
        component.operation();
    }
}

//具体装饰器 A
static class ConcreteDecoratorA extends Decorator {
    public ConcreteDecoratorA(Component component) {
        super(component);
    }
    private void operationFirst(){ } //在调用父类的 operation 方法之前需要执行的操作
    private void operationLast(){ }  //在调用父类的 operation 方法之后需要执行的操作
    public void operation() {
        //调用父类的方法，可以在调用前后执行一些附加动作
        operationFirst();              //添加的功能
         //这里可以选择性地调用父类的方法，
         //如果不调用，则相当于完全改写了方法，实现了新的功能
        super.operation();
        operationLast();               //添加的功能
    }
}

//具体装饰器 B
static class ConcreteDecoratorB extends Decorator {
    public ConcreteDecoratorB(Component component) {
        super(component);
    }
    private void operationFirst(){ } //在调用父类的 operation 方法之前需要执行的操作
```

```
        private void operationLast(){ }  //在调用父类的 operation 方法之后需要执行的操作
        public void operation() {
            //调用父类的方法，可以在调用前后执行一些附加动作
            operationFirst();               //添加的功能
            super.operation();
            operationLast();                //添加的功能
        }
    }
}
```

13.2　使用装饰器模式解决实际问题

13.2.1　使用装饰器模式解决煎饼“加码”问题

下面用代码来模拟给煎饼“加码”的业务场景，先来看不用装饰器模式的情况。首先创建一个煎饼 Battercake 类。

```
public class Battercake {

    protected String getMsg(){
        return "煎饼";
    }

    public int getPrice(){
        return 5;
    }

}
```

然后创建一个加鸡蛋的煎饼 BattercakeWithEgg 类。

```
public class BattercakeWithEgg extends Battercake{
    @Override
    protected String getMsg() {
        return super.getMsg() + "+1 个鸡蛋";
    }

    @Override
    //加 1 个鸡蛋加 1 元钱
    public int getPrice() {
        return super.getPrice() + 1;
    }
}
```

再创建一个既加鸡蛋又加香肠的 BattercakeWithEggAndSausage 类。

```
public class BattercakeWithEggAndSausage extends BattercakeWithEgg{
    @Override
    protected String getMsg() {
        return super.getMsg() + "+1 根香肠";
    }

    @Override
    //加 1 根香肠加 2 元钱
    public int getPrice() {
        return super.getPrice() + 2;
    }
}
```

最后编写客户端测试代码。

```
public static void main(String[] args) {

        Battercake battercake = new Battercake();
        System.out.println(battercake.getMsg() + ",总价格：" + battercake.getPrice());

        Battercake battercakeWithEgg = new BattercakeWithEgg();
        System.out.println(battercakeWithEgg.getMsg() + ",总价格：" +
                battercakeWithEgg.getPrice());

        Battercake battercakeWithEggAndSausage = new BattercakeWithEggAndSausage();
        System.out.println(battercakeWithEggAndSausage.getMsg() + ",总价格：" +
                battercakeWithEggAndSausage.getPrice());

    }
```

运行结果如下图所示。

```
D:\Java\jdk1.8.0_151\bin\java ...
煎饼,总价格：5
煎饼+1个鸡蛋,总价格：6
煎饼+1个鸡蛋+1根香肠,总价格：8
```

运行结果没有问题。但是，如果用户需要一个加 2 个鸡蛋和 1 根香肠的煎饼，则用现在的类结构是创建不出来的，也无法自动计算出价格，除非再创建一个类做定制。如果需求再变，那么一直加定制显然是不科学的。

下面用装饰器模式来解决上面的问题。首先创建一个煎饼的抽象 Battercake 类。

```
public abstract class Battercake {
    protected abstract String getMsg();
    protected abstract int getPrice();
}
```

创建一个基本的煎饼（或者叫基础套餐）BaseBattercake。

```
public class BaseBattercake extends Battercake {
    protected String getMsg(){
        return "煎饼";
    }

    public int getPrice(){ return 5;  }
}
```

然后创建一个扩展套餐的抽象装饰器 BattercakeDecotator 类。

```
public abstract class BattercakeDecorator extends Battercake {
    //静态代理，委派
    private Battercake battercake;

    public BattercakeDecorator(Battercake battercake) {
        this.battercake = battercake;
    }
    protected abstract void doSomething();

    @Override
    protected String getMsg() {
        return this.battercake.getMsg();
    }
    @Override
    protected int getPrice() {
        return this.battercake.getPrice();
    }
}
```

接着创建鸡蛋装饰器 EggDecorator 类。

```
public class EggDecorator extends BattercakeDecorator {
    public EggDecorator(Battercake battercake) {
        super(battercake);
    }

    protected void doSomething() {}

    @Override
    protected String getMsg() {
        return super.getMsg() + "+1 个鸡蛋";
    }

    @Override
    protected int getPrice() {
```

```
        return super.getPrice() + 1;
    }
}
```

创建香肠装饰器 SausageDecorator 类。

```
public class SausageDecorator extends BattercakeDecorator {
    public SausageDecorator(Battercake battercake) {
        super(battercake);
    }

    protected void doSomething() {}

    @Override
    protected String getMsg() {
        return super.getMsg() + "+1 根香肠";
    }
    @Override
    protected int getPrice() {
        return super.getPrice() + 2;
    }
}
```

再编写客户端测试代码。

```
public class BattercakeTest {
    public static void main(String[] args) {
        Battercake battercake;
        //买一个煎饼
        battercake = new BaseBattercake();
        //煎饼有点小，想再加 1 个鸡蛋
        battercake = new EggDecorator(battercake);
        //再加 1 个鸡蛋
        battercake = new EggDecorator(battercake);
        //很饿，再加 1 根香肠
        battercake = new SausageDecorator(battercake);

        //与静态代理的最大区别就是职责不同
        //静态代理不一定要满足 is-a 的关系
        //静态代理会做功能增强，同一个职责变得不一样

        //装饰器更多考虑的是扩展
        System.out.println(battercake.getMsg() + ",总价: " + battercake.getPrice());
    }
}
```

运行结果如下图所示。

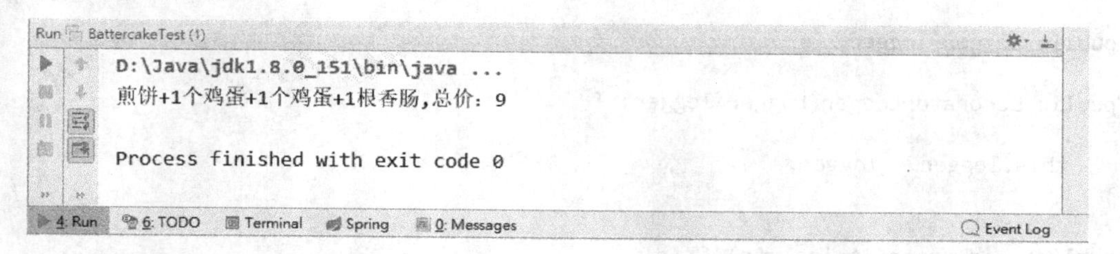

最后来看类图，如下图所示。

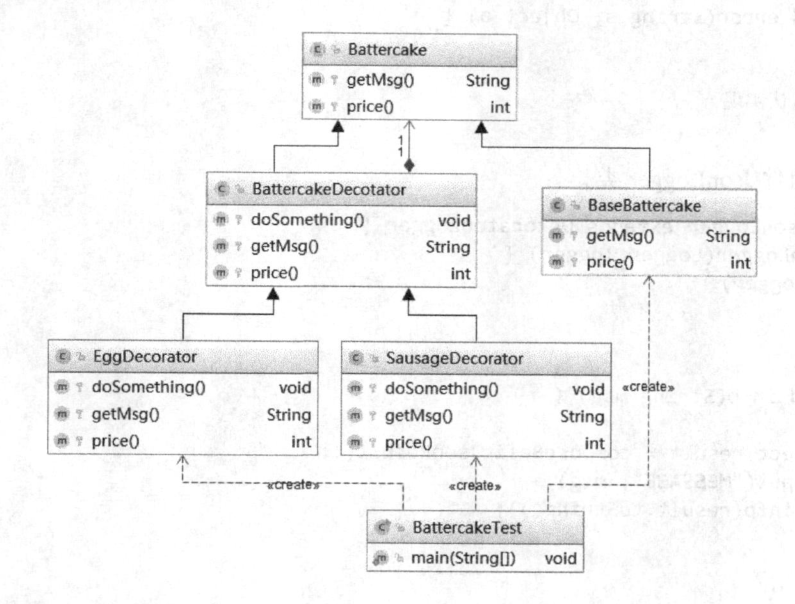

13.2.2　使用装饰器模式扩展日志格式输出

为了加深印象，我们再来看一个应用场景。需求大致是这样的，系统采用的是 SLS 服务监控项目日志，以 JSON 格式解析，因此需要将项目中的日志封装成 JSON 格式再打印。现有的日志体系采用 Log4j + Slf4j 框架搭建而成。客户端调用如下。

```
private static final Logger logger = LoggerFactory.getLogger(Component.class);
logger.error(string);
```

这样打印出来的是毫无规则的一行行字符串。当考虑将其转换成 JSON 格式时，笔者采用装饰器模式。目前有的是统一接口 Logger 和其具体实现类，笔者要加的就是一个装饰类和真正封装成 JSON 格式的装饰产品类。创建装饰器类 DecoratorLogger。

```
public class DecoratorLogger implements Logger {

    public Logger logger;

    public DecoratorLogger(Logger logger) {

        this.logger = logger;
    }

    public void error(String str) {}

    public void error(String s, Object o) {

    }
    //省略其他默认实现
}
```

创建具体组件 JsonLogger 类。

```
public class JsonLogger extends DecoratorLogger {
    public JsonLogger(Logger logger) {
        super(logger);
    }

    @Override
    public void info(String msg) {

        JSONObject result = composeBasicJsonResult();
        result.put("MESSAGE", msg);
        logger.info(result.toString());
    }

    @Override
    public void error(String msg) {

        JSONObject result = composeBasicJsonResult();
        result.put("MESSAGE", msg);
        logger.error(result.toString());
    }

    public void error(Exception e) {

        JSONObject result = composeBasicJsonResult();
        result.put("EXCEPTION", e.getClass().getName());
        String exceptionStackTrace = Arrays.toString(e.getStackTrace());
        result.put("STACKTRACE", exceptionStackTrace);
        logger.error(result.toString());
```

```
    }

    private JSONObject composeBasicJsonResult() {
        //拼装了一些运行时的信息
        return new JSONObject();
    }
}
```

可以看到，在 JsonLogger 中，对于 Logger 的各种接口，我们都用 JsonObject 对象进行一层封装。在打印的时候，最终还是调用原生接口 logger.error(string)，只是这个 String 参数已经被装饰过了。如果有额外的需求，则可以再写一个函数去实现。比如 error(Exception e)，只传入一个异常对象，这样在调用时就非常方便。

另外，为了在新老交替的过程中尽量不改变太多代码和使用方式，笔者又在 JsonLogger 中加入了一个内部的工厂类 JsonLoggerFactory（这个类转移到 DecoratorLogger 中可能更好一些）。它包含一个静态方法，用于提供对应的 JsonLogger 实例。最终在新的日志体系中，使用方式如下。

```
private static final Logger logger = JsonLoggerFactory.getLogger(Client.class);
public static void main(String[] args) {

    logger.error("错误信息");
}
```

对于客户端而言，唯一与原先不同的地方就是将 LoggerFactory 改为 JsonLoggerFactory 即可，这样的实现，也会更快更方便地被其他开发者接受和习惯。最后看如下图所示的类图。

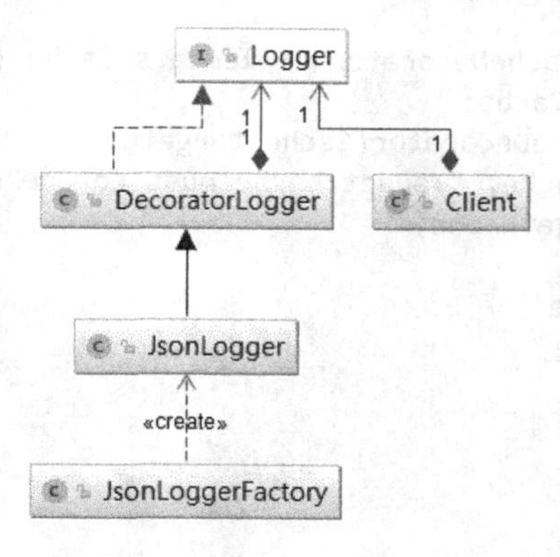

装饰器模式最本质的特征是将原有类的附加功能抽离出来，简化原有类的逻辑。通过这样两个案例，我们可以总结出来，其实抽象的装饰器是可有可无的，具体可以根据业务模型来选择。

13.3 装饰器模式在框架源码中的应用

13.3.1 装饰器模式在 JDK 源码中的应用

装饰器模式在源码中应用得非常多，在 JDK 中体现最明显的类就是与 I/O 相关的类，如 BufferedReader、InputStream、OutputStream，看一下常用的 InputStream 的类图，如下图所示。

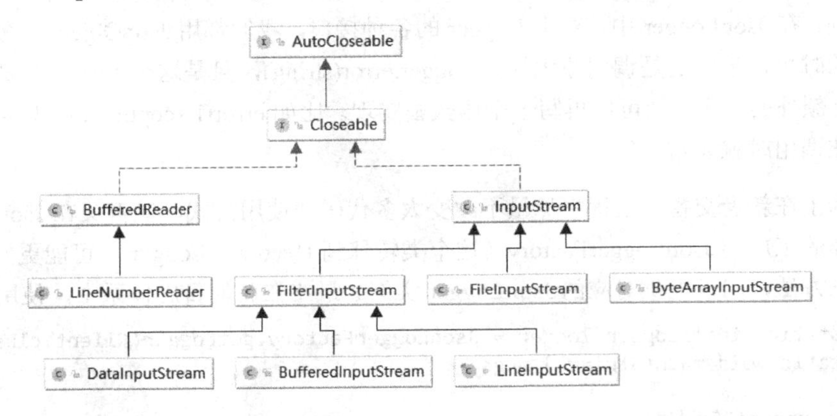

13.3.2 装饰器模式在 Spring 源码中的应用

在 Spring 中，我们可以尝试理解一下 TransactionAwareCacheDecorator 类，这个类主要用来处理事务缓存，代码如下。

```
public class TransactionAwareCacheDecorator implements Cache {
    private final Cache targetCache;
    public TransactionAwareCacheDecorator(Cache targetCache) {
        Assert.notNull(targetCache, "Target Cache must not be null");
        this.targetCache = targetCache;
    }
    public Cache getTargetCache() {
        return this.targetCache;

    }
    ...
}
```

TransactionAwareCacheDecorator 就是对 Cache 的一个包装。再来看一个 MVC 中的装饰器模式 HttpHeadResponseDecorator 类。

```
public class HttpHeadResponseDecorator extends ServerHttpResponseDecorator {
```

```
    public HttpHeadResponseDecorator(ServerHttpResponse delegate) {
        super(delegate);
    }
    ...
}
```

13.3.3 装饰器模式在 MyBatis 源码中的应用

最后，来看 MyBatis 中一段处理缓存的设计 org.apache.ibatis.cache.Cache 类，找到它的包定位，如下图所示。

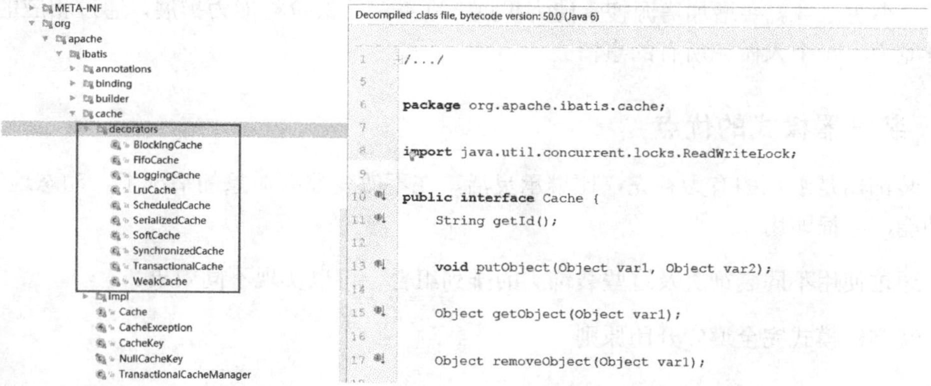

```
META-INF
org
  apache
    ibatis
      annotations
      binding
      builder
      cache
        decorators
          BlockingCache
          FifoCache
          LoggingCache
          LruCache
          ScheduledCache
          SerializedCache
          SoftCache
          SynchronizedCache
          TransactionalCache
          WeakCache
        impl
        Cache
        CacheException
        CacheKey
        NullCacheKey
        TransactionalCacheManager
```

Decompiled .class file, bytecode version: 50.0 (Java 6)

```
1   /.../
5
6   package org.apache.ibatis.cache;
7
8   import java.util.concurrent.locks.ReadWriteLock;
9
10  public interface Cache {
11      String getId();
12
13      void putObject(Object var1, Object var2);
14
15      Object getObject(Object var1);
16
17      Object removeObject(Object var1);
```

从名字上来看，其实更容易理解。比如，FifoCache（先入先出算法的缓存）、LruCache（最近最少使用的缓存）、TransactionlCache（事务相关的缓存）都采用装饰器模式。

13.4 装饰器模式扩展

13.4.1 装饰器模式与代理模式的区别

从代理模式的 UML 类图和通用代码实现上看，代理模式与装饰器模式几乎一模一样。代理模式的 Subject 对应装饰器模式的 Component，代理模式的 RealSubject 对应装饰器模式的 Concrete Component，代理模式的 Proxy 对应装饰器模式的 Decorator。确实，从代码实现上看，代理模式的确与装饰器模式是一样的（其实装饰器模式就是代理模式的一个特殊应用），但是这两种设计模式所面向的功能扩展面是不一样的。

装饰器模式强调自身功能的扩展。Decorator 所做的就是增强 Concrete Component 的功能（也有可能减弱功能），主体对象为 Concrete Component，着重类功能的变化。

代理模式强调对代理过程的控制。Proxy 完全掌握对 RealSubject 的访问控制，因此，Proxy 可以决定对 RealSubject 进行功能扩展、功能缩减甚至功能散失（不调用 RealSubject 方法），主体对象为 Proxy。

简单来讲，假设现在小明想租房，那么势必会有一些事务发生：房源搜索、联系房东谈价格等。

假设按照代理模式进行思考，那么小明只需找到一个房产中介，让他去做房源搜索、联系房东谈价格这些事情，小明只需等待通知然后付中介费就行了。

而如果采用装饰器模式进行思考，因为装饰器模式强调的是自身功能扩展，也就是说，如果要找房子，小明自身就要增加房源搜索能力扩展、联系房东谈价格能力扩展，通过相应的装饰器，提升自身能力，一个人做完所有的事情。

13.4.2 装饰器模式的优点

（1）装饰器是继承的有力补充，比继承灵活，在不改变原有对象的情况下，动态地给一个对象扩展功能，即插即用。

（2）通过使用不同装饰类及这些装饰类的排列组合，可以实现不同效果。

（3）装饰器模式完全遵守开闭原则。

13.4.3 装饰器模式的缺点

（1）会出现更多的代码、更多的类，增加程序的复杂性。

（2）动态装饰在多层装饰时会更复杂。

14

第 14 章 享元模式

14.1 享元模式概述

面向对象技术可以很好地解决一些灵活性或可扩展性问题，但在很多情况下需要在系统中增加类和对象的个数。当对象数量太多时，将导致运行代价过高，带来性能下降等问题。享元模式正是为解决这一类问题而诞生的。

14.1.1 享元模式的定义

享元模式（Flyweight Pattern）又叫作轻量级模式，是对象池的一种实现。类似线程池，线程池可以避免不停地创建和销毁多个对象，消耗性能。享元模式提供了减少对象数量从而改善应用所需的对象结构的方式。其宗旨是共享细粒度对象，将多个对同一对象的访问集中起来，不必为每个访问者都创建一个单独的对象，以此来降低内存的消耗，属于结构型设计模式。

原文：Use sharing to support large numbers of fine-grained objects efficiently.

享元模式把一个对象的状态分成内部状态和外部状态，内部状态是不变的，外部状态是变化的；然后通过共享不变的部分，达到减少对象数量并节约内存的目的。

享元模式的本质是缓存共享对象，降低内存消耗。

14.1.2 享元模式的应用场景

在生活中，享元模式非常常见，比如各中介机构的房源共享，再比如全国社保联网。

当系统中多处需要同一组信息时，可以把这些信息封装到一个对象中，然后对该对象进行缓存，这样，一个对象就可以提供给多处需要使用的地方，避免大量同一对象的多次创建，降低大量内存空间的消耗。

享元模式其实是工厂方法模式的一个改进机制，享元模式同样要求创建一个或一组对象，并且就是通过工厂方法模式生成对象的，只不过享元模式为工厂方法模式增加了缓存这一功能。主要应用场景如下。

（1）常应用于系统底层的开发，以便解决系统的性能问题。

（2）系统有大量相似对象、需要缓冲池的场景。

14.1.3 享元模式的 UML 类图

享元模式的 UML 类图如下。

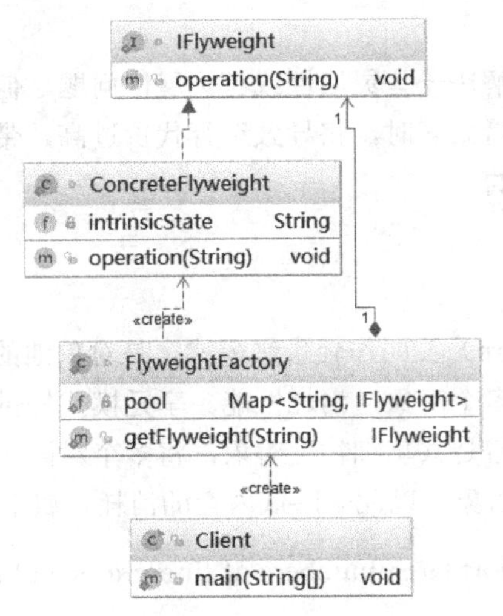

由上图可以看到，享元模式主要包含 3 个角色。

（1）抽象享元角色（IFlyweight）：享元对象抽象基类或者接口，同时定义出对象的外部状态和内部状态的接口或实现。

（2）具体享元角色（ConcreteFlyweight）：实现抽象角色定义的业务。该角色的内部状态处理应该与环境无关，不会出现一个操作改变内部状态、同时修改了外部状态的情况。

（3）享元工厂（FlyweightFactory）：负责管理享元对象池和创建享元对象。

14.1.4　享元模式的通用写法

下面是享元模式的通用写法。

```
public class Client {
    public static void main(String[] args) {
        FlyweightFactory flyweightFactory = new FlyweightFactory();
        IFlyweight flyweight1 = flyweightFactory.getFlyweight("aa");
        IFlyweight flyweight2 = flyweightFactory.getFlyweight("bb");
        flyweight1.operation("a");
        flyweight2.operation("b");
    }

    interface IFlyweight {
        void operation(String extrinsicState);
    }

    //具体享元角色
    static class ConcreteFlyweight implements IFlyweight {
        private String intrinsicState;

        public ConcreteFlyweight(String intrinsicState) {
            this.intrinsicState = intrinsicState;
        }


        public void operation(String extrinsicState) {
            System.out.println("Object address: " + System.identityHashCode(this));
            System.out.println("IntrinsicState: " + this.intrinsicState);
            System.out.println("ExtrinsicState: " + extrinsicState);
        }
    }

    //享元工厂
    static class FlyweightFactory {

        private static Map<String, IFlyweight> pool = new HashMap<String, IFlyweight>();

        //因为内部状态具备不变性，所以作为缓存的键
```

```
        public static IFlyweight getFlyweight(String intrinsicState) {
            if (!pool.containsKey(intrinsicState)) {
                IFlyweight flyweight = new ConcreteFlyweight(intrinsicState);
                pool.put(intrinsicState, flyweight);
            }
            return pool.get(intrinsicState);
        }
    }
}
```

14.2 使用享元模式解决实际问题

14.2.1 使用享元模式实现资源共享池

举个例子，每年春节为了买到一张回家的火车票，大家都要大费周章。为了解决这一问题，12306 网站提供了自动查票的功能。如果开启自动查票功能，则系统会将我们填写的信息缓存起来，然后定时查询余票信息。在买票的时候，我们肯定要查询一下有没有我们需要的车票，假设一张火车票包含出发站、目的站、价格、座位类别等信息。现在要求编写火车票查询模拟代码，可以通过出发站、目的站查到相关票的信息。

比如要求通过出发站、目的站查询火车票的相关信息，那么只需构建出火车票类对象，向用户提供一个查询出发站、目的站的接口进行查询即可。

首先创建 ITicket 接口。

```
public interface ITicket {
    void showInfo(String bunk);
}
```

然后创建 TrainTicket 接口。

```
public class TrainTicket implements ITicket {
    private String from;
    private String to;
    private int price;

    public TrainTicket(String from, String to) {
        this.from = from;
        this.to = to;
    }

    public void showInfo(String bunk) {
```

```
        this.price = new Random().nextInt(500);
        System.out.println(String.format("%s->%s: %s 价格: %s 元", this.from, this.to, bunk,
this.price));
    }
}
```

接着创建 TicketFactory 类。

```
public static class TicketFactory {
    public static ITicket queryTicket(String from, String to) {
        return new TrainTicket(from, to);
    }
}
```

最后编写客户端代码如下。

```
public static void main(String[] args) {
       ITicket ticket = TicketFactory.queryTicket("深圳北", "潮汕");
       ticket.showInfo("硬座");
}
```

由上面代码可以知道，当客户端进行查询时，系统通过 TicketFactory 直接创建一个火车票对象，但是这样做的话，当某个瞬间如果有大量用户查询同一张票的信息时，系统就会创建出大量该火车票对象，内存压力骤增。其实更好的做法应该是缓存该火车票对象，然后提供给其他查询请求复用，这样一个对象就足以支撑数以千计的查询请求，内存完全无压力，使用享元模式可以很好地解决这个问题。我们继续优化代码，只需在 TicketFactory 类中进行更改，增加缓存机制。

```
class TicketFactory {
    private static Map<String, ITicket> sTicketPool = new ConcurrentHashMap<String,ITicket>();

    public static ITicket queryTicket(String from, String to) {
        String key = from + "->" + to;
        if (TicketFactory.sTicketPool.containsKey(key)) {
            System.out.println("使用缓存: " + key);
            return TicketFactory.sTicketPool.get(key);
        }
        System.out.println("首次查询，创建对象: " + key);
        ITicket ticket = new TrainTicket(from, to);
        TicketFactory.sTicketPool.put(key, ticket);
        return ticket;
    }
}
```

运行结果如下图所示。

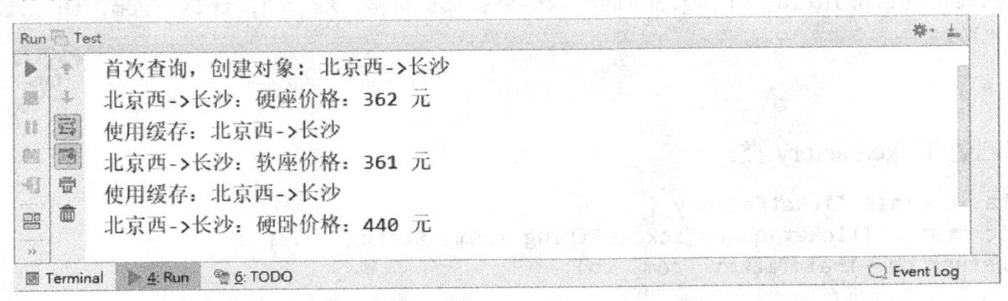

可以看到，除了第一次查询创建对象，后续查询相同车次票的信息都使用缓存对象，不需要创建新对象，其类图如下。

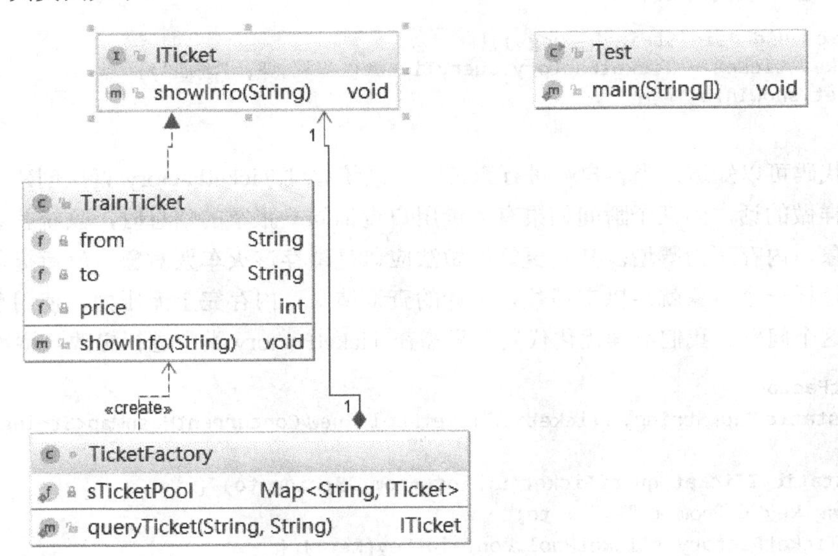

其中，ITicket 是抽象享元角色，TrainTicket 是具体享元角色，TicketFactory 是享元工厂。有些小伙伴一定会有疑惑了，这不就是注册式单例模式吗？对，这就是注册式单例模式。虽然在结构上很像，但是享元模式的重点在结构上，而不在创建对象上。后面结合享元模式在 JDK 源码中的应用，大家应该就能彻底明白了。

14.2.2 使用享元模式实现数据库连接池

再举个例子，我们经常使用的数据库连接池，因为使用 Connection 对象时主要性能消耗在建立连接和关闭连接的时候，为了提高 Connection 对象在调用时的性能，将 Connection 对象在调用前创建好并缓存起来，在用的时候直接从缓存中取值，用完后再放回去，达到资源重复利用的目的，代码如下。

```
public class ConnectionPool {

    private Vector<Connection> pool;

    private String url = "jdbc:mysql://localhost:3306/test";
    private String username = "root";
    private String password = "root";
    private String driverClassName = "com.mysql.jdbc.Driver";
    private int poolSize = 100;

    public ConnectionPool() {
        pool = new Vector<Connection>(poolSize);

        try{
            Class.forName(driverClassName);
            for (int i = 0; i < poolSize; i++) {
                Connection conn = DriverManager.getConnection(url,username,password);
                pool.add(conn);
            }
        }catch (Exception e){
            e.printStackTrace();
        }

    }

    public synchronized Connection getConnection(){
        if(pool.size() > 0){
            Connection conn = pool.get(0);
            pool.remove(conn);
            return conn;
        }
        return null;
    }

    public synchronized void release(Connection conn){
        pool.add(conn);
    }
}
```

这样的连接池普遍应用于开源框架，可以有效提升底层的运行性能。

14.3 享元模式在框架源码中的应用

14.3.1 享元模式在 JDK 源码中的应用

1. String 中的享元模式

Java 中将 String 类定义为由 final 修饰的（不可改变的），JVM 中字符串一般被保存在字符串常量池中，Java 会确保一个字符串在常量池中只有一个“复制”，这个字符串常量池在 JDK 6.0 以前是位于常量池中的，位于永久代；而在 JDK 7.0 中，JVM 将其从永久代拿出来放置于堆中。

我们做一个测试，代码如下。

```
public static void main(String[] args) {
    String s1 = "hello";
    String s2 = "hello";
    String s3 = "he" + "llo";
    String s4 = "hel" + new String("lo");
    String s5 = new String("hello");
    String s6 = s5.intern();
    String s7 = "h";
    String s8 = "ello";
    String s9 = s7 + s8;
    System.out.println(s1==s2);//true
    System.out.println(s1==s3);//true
    System.out.println(s1==s4);//false
    System.out.println(s1==s9);//false
    System.out.println(s4==s5);//false
    System.out.println(s1==s6);//true
}
```

String 类是由 final 修饰的，当以字面量的形式创建 String 变量时，JVM 会在编译期间就把该字面量"hello"放到字符串常量池中，在 Java 启动的时候就已经加载到内存中了。这个字符串常量池的特点就是有且只有一份相同的字面量。如果有其他相同的字面量，则 JVM 返回这个字面量的引用；如果没有相同的字面量，则在字符串常量池中创建这个字面量并返回它的引用。

由于 s2 指向的字面量"hello"在常量池中已经存在（s1 先于 s2），于是 JVM 就返回这个字面量绑定的引用，所以 s1==s2。

s3 中字面量的拼接其实就是"hello"，JVM 在编译期间就已经对它进行了优化，所以 s1 和 s3 也是相等的。

s4 中的 new String("lo")生成了两个对象：lo 和 new String("lo")。lo 存在于字符串常量池中，new String("lo")存在于堆中，String s4 = "hel" + new String("lo")实质上是两个对象的相加，编译器不会进行优化，相加的结果存在于堆中，而 s1 存在于字符串常量池中，当然不相等。s1==s9 的原理也一样。

s4 和 s5 的结果都在堆中，不用说，肯定不相等。

s5.intern()方法能使一个位于堆中的字符串在运行期间动态地加入字符串常量池（字符串常量池的内容是在程序启动的时候就已经加载好了的）。如果字符串常量池中有该对象对应的字面量，则返回该字面量在字符串常量池中的引用；否则，复制一份该字面量到字符串常量池并返回它的引用。因此 s1==s6 输出 true。

2. Integer 中的享元模式

再举一个例子，大家都非常熟悉的对象 Integer 也用到了享元模式，其中暗藏玄机，来看代码。

```
public static void main(String[] args) {
    Integer a = Integer.valueOf(100);
    Integer b = 100;

    Integer c = Integer.valueOf(1000);
    Integer d = 1000;

    System.out.println("a==b:" + (a==b));
    System.out.println("c==d:" + (c==d));
}
```

大家猜它的运行结果是什么？在运行完程序后，我们才发现有些不对，得到了一个意想不到的运行结果，如下图所示。

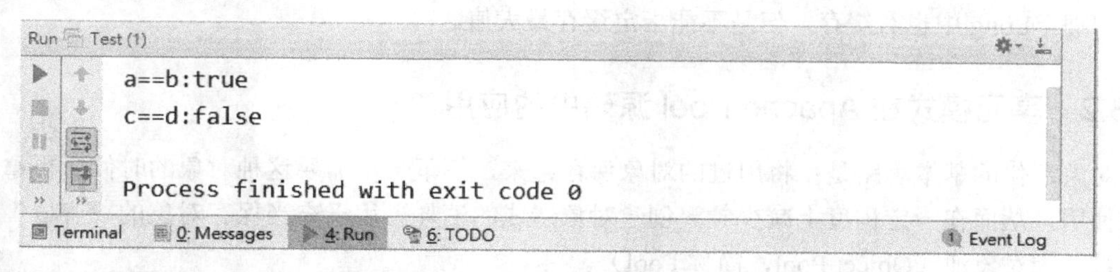

之所以得到这样的结果，是因为 Integer 用到了享元模式，来看 Integer 的源码。

```
public final class Integer extends Number implements Comparable<Integer> {
...
    public static Integer valueOf(int i) {
        if (i >= IntegerCache.low && i <= IntegerCache.high)
            return IntegerCache.cache[i + (-IntegerCache.low)];
```

```
        return new Integer(i);
}
...
}
```

由上可知，Integer 源码中的 valueOf()方法做了一个条件判断，如果目标值在-128～127，则直接从缓存中取值，否则新建对象。JDK 为何要这样做呢？因为-128～127 的数据在 int 范围内是使用最频繁的，为了减少频繁创建对象带来的内存消耗，这里就用到了享元模式，以提高性能。

3. Long 中的享元模式

和 Integer 类似，Long 源码也用到了享元模式，将-128～127 的值缓存起来，源码如下。

```
public final class Long extends Number implements Comparable<Long> {
    public static Long valueOf(long var0) {
        return var0 >= -128L && var0 <= 127L ? Long.LongCache.cache[(int)var0 + 128] : new
Long(var0);
    }
    private static class LongCache {
        private LongCache(){}

        static final Long cache[] = new Long[-(-128) + 127 + 1];

        static {
            for(int i = 0; i < cache.length; i++)
                cache[i] = new Long(i - 128);
        }
    }
    //...
}
```

同理，Long 中也有缓存，但是不能指定缓存最大值。

14.3.2　享元模式在 Apache Pool 源码中的应用

对象池化的基本思路是：将用过的对象保存起来，等下一次需要这种对象的时候，再拿出来重复使用，从而在一定程度上减少频繁创建对象造成的消耗。用于充当保存对象的“容器”的对象，被称为对象池（Object Pool，简称 Pool）。

Apache Pool 实现了对象池的功能，定义了对象的生成、销毁、激活、钝化等操作及其状态转换，并提供几个默认的对象池实现，有如下几个重要的角色。

Pooled Object（池化对象）：用于封装对象（例如，线程、数据库连接和 TCP 连接），将其包裹成可被对象池管理的对象。

Pooled Object Factory（池化对象工厂）：定义了操作 Pooled Object 实例生命周期的一些方法，Pooled Object Factory 必须实现线程安全。

Object Pool（对象池）：Object Pool 负责管理 Pooled Object，例如，借出对象、返回对象、校验对象、有多少激活对象和有多少空闲对象。

在类中定义了一个缓存池化对象的容器，源码如下。

```
public class GenericObjectPool<T> extends BaseGenericObjectPool<T> implements ObjectPool<T>,
GenericObjectPoolMXBean, UsageTracking<T> {
    // 定义池化对象工厂
    private final PooledObjectFactory<T> factory;
    // 缓存池化对象
    private final Map<IdentityWrapper<T>, PooledObject<T>> allObjects;

    ...
}
```

这里不再分析源码中具体的创建逻辑。

14.4　享元模式扩展

14.4.1　享元模式的内部状态和外部状态

享元模式的定义提出了两个要求：细粒度和共享对象。因为要求细粒度，所以不可避免地会使对象数量多且性质相近，此时我们就将这些对象的信息分为两个部分：内部状态和外部状态。

内部状态指对象共享出来的信息，存储在享元对象内部，并且不会随环境的改变而改变；外部状态指对象得以依赖的一个标记，随环境的改变而改变，不可共享。

比如，连接池中的连接对象，保存在连接对象中的用户名、密码、连接 URL 等信息，在创建对象的时候就设置好了，不会随环境的改变而改变，这些为内部状态。而当每个连接要被回收利用时，我们需要将它标记为可用状态，这些为外部状态。

14.4.2 享元模式的优点

（1）减少对象的创建，降低内存中对象的数量，降低系统的内存，提高效率。

（2）减少内存之外的其他资源占用。

14.4.3 享元模式的缺点

（1）关注内、外部状态，关注线程安全问题。

（2）使系统、程序的逻辑复杂化。

15

第 15 章 组合模式

15.1 组合模式概述

我们知道古代的皇帝想要管理国家，是不可能直接管理到每一个老百姓的，因此设置了很多机构，比如三省六部，这些机构下面又有很多小的组织。这些组织共同管理着国家。再比如一个大公司，下面有很多个部门，每一个部门下面又都有很多个小部门。这就是组合模式。

15.1.1 组合模式的定义

组合模式（Composite Pattern）又叫作整体-部分（Part-Whole）模式，它的宗旨是通过将单个对象（叶子节点）和组合对象（树枝节点）用相同的接口进行表示，使得客户对单个对象和组合对象的使用具有一致性，属于结构型设计模式。

原文：Compose objects into tree structures to represent part-whole hierarchies. Composite lets clients treat individual objects and compositions of objects uniformly.

组合模式一般用来描述整体与部分的关系，它将对象组织到树形结构中，顶层的节点被称为根节点，根节点下面可以包含树枝节点和叶子节点，树枝节点下面又可以包含树枝节点和叶子节点，树形结构图如下。

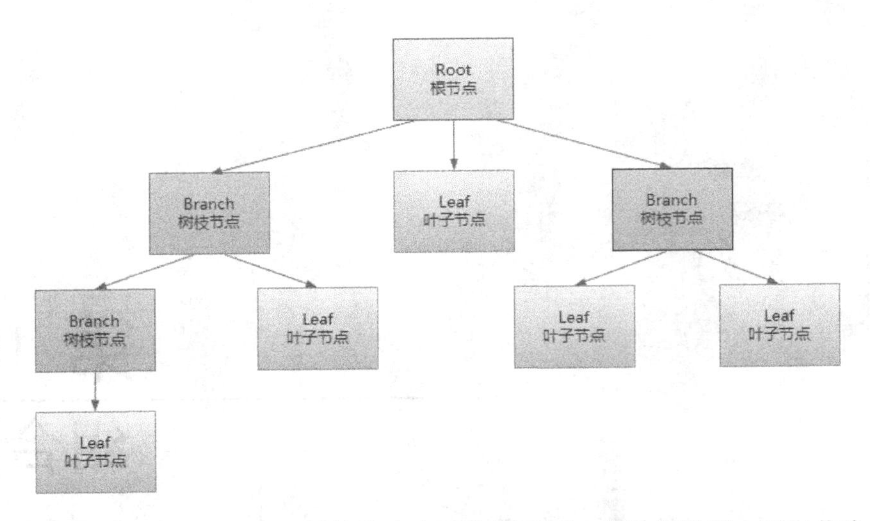

由上图可以看出，其实根节点和树枝节点本质上属于同一种数据类型，可以作为容器使用；而叶子节点与树枝节点在语义上不属于同一种类型。但是在组合模式中，会把树枝节点和叶子节点看作属于同一种数据类型（用同一接口定义），让它们具备一致行为。这样，在组合模式中，整个树形结构中的对象都属于同一种类型，带来的好处就是用户不需要辨别是树枝节点还是叶子节点，可以直接进行操作，给用户的使用带来极大的便利。

15.1.2 组合模式的应用场景

在生活中，组合模式非常常见，比如，树形菜单、公司组织架构和操作系统目录结构等，如下图所示。

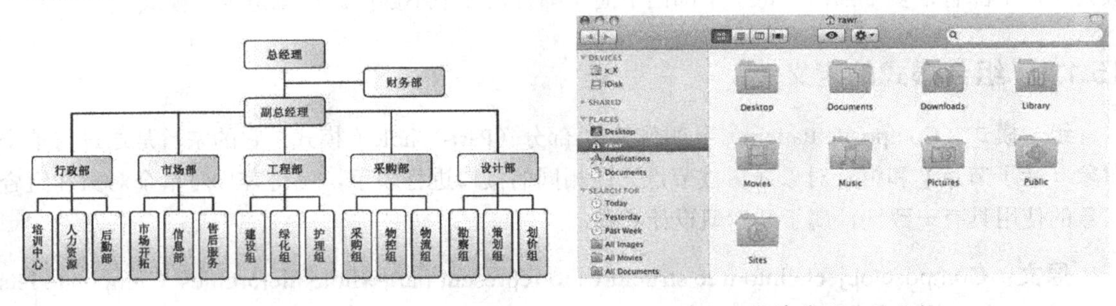

公司组织架构　　　　操作系统目录结构

当子系统与其内各个对象层次呈树形结构时，可以使用组合模式让子系统内各个对象层次的行为操作具备一致性。当客户端使用该子系统内任意一个对象时，不用进行区分，直接使用通用操作即可，非常便捷。

注：如果树形结构系统不使用组合模式进行架构，则按照正常的思维逻辑，对该系统进行职责分析。如上面的树形结构图所示，该系统具备两种对象层次类型：树枝节点和叶子节点。那么我们就需要构造两种对应的类型，由于树枝节点具备容器功能，因此树枝节点类内部需维护多个集合存储其他对象层次，例如 List<Composite>,List<Leaf>。如果当前系统对象层次更复杂，那么树枝节点就又要增加对应的层次集合，这给树枝节点的构建带来了巨大的复杂性、臃肿性及不可扩展性。同时，当客户端访问该系统层次时，还需进行层次区分，这样才能使用对应的行为，给客户端的使用也带来了巨大的复杂性。而如果使用组合模式构建该系统，由于组合模式抽取了系统各个层次的共性行为，具体层次只要按需实现所需行为即可，这样子系统各个层次就都属于同一种类型，所以树枝节点只需维护一个集合（List<Component>）即可存储系统所有层次的内容，并且客户端也不需要区分该系统各个层次的对象，对内系统架构简洁优雅，对外接口精简易用。

总结一下，组合模式主要有以下应用场景。

（1）希望客户端可以忽略组合对象与单个对象的差异。

（2）对象层次具备整体和部分，呈树形结构。

15.1.3　透明组合模式的 UML 类图及通用写法

透明组合模式是把所有公共方法都定义在 Component 中，这样做的好处是客户端无须分辨叶子节点和树枝节点，它们具备完全一致的接口。其 UML 类图如下。

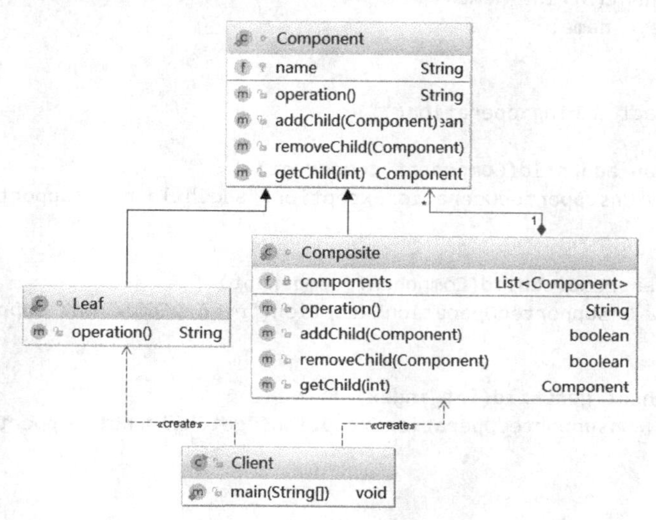

透明组合模式的通用写法如下。

```
public class Client {
    public static void main(String[] args) {
        //创建一个根节点
        Component root = new Composite("root");
        //创建一个树枝节点
        Component branchA = new Composite("---branchA");
        Component branchB = new Composite("------branchB");
        //创建一个叶子节点
        Component leafA = new Leaf("------leafA");
        Component leafB = new Leaf("---------leafB");
        Component leafC = new Leaf("---leafC");

        root.addChild(branchA);
        root.addChild(leafC);
        branchA.addChild(leafA);
        branchA.addChild(branchB);
        branchB.addChild(leafB);

        String result = root.operation();
        System.out.println(result);

    }

    static abstract class Component {
        protected String name;

        public Component(String name) {
            this.name = name;
        }

        public abstract String operation();

        public boolean addChild(Component component) {
            throw new UnsupportedOperationException("addChild not supported!");
        }

        public boolean removeChild(Component component) {
            throw new UnsupportedOperationException("removeChild not supported!");
        }

        public Component getChild(int index) {
            throw new UnsupportedOperationException("getChild not supported!");
        }
    }
```

```
    //树枝节点
    static class Composite extends Component {
        private List<Component> components;

        public Composite(String name) {
            super(name);
            this.components = new ArrayList<Component>();
        }

        @Override
        public String operation() {
            StringBuilder builder = new StringBuilder(this.name);
            for (Component component : this.components) {
                builder.append("\n");
                builder.append(component.operation());
            }
            return builder.toString();
        }

        public boolean addChild(Component component) {
            throw new UnsupportedOperationException("addChild not supported!");
        }

        public boolean removeChild(Component component) {
            throw new UnsupportedOperationException("removeChild not supported!");
        }

        public Component getChild(int index) {
            throw new UnsupportedOperationException("getChild not supported!");
        }
    }

    //叶子节点
    static class Leaf extends Component {

        public Leaf(String name) {
            super(name);
        }

        @Override
        public String operation() {
            return this.name;
        }
    }

}
```

15.1.4 安全组合模式的 UML 类图及通用写法

安全组合模式只规定系统各个层次的最基础的一致行为，而把组合（节点）本身的方法（管理子类对象的添加、删除等）放到自身当中。其 UML 类图如下。

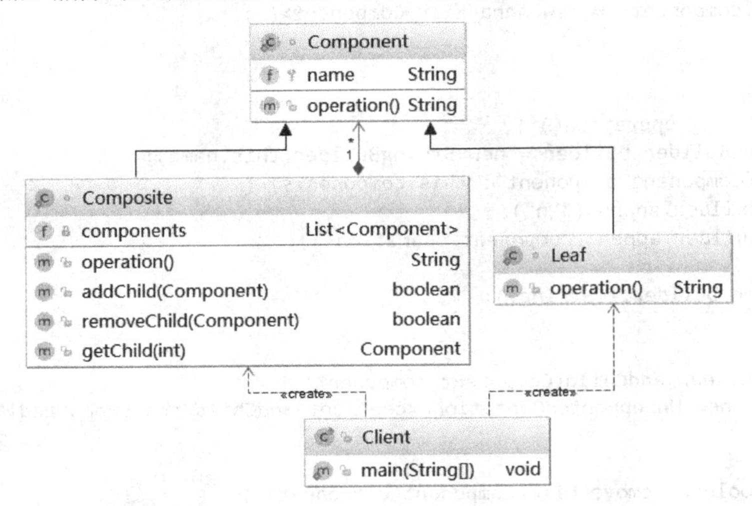

安全组合模式的通用代码相对透明组合模式而言，需要进行如下修改。

首先修改 Component 代码，只保留各层次的公共行为。

```
//抽象根节点
static abstract class Component {
    protected String name;

    public Component(String name) {
        this.name = name;
    }

    public abstract String operation();
}
```

然后修改客户端代码，将树枝节点类型更改为 Composite 类型，以便获取管理子类操作的方法。

```
class Client {
    public static void main(String[] args) {
        //创建一个根节点
        Composite root = new Composite("root");
        //创建一个树枝节点
```

```
        Composite branchA = new Composite("---branchA");
        Composite branchB = new Composite("------branchB");
        //创建一个叶子节点
        Component leafA = new Leaf("------leafA");
        Component leafB = new Leaf("---------leafB");
        Component leafC = new Leaf("---leafC");

        root.addChild(branchA);
        root.addChild(leafC);
        branchA.addChild(leafA);
        branchA.addChild(branchB);
        branchB.addChild(leafB);

        String result = root.operation();
        System.out.println(result);
    }
}
```

15.2　使用组合模式解决实际问题

15.2.1　使用透明组合模式实现课程目录结构

以咕泡学院的课程为例，我们设计一个课程的关系结构。比如，我们有 Java 入门课程、人工智能课程、Java 设计模式、源码分析、软技能等，而 Java 设计模式、源码分析、软技能又属于 Java 架构师系列课程包，每个课程的定价都不一样。但是，这些课程不论怎么组合，都有一些共性，而且是整体和部分的关系，所以可以用组合模式来设计。首先创建一个顶层的抽象组件 CourseComponent 类。

```
public abstract class CourseComponent {

    public void addChild(CourseComponent catalogComponent){
        throw new UnsupportedOperationException("不支持添加操作");
    }

    public void removeChild(CourseComponent catalogComponent){
        throw new UnsupportedOperationException("不支持删除操作");
    }


    public String getName(CourseComponent catalogComponent){
        throw new UnsupportedOperationException("不支持获取名称操作");
    }
```

```
    public double getPrice(CourseComponent catalogComponent){
        throw new UnsupportedOperationException("不支持获取价格操作");
    }


    public void print(){
        throw new UnsupportedOperationException("不支持打印操作");
    }

}
```

把所有可能用到的方法都定义到这个顶层的抽象组件中，但是不写任何逻辑处理的代码，而是直接抛异常。这里，有些小伙伴会有疑惑，为什么不用抽象方法？因为用了抽象方法，其子类就必须实现，这样便体现不出各子类的细微差异。所以子类继承此抽象类后，只需要重写有差异的方法覆盖父类的方法即可。

然后分别创建课程 Course 类和课程包 CoursePackage 类。创建 Course 类的代码如下。

```
public class Course extends CourseComponent {
    private String name;
    private double price;

    public Course(String name, double price) {
        this.name = name;
        this.price = price;
    }

    @Override
    public String getName(CourseComponent catalogComponent) {
        return this.name;
    }

    @Override
    public double getPrice(CourseComponent catalogComponent) {
        return this.price;
    }

    @Override
    public void print() {
        System.out.println(name + " (¥" + price + "元)");
    }

}
```

创建 CoursePackage 类的代码如下。

```
public class CoursePackage extends CourseComponent {
    private List<CourseComponent> items = new ArrayList<CourseComponent>();
    private String name;
    private Integer level;


    public CoursePackage(String name, Integer level) {
        this.name = name;
        this.level = level;
    }

    @Override
    public void addChild(CourseComponent catalogComponent) {
        items.add(catalogComponent);
    }

    @Override
    public String getName(CourseComponent catalogComponent) {
        return this.name;
    }

    @Override
    public void removeChild(CourseComponent catalogComponent) {
        items.remove(catalogComponent);
    }

    @Override
    public void print() {
        System.out.println(this.name);

        for(CourseComponent catalogComponent : items){
            //控制显示格式
            if(this.level != null){
                for(int  i = 0; i < this.level; i ++){
                    //打印空格控制格式
                    System.out.print("  ");
                }
                for(int  i = 0; i < this.level; i ++){
                    //每一行开始打印一个+号
                    if(i == 0){ System.out.print("+"); }
                    System.out.print("-");
                }
            }
```

```
            //打印标题
            catalogComponent.print();
        }
    }

}
```

最后编写客户端测试代码。

```
public static void main(String[] args) {

    System.out.println("============透明组合模式===========");

    CourseComponent javaBase = new Course("Java 入门课程",8280);
    CourseComponent ai = new Course("人工智能",5000);

    CourseComponent packageCourse = new CoursePackage("Java 架构师课程",2);

    CourseComponent design = new Course("Java 设计模式",1500);
    CourseComponent source = new Course("源码分析",2000);
    CourseComponent softSkill = new Course("软技能",3000);

    packageCourse.addChild(design);
    packageCourse.addChild(source);
    packageCourse.addChild(softSkill);

    CourseComponent catalog = new CoursePackage("课程主目录",1);
    catalog.addChild(javaBase);
    catalog.addChild(ai);
    catalog.addChild(packageCourse);

    catalog.print();

}
```

运行结果如下图所示。

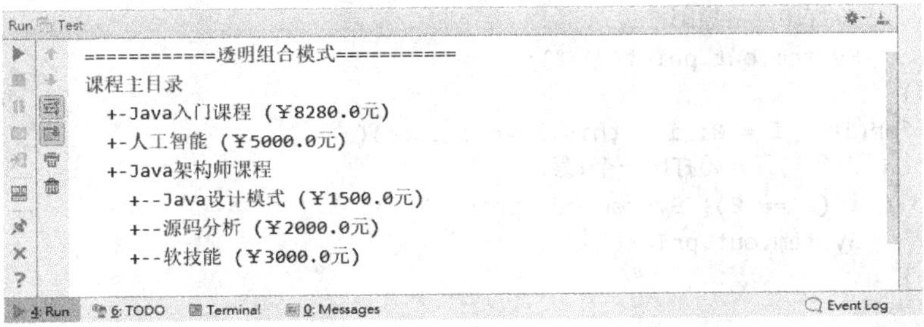

```
============透明组合模式===========
课程主目录
  +-Java入门课程 (￥8280.0元)
  +-人工智能 (￥5000.0元)
  +-Java架构师课程
    +--Java设计模式 (￥1500.0元)
    +--源码分析 (￥2000.0元)
    +--软技能 (￥3000.0元)
```

透明组合模式把所有公共方法都定义在 Component 中，这样客户端就不需要区分操作对象是叶子节点还是树枝节点；但是，叶子节点会继承一些它不需要（管理子类操作的方法）的方法，这与设计模式的接口隔离原则相违背。

15.2.2　使用安全组合模式实现无限级文件系统

再举一个程序员更熟悉的例子。对于程序员来说，计算机是每天都要接触的。计算机的文件系统其实就是一个典型的树形结构，目录包含文件夹和文件，文件夹里面又可以包含文件夹和文件。下面用代码来实现一个目录系统。

文件系统有两个大的层次：文件夹和文件。其中，文件夹能容纳其他层次，为树枝节点；文件是最小单位，为叶子节点。由于目录系统层次较少，且树枝节点（文件夹）结构相对稳定，而文件其实可以有很多类型，所以我们选择使用安全组合模式来实现目录系统，可以避免为叶子节点类型（文件）引入冗余方法。首先创建顶层的抽象组件 Directory 类。

```
public abstract class Directory {

    protected String name;

    public Directory(String name) {
        this.name = name;
    }

    public abstract void show();

}
```

然后分别创建 File 类和 Folder 类。创建 File 类的代码如下。

```
public class File extends Directory {

    public File(String name) {
        super(name);
    }

    @Override
    public void show() {
        System.out.println(this.name);
    }

}
```

创建 Folder 类的代码如下。

```java
import java.util.ArrayList;
import java.util.List;

public class Folder extends Directory {
    private List<Directory> dirs;

    private Integer level;

    public Folder(String name,Integer level) {
        super(name);
        this.level = level;
        this.dirs = new ArrayList<Directory>();
    }

    @Override
    public void show() {
        System.out.println(this.name);
        for (Directory dir : this.dirs) {
            //控制显示格式
            if(this.level != null){
                for(int  i = 0; i < this.level; i ++){
                    //打印空格控制格式
                    System.out.print("  ");
                }
                for(int  i = 0; i < this.level; i ++){
                    //每一行开始打印一个+号
                    if(i == 0){ System.out.print("+"); }
                    System.out.print("-");
                }
            }
            //打印名称
            dir.show();
        }
    }

    public boolean add(Directory dir) {
        return this.dirs.add(dir);
    }

    public boolean remove(Directory dir) {
        return this.dirs.remove(dir);
    }

    public Directory get(int index) {
        return this.dirs.get(index);
    }
```

```
    public void list(){
        for (Directory dir : this.dirs) {
            System.out.println(dir.name);
        }
    }

}
```

注意，Folder 类不仅覆盖了顶层的 show()方法，还增加了 list()方法。

最后编写客户端测试代码。

```
    public static void main(String[] args) {

        System.out.println("============安全组合模式===========");

        File qq = new File("QQ.exe");
        File wx = new File("微信.exe");

        Folder office = new Folder("办公软件",2);

        File word = new File("Word.exe");
        File ppt = new File("PowerPoint.exe");
        File excel = new File("Excel.exe");

        office.add(word);
        office.add(ppt);
        office.add(excel);

        Folder wps = new Folder("金山软件",3);
        wps.add(new File("WPS.exe"));
        office.add(wps);

        Folder root = new Folder("根目录",1);
        root.add(qq);
        root.add(wx);
        root.add(office);

        System.out.println("----------show()方法效果-----------");
        root.show();

        System.out.println("----------list()方法效果-----------");
        root.list();

}
```

运行结果如下图所示。

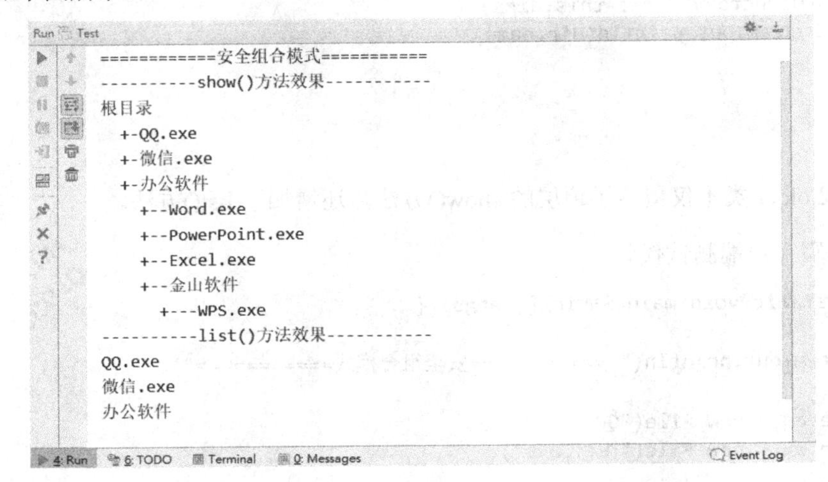

安全组合模式的好处是接口定义职责清晰，符合设计模式的单一职责原则和接口隔离原则；缺点是用户需要区分树枝节点和叶子节点，这样才能正确处理各个层次的操作，客户端无法依赖抽象接口（Component），违背了设计模式的依赖倒置原则。

15.3 组合模式在框架源码中的应用

15.3.1 组合模式在 JDK 源码中的应用

首先来看一个非常熟悉的 HashMap，它有一个 putAll()方法。

```
public class HashMap<K,V> extends AbstractMap<K,V>
implements Map<K,V>, Cloneable, Serializable {
...
    public void putAll(Map<? extends K, ? extends V> m) {
        putMapEntries(m, true);
    }
...
final void putMapEntries(Map<? extends K, ? extends V> m, boolean evict) {
        int s = m.size();
        if (s > 0) {
            if (table == null) { // pre-size
                float ft = ((float)s / loadFactor) + 1.0F;
                int t = ((ft < (float)MAXIMUM_CAPACITY) ?
                         (int)ft : MAXIMUM_CAPACITY);
                if (t > threshold)
```

```
                threshold = tableSizeFor(t);
        }
        else if (s > threshold)
            resize();
        for (Map.Entry<? extends K, ? extends V> e : m.entrySet()) {
            K key = e.getKey();
            V value = e.getValue();
            putVal(hash(key), key, value, false, evict);
        }
    }
}
...
}
```

我们看到 putAll()方法传入的是 Map 对象，Map 就是一个抽象构件（这个构件只支持键值对的存储格式），而 HashMap 是一个中间构件，HashMap 中的 Node 节点就是叶子节点。说到中间构件就会有规定的存储方式。HashMap 中的存储使用一个静态内部类的数组 Node<K,V>[] tab，源码如下。

```
static class Node<K,V> implements Map.Entry<K,V> {
        final int hash;
        final K key;
        V value;
        Node<K,V> next;

        Node(int hash, K key, V value, Node<K,V> next) {
            this.hash = hash;
            this.key = key;
            this.value = value;
            this.next = next;
        }
          ...
        }

...

    final V putVal(int hash, K key, V value, boolean onlyIfAbsent,
                   boolean evict) {
        Node<K,V>[] tab; Node<K,V> p; int n, i;
        if ((tab = table) == null || (n = tab.length) == 0)
            n = (tab = resize()).length;
        if ((p = tab[i = (n - 1) & hash]) == null)
            tab[i] = newNode(hash, key, value, null);
        else {
            Node<K,V> e; K k;
            if (p.hash == hash &&
```

```
            ((k = p.key) == key || (key != null && key.equals(k))))
            e = p;
        else if (p instanceof TreeNode)
            e = ((TreeNode<K,V>)p).putTreeVal(this, tab, hash, key, value);
        else {
            for (int binCount = 0; ; ++binCount) {
                if ((e = p.next) == null) {
                    p.next = newNode(hash, key, value, null);
                    if (binCount >= TREEIFY_THRESHOLD - 1) // -1 for 1st
                        treeifyBin(tab, hash);
                    break;
                }
                if (e.hash == hash &&
                    ((k = e.key) == key || (key != null && key.equals(k))))
                    break;
                p = e;
            }
        }
        if (e != null) { // existing mapping for key
            V oldValue = e.value;
            if (!onlyIfAbsent || oldValue == null)
                e.value = value;
            afterNodeAccess(e);
            return oldValue;
        }
    }
    ++modCount;
    if (++size > threshold)
        resize();
    afterNodeInsertion(evict);
    return null;
}
...
```

同理，我们常用的 ArrayList 对象也有 addAll()方法，其参数也是 ArrayList 的父类 Collection，源码如下。

```
public class ArrayList<E> extends AbstractList<E>
        implements List<E>, RandomAccess, Cloneable, java.io.Serializable
{
...
    public boolean addAll(Collection<? extends E> c) {
        Object[] a = c.toArray();
        int numNew = a.length;
```

```
        ensureCapacityInternal(size + numNew);  // Increments modCount
        System.arraycopy(a, 0, elementData, size, numNew);
        size += numNew;
        return numNew != 0;
    }
    ...
}
```

组合对象和被组合对象都应该有统一的接口实现或者统一的抽象父类。

15.3.2　组合模式在 MyBatis 源码中的应用

我们再看一个开源框架中非常经典的案例，MyBatis 解析各种 Mapping 文件中的 SQL 语句时，设计了一个非常关键的类叫作 SqlNode，XML 中的每一个 Node 都会被解析为一个 SqlNode 对象，最后把所有 SqlNode 都拼装到一起，就成为一条完整的 SQL 语句，它的顶层设计非常简单，源码如下。

```
public interface SqlNode {
  boolean apply(DynamicContext context);
}
```

apply()方法会根据传入的参数 context，解析该 SqlNode 记录的 SQL 语句片段，并调用 DynamicContext.appendSql()方法将解析后的 SQL 语句片段追加到 DynamicContext 的 sqlBuilder 中保存。当所有 SqlNode 都完成解析时，可以通过 DynamicContext.getSql()获取一条完成的 SQL 语句。对具体源码实现感兴趣的小伙伴可以去研究一下，这里给大家展示一下类图，如下图所示。

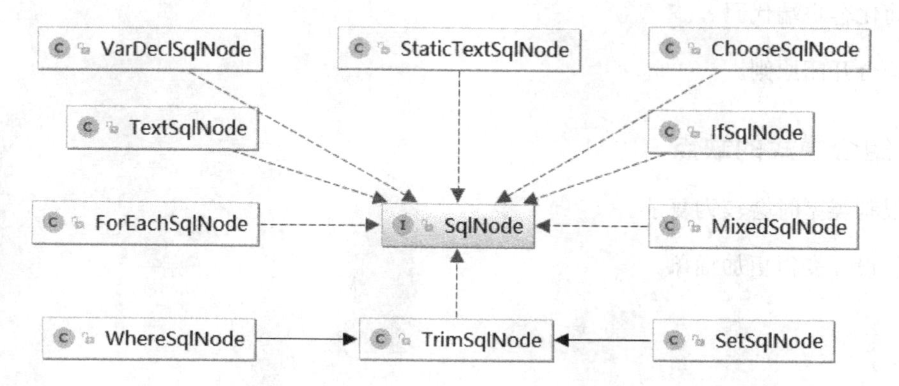

15.4 组合模式扩展

很多小伙伴肯定还有疑问，既然组合模式会被分为两种实现，那么肯定不同的场景应该使用不同的实现方式，即具体情况具体分析。透明组合模式将公共接口封装到抽象根节点（Component）中，那么系统所有节点就都具备一致行为，因此，如果系统绝大多数层次都具备相同的公共行为，则采用透明组合模式也许会更好（代价是为剩下的少数层次节点引入不需要的方法）；而如果系统各个层次的差异性行为较多或者树枝节点层次相对稳定（健壮），则采用安全组合模式。

> 注：设计模式的出现并不是说我们要写的代码一定要遵循设计模式所要求的方方面面，这是不现实也是不可能的。设计模式的出现，其实只是强调好的代码所具备的一些特征（遵循七大设计原则），这些特征对于项目开发是具备积极效应的，但不是说我们每实现一个类就一定要全部满足设计模式的要求，如果真的完全满足设计模式的要求，反而可能存在过度设计的嫌疑。同时，23 种设计模式其实都是严格遵循七大设计原则的，只是不同的模式在不同的场景中会更加适用。对设计模式的理解应该重于意而不是形。当真正编码时，经常使用的是某种设计模式的变形体，真正切合项目的设计模式才是正确的设计模式。

15.4.1 组合模式的优点

（1）清楚地定义各层次的复杂对象，表示对象的全部或部分层次。

（2）让客户端忽略了层次的差异，方便对整个层次结构进行控制。

（3）简化客户端代码。

（4）符合开闭原则。

15.4.2 组合模式的缺点

（1）限制类型时会较为复杂。

（2）使设计变得更加抽象。

16 第 16 章 适配器模式

16.1 适配器模式概述

16.1.1 适配器模式的定义

适配器模式（Adapter Pattern）又叫作变压器模式，它的功能是将一个类的接口变成客户端所期望的另一种接口，从而使原本因接口不匹配而导致无法在一起工作的两个类能够一起工作，属于结构型设计模式。

原文：Convert the interface of a class into another interface clients expect. Adapter lets classes work together that couldn't otherwise because of incompatible interfaces.

也就是说，当前系统存在两种接口 A 和 B，客户端只支持访问 A 接口，但是当前系统没有 A 接口对象，有 B 接口对象，而客户端无法识别 B 接口，因此需要通过一个适配器 C，将 B 接口内容转换成 A 接口，从而使得客户端能够从 A 接口获取 B 接口的内容。

在软件开发中，基本上任何问题都可以通过增加一个中间层来解决。适配器模式其实就是一个中间层。综上，适配器模式起着转化/委托的作用，将一种接口转化为另一种符合需求的接口。

16.1.2 适配器模式的应用场景

在生活中，适配器模式有非常多应用场景，例如，电源插转换头、手机充电转换头、显示器转接头等。

适配器模式提供一个适配器，将当前系统存在的一个对象转化为客户端能够访问的接口对象。适配器模式适用于以下业务场景。

（1）已经存在的类，它的方法和需求不匹配（方法结果相同或相似）的情况。

（2）适配器模式不是软件设计阶段考虑的设计模式，是随着软件维护，由于不同产品、不同厂家造成功能类似而接口不相同情况下的解决方案，有种亡羊补牢的感觉。

适配器模式有 3 种形式：类适配器、对象适配器、接口适配器。适配器模式一般包含 3 个角色。

（1）目标角色（ITarget）：也就是我们期望的接口。

（2）源角色（Adaptee）：存在于系统中，是指内容满足客户需求（需转换）但接口不匹配的接口实例。

（3）适配器（Adapter）：将 Adaptee 转化为目标角色 ITarget 的类实例。

适配器模式各角色之间的关系如下。

假设在当前系统中，客户端需要访问的是 ITarget 接口，但 ITarget 接口没有一个实例符合需求，而 Adaptee 实例符合需求，但是客户端无法直接使用 Adaptee（接口不兼容）；因此，需要一个 Adapter 来进行中转，使 Adaptee 能转化为 ITarget 接口的形式。

16.1.3 类适配器的 UML 类图及通用写法

类适配器的原理就是通过继承来实现适配器功能。具体做法是，让 Adapter 实现 ITarget 接口，并且继承 Adaptee，这样 Adapter 就具备了 ITarget 和 Adaptee 的特性，可以将两者进行转化。其 UML 类图如下。

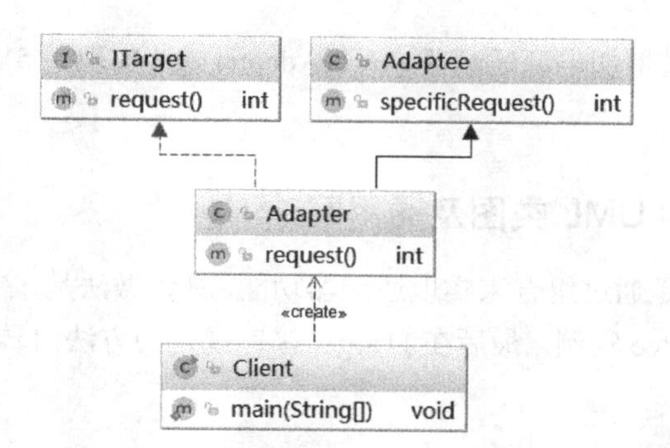

我们用一个示例进行讲解。在中国，民用电都是 220V 交流电，但手机使用的锂电池使用的是 5V 直流电。因此，当手机充电时，就需要使用电源适配器来进行转换。我们用代码来还原这个生活场景，创建 Adaptee 角色，需要被转换的对象 AC220 类，表示 220V 交流电。

```
public class AC220 {
    public int outputAC220V(){
        int output = 220;
        System.out.println("输出电压" + output + "V");
        return output;
    }
}
```

创建 ITarget 角色 DC5 接口，表示 5V 直流电。

```
public interface DC5 {
    int output5V();
}
```

创建 Adapter 角色电源适配器 PowerAdapter 类。

```
public class PowerAdapter extends AC220 implements DC5 {
    public int output5V() {
        int adapterInput = super.outputAC220V();
        int adapterOutput = adapterInput / 44;
        System.out.println("使用 Adapter 输入 AC" + adapterInput + "V,输出 DC" + adapterOutput +
"V");
        return adapterOutput;
    }
}
```

编写客户端测试代码。

```
    public static void main(String[] args) {
        DC5 adapter = new PowerAdapter();
        adapter.output5V();
    }
```

在上面的案例中，通过增加电源适配器 PowerAdapter 类，实现了 5V 直流电和 220V 交流电的兼容。

16.1.4 对象适配器的 UML 类图及通用写法

对象适配器的原理就是通过组合来实现适配器功能。具体做法是，首先让 Adapter 实现 ITarget 接口，然后内部持有 Adaptee 实例，最后在 ITarget 接口规定的方法内转换 Adaptee。其 UML 类图如下。

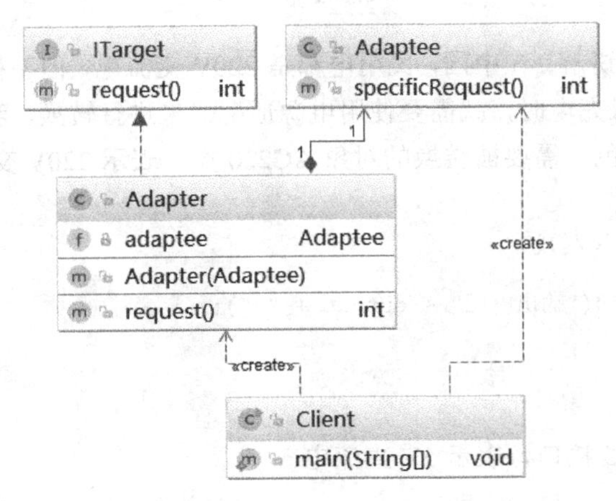

接前面的示例，代码只需更改 Adapter 实现，其他与类适配器一致。

```
public class PowerAdapter implements DC5 {
    private AC220 ac220;

    public PowerAdapter(AC220 ac220) {
        this.ac220 = ac220;
    }

    public int output5V() {
        int adapterInput = ac220.outputAC220V();
        int adapterOutput = adapterInput / 44;
        System.out.println("使用 Adapter 输入 AC" + adapterInput + "V,输出 DC" + adapterOutput +
"V");
        return adapterOutput;
    }
}
```

16.1.5　接口适配器的 UML 类图及通用写法

接口适配器的关注点与类适配器、对象适配器的关注点不太一样，类适配器和对象适配器着重于将系统存在的一个角色（Adaptee）转化成目标接口（ITarget）所需的内容，而接口适配器的使用场景是当接口的方法过多时，如果直接实现接口，则类会多出许多空实现的方法，显得很臃肿。此时，使用接口适配器就能只实现我们需要的接口方法，使目标更清晰。其 UML 类图如下。

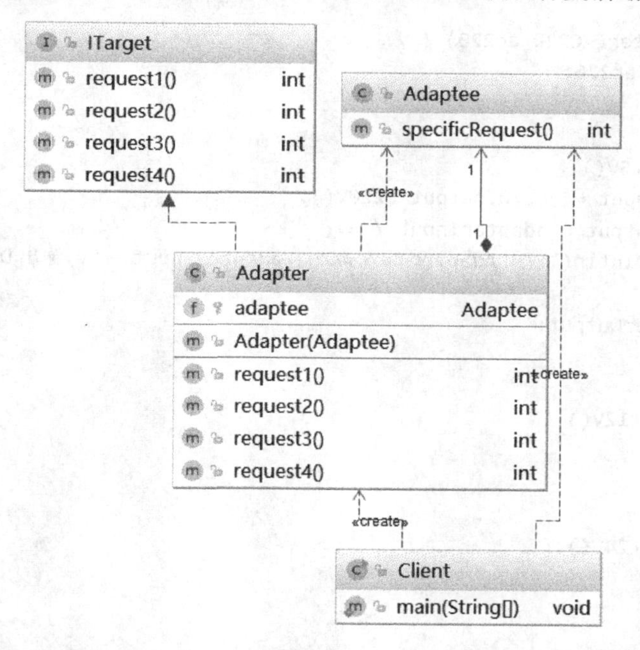

接口适配器的主要原理就是使用抽象类实现接口，并且空实现众多。我们来看接口适配器的源码实现，首先创建 ITarget 角色 DC 类。

```
public interface DC {
    int output5V();
    int output12V();
    int output24V();
    int output36V();
}
```

然后创建 Adaptee 角色 AC220 类。

```
public class AC220 {
    public int outputAC220V(){
        int output = 220;
```

```
        System.out.println("输出电压" + 220 + "V");
        return output;
    }
}
```

最后创建 Adapter 角色 PowerAdapter 类。

```
public class PowerAdapter implements DC {
    private AC220 ac220;

    public PowerAdapter(AC220 ac220) {
        this.ac220 = ac220;
    }

    public int output5V() {
        int adapterInput = ac220.outputAC220V();
        int adapterOutput = adapterInput / 44;
        System.out.println("使用 Adapter 输入 AC" + adapterInput + "V,输出 DC" + adapterOutput +
"V");
        return adapterOutput;
    }

    public int output12V() {
        return 0;
    }

    public int output24V() {
        return 0;
    }

    public int output36V() {
        return 0;
    }
}
```

编写客户端测试代码。

```
public class Test {
    public static void main(String[] args) {
        DC adapter = new PowerAdapter(new AC220());
        adapter.output5V();
    }
}
```

16.2　使用适配器模式解决实际问题

16.2.1　使用类适配器重构第三方登录自由适配

我们使用类适配器来实现一个实际的业务场景，解决实际问题。年纪稍微大一点的小伙伴一定经历过这样的过程。很早以前开发的老系统应该都有登录接口，但是随着业务的发展和社会的进步，单纯地依赖用户名和密码登录显然不能满足用户需求。现在，大部分系统都已经支持多种登录方式，如 QQ 登录、微信登录、手机登录、微博登录等，同时保留用户名和密码登录的方式。虽然登录形式丰富，但是登录后的处理逻辑可以不必改，都是将登录状态保存到 Session，遵循开闭原则。首先创建统一的返回结果 ResultMsg 类。

```
public class ResultMsg {

    private int code;
    private String msg;
    private Object data;

    public ResultMsg(int code, String msg, Object data) {
        this.code = code;
        this.msg = msg;
        this.data = data;
    }

    public int getCode() {
        return code;
    }

    public void setCode(int code) {
        this.code = code;
    }

    public String getMsg() {
        return msg;
    }

    public void setMsg(String msg) {
        this.msg = msg;
    }

    public Object getData() {
        return data;
    }
```

```
    public void setData(Object data) {
        this.data = data;
    }
}
```

假设在老系统中，处理登录逻辑的代码在 PassportService 类中。

```
public class PassportService {

    /**
     * 注册方法
     * @param username
     * @param password
     * @return
     */
    public ResultMsg regist(String username,String password){
        return  new ResultMsg(200,"注册成功",new Member());
    }


    /**
     * 登录方法
     * @param username
     * @param password
     * @return
     */
    public ResultMsg login(String username,String password){
        return null;
    }

}
```

为了遵循开闭原则，不修改老系统的代码。下面开启代码重构之路，创建 Member 类。

```
public class Member {

    private String username;
    private String password;
    private String mid;
    private String info;

    public String getUsername() {
        return username;
    }

    public void setUsername(String username) {
        this.username = username;
```

```
    }

    public String getPassword() {
        return password;
    }

    public void setPassword(String password) {
        this.password = password;
    }

    public String getMid() {
        return mid;
    }

    public void setMid(String mid) {
        this.mid = mid;
    }

    public String getInfo() {
        return info;
    }

    public void setInfo(String info) {
        this.info = info;
    }
}
```

我们也不改动运行非常稳定的代码，创建 ITarget 角色 IPassportForThird 接口。

```
public interface IPassportForThird {

    ResultMsg loginForQQ(String openId);

    ResultMsg loginForWechat(String openId);

    ResultMsg loginForToken(String token);

    ResultMsg loginForTelphone(String phone,String code);

}
```

创建 Adapter 角色实现兼容，创建一个新的类 PassportForThirdAdapter，继承原来的逻辑。

```
public class PassportForThirdAdapter extends PassportService implements IPassportForThird {

    public ResultMsg loginForQQ(String openId) {
        return loginForRegist(openId,null);
    }
```

```
    public ResultMsg loginForWechat(String openId) {
        return loginForRegist(openId,null);
    }

    public ResultMsg loginForToken(String token) {
        return loginForRegist(token,null);
    }

    public ResultMsg loginForTelphone(String phone, String code) {
        return loginForRegist(phone,null);
    }

    private ResultMsg loginForRegist(String username,String password){
        if(null == password){
            password = "THIRD_EMPTY";
        }
        super.regist(username,password);
        return super.login(username,password);
    }
}
```

客户端测试代码如下。

```
public static void main(String[] args) {
        PassportForThirdAdapter adapter = new PassportForThirdAdapter();
        adapter.login("tom","123456");
        adapter.loginForQQ("sjooguwoersdfjhasjfsa");
        adapter.loginForWechat("slfsjoljsdo8234ssdfs");
    }
```

16.2.2 使用接口适配器优化代码

通过这么一个简单的适配动作，我们完成了代码兼容。当然，代码还可以更加优雅，根据不同的登录方式，创建不同的 Adapter。首先创建 LoginAdapter 接口。

```
public interface ILoginAdapter {
    boolean support(Object object);
    ResultMsg login(String id,Object adapter);
}
```

然后创建一个抽象类 AbstractAdapter 继承 PassportService 原有的功能，同时实现 ILoginAdapter 接口，再分别实现不同的登录适配。QQ 登录 LoginForQQAdapter 如下。

```
public class LoginForQQAdapter extends AbstractAdapter{
    public boolean support(Object adapter) {
```

```
        return adapter instanceof LoginForQQAdapter;
    }

    public ResultMsg login(String id, Object adapter) {
        if(!support(adapter)){return null;}
        //accesseToken
        //time
        return super.loginForRegist(id,null);

    }

}
```

手机登录 LoginForTelAdapter 如下。

```
public class LoginForTelAdapter extends AbstractAdapter{
    public boolean support(Object adapter) {
        return adapter instanceof LoginForTelAdapter;
    }

    public ResultMsg login(String id, Object adapter) {
        return super.loginForRegist(id,null);
    }
}
```

Token 自动登录 LoginForTokenAdapter 如下。

```
public class LoginForTokenAdapter extends AbstractAdapter {
    public boolean support(Object adapter) {
        return adapter instanceof LoginForTokenAdapter;
    }

    public ResultMsg login(String id, Object adapter) {
        return super.loginForRegist(id,null);
    }
}
```

微信登录 LoginForWechatAdapter 如下。

```
public class LoginForWechatAdapter extends AbstractAdapter{
    public boolean support(Object adapter) {
        return adapter instanceof LoginForWechatAdapter;
    }

    public ResultMsg login(String id, Object adapter) {
        return super.loginForRegist(id,null);
    }
}
```

接着创建适配器 PassportForThirdAdapter 类，实现目标接口 IPassportForThird 完成兼容。

```
public class PassportForThirdAdapter implements IPassportForThird {

    public ResultMsg loginForQQ(String openId) {
        return processLogin(openId, LoginForQQAdapter.class);
    }

    public ResultMsg loginForWechat(String openId) {

        return processLogin(openId, LoginForWechatAdapter.class);

    }

    public ResultMsg loginForToken(String token) {

        return processLogin(token, LoginForTokenAdapter.class);
    }

    public ResultMsg loginForTelphone(String phone, String code) {
        return processLogin(phone, LoginForTelAdapter.class);
    }


    private ResultMsg processLogin(String id,Class<? extends ILoginAdapter> clazz){
        try {
            ILoginAdapter adapter = clazz.newInstance();
            if (adapter.support(adapter)){
                return adapter.login(id,adapter);
            }
        } catch (Exception e) {
            e.printStackTrace();
        }
        return null;
    }

}
```

客户端测试代码如下。

```
public static void main(String[] args) {
    IPassportForThird adapter = new PassportForThirdAdapter();
    adapter.loginForQQ("sdfasdfasfasfas");
}
```

最后来看如下图所示的类图。

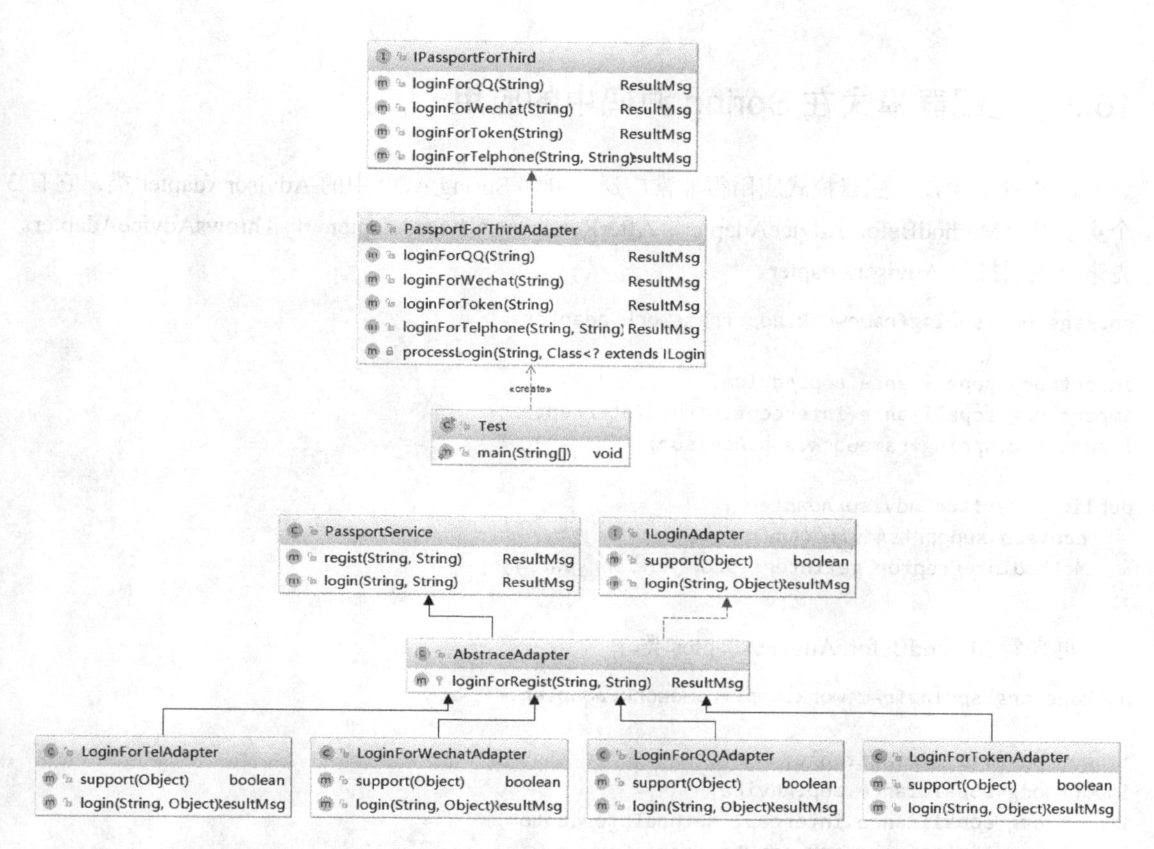

至此，在遵循开闭原则的前提下，我们完整地实现了一个兼容多平台登录的业务场景。当然，目前的这个设计并不完美，仅供参考，感兴趣的小伙伴可以继续完善这段代码。例如适配器类中的参数类型目前设置为 String，改为 Object[]应该更合理。

学习到这里，小伙伴们可能会有一个疑问：适配器模式与策略模式好像区别不大？笔者要强调一下，适配器模式主要解决的是功能兼容问题，单场景适配大家可能不会和策略模式对比，但复杂场景适配大家就很容易混淆。其实，大家有没有发现一个细节，笔者给每个适配器类都加上了一个 support()方法，用来判断是否兼容，support()方法的参数类型也是 Object，而 support()来自接口。适配器类的实现逻辑并不依赖接口，完全可以将 ILoginAdapter 接口去掉。而加上接口，只是为了代码规范。上面的代码可以说是策略模式、简单工厂模式和适配器模式的综合运用。

16.3 适配器模式在 Spring 源码中的应用

在 Spring 中，适配器模式应用得非常广泛，例如 Spring AOP 中的 AdvisorAdapter 类，它有 3 个实现类：MethodBeforeAdviceAdapter、AfterReturningAdviceAdapter 和 ThrowsAdviceAdapter。先来看顶层接口 AdvisorAdapter。

```
package org.springframework.aop.framework.adapter;

import org.aopalliance.aop.Advice;
import org.aopalliance.intercept.MethodInterceptor;
import org.springframework.aop.Advisor;

public interface AdvisorAdapter {
    boolean supportsAdvice(Advice var1);
    MethodInterceptor getInterceptor(Advisor var1);
}
```

再来看 MethodBeforeAdviceAdapter 类。

```
package org.springframework.aop.framework.adapter;

import java.io.Serializable;
import org.aopalliance.aop.Advice;
import org.aopalliance.intercept.MethodInterceptor;
import org.springframework.aop.Advisor;
import org.springframework.aop.MethodBeforeAdvice;

class MethodBeforeAdviceAdapter implements AdvisorAdapter, Serializable {
    MethodBeforeAdviceAdapter() {
    }

    public boolean supportsAdvice(Advice advice) {
        return advice instanceof MethodBeforeAdvice;
    }
    public MethodInterceptor getInterceptor(Advisor advisor) {
        MethodBeforeAdvice advice = (MethodBeforeAdvice)advisor.getAdvice();
        return new MethodBeforeAdviceInterceptor(advice);
    }
}
```

这里就不把其他两个类的代码贴出来了。Spring 会根据不同的 AOP 配置来确定使用对应的 Advice，与策略模式不同的是，一个方法可以同时拥有多个 Advice。

再来看一个 Spring MVC 中的 HandlerAdapter 类，它也有多个子类，类图如下。

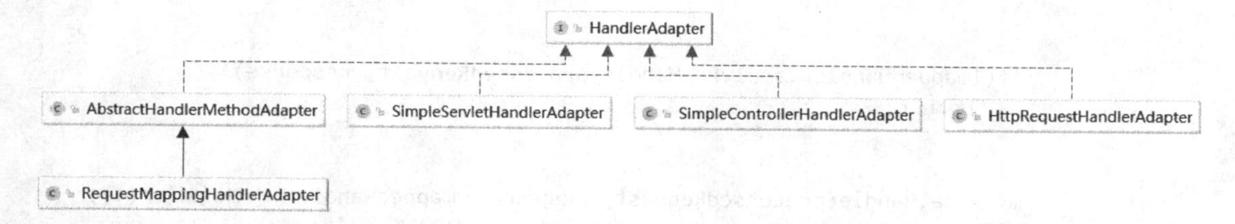

其适配调用的关键代码还是在 DispatcherServlet 的 doDispatch()方法中，代码如下。

```
protected void doDispatch(HttpServletRequest request, HttpServletResponse response) throws
Exception {
    HttpServletRequest processedRequest = request;
    HandlerExecutionChain mappedHandler = null;
    boolean multipartRequestParsed = false;
    WebAsyncManager asyncManager = WebAsyncUtils.getAsyncManager(request);

    try {
        try {
            ModelAndView mv = null;
            Object dispatchException = null;

            try {
                processedRequest = this.checkMultipart(request);
                multipartRequestParsed = processedRequest != request;
                mappedHandler = this.getHandler(processedRequest);
                if(mappedHandler == null) {
                    this.noHandlerFound(processedRequest, response);
                    return;
                }

                HandlerAdapter ha = this.getHandlerAdapter(mappedHandler.getHandler());
                String method = request.getMethod();
                boolean isGet = "GET".equals(method);
                if(isGet || "HEAD".equals(method)) {
                    long lastModified = ha.getLastModified(request, mappedHandler.getHandler());
                    if(this.logger.isDebugEnabled()) {
                        this.logger.debug("Last-Modified value for [" + getRequestUri(request)
                            + "] is: " + lastModified);
                    }
```

```
                if((new ServletWebRequest(request, response)).checkNotModified(lastModified)
                  && isGet) {
                    return;
                }
            }

            if(!mappedHandler.applyPreHandle(processedRequest, response)) {
                return;
            }

            mv = ha.handle(processedRequest, response, mappedHandler.getHandler());
            if(asyncManager.isConcurrentHandlingStarted()) {
                return;
            }

            this.applyDefaultViewName(processedRequest, mv);
            mappedHandler.applyPostHandle(processedRequest, response, mv);
        } catch (Exception var20) {
            dispatchException = var20;
        } catch (Throwable var21) {
            dispatchException = new NestedServletException("Handler dispatch failed", var21);
        }

        this.processDispatchResult(processedRequest, response, mappedHandler, mv, (Exception)
         dispatchException);
    } catch (Exception var22) {
        this.triggerAfterCompletion(processedRequest, response, mappedHandler, var22);
    } catch (Throwable var23) {
        this.triggerAfterCompletion(processedRequest, response, mappedHandler, new
         NestedServletException("Handler processing failed", var23));
    }

} finally {
    if(asyncManager.isConcurrentHandlingStarted()) {
        if(mappedHandler != null) {
            mappedHandler.applyAfterConcurrentHandlingStarted(processedRequest, response);
        }
    } else if(multipartRequestParsed) {
        this.cleanupMultipart(processedRequest);
    }
}
}
```

在 doDispatch()方法中调用了 getHandlerAdapter()方法，代码如下。

```
protected HandlerAdapter getHandlerAdapter(Object handler) throws ServletException {
    if(this.handlerAdapters != null) {
        Iterator var2 = this.handlerAdapters.iterator();

        while(var2.hasNext()) {
            HandlerAdapter ha = (HandlerAdapter)var2.next();
            if(this.logger.isTraceEnabled()) {
                this.logger.trace("Testing handler adapter [" + ha + "]");
            }

            if(ha.supports(handler)) {
                return ha;
            }
        }
    }

    throw new ServletException("No adapter for handler [" + handler + "]: The DispatcherServlet
configuration needs to include a HandlerAdapter that supports this handler");
}
```

在 getHandlerAdapter()方法中循环调用 supports()方法来判断是否兼容，循环迭代集合中的 Adapter 在初始化时早已被赋值。

16.4 适配器模式扩展

16.4.1 适配器模式与装饰器模式的区别

适配器模式和装饰器模式都是包装器模式（Wrapper Pattern），装饰器模式其实就是一种特殊的代理模式。

对比维度	适配器模式	装饰器模式
形式	没有层级关系	一种非常特别的代理模式，有层级关系
定义	适配器和被适配者没有必然的联系，通常采用继承或代理的形式进行包装	装饰器和被装饰者都实现同一个接口，主要目的是扩展之后依旧保留OOP关系
关系	满足has-a的关系	满足is-a的关系
功能	注重兼容、转换	注重覆盖、扩展
设计	后置考虑	前置考虑

16.4.2 适配器模式的优点

（1）能提高类的透明性和复用，但现有的类复用不需要改变。

（2）适配器类和原角色类解耦，提高程序的扩展性。

（3）在很多业务场景中符合开闭原则。

16.4.3 适配器模式的缺点

（1）适配器编写过程需要结合业务场景全面考虑，可能会增加系统的复杂性。

（2）增加代码阅读难度，降低代码可读性，过多使用适配器会使系统代码变得凌乱。

17

第 17 章 桥接模式

17.1 桥接模式概述

17.1.1 桥接模式的定义

桥接模式（Bridge Pattern）又叫作桥梁模式、接口（Interface）模式或柄体（Handle and Body）模式，指将抽象部分与具体实现部分分离，使它们都可以独立地变化，属于结构型设计模式。

原文：Decouple an abstraction from its implementation so that the two can vary independently.

桥接模式的主要目的是通过组合的方式建立两个类之间的联系，而不是继承，但又类似多重继承方案。但是多重继承方案往往违背了类的单一职责原则，其复用性比较差，桥接模式是比多重继承方案更好的替代方案。桥接模式的核心在于把抽象与实现解耦。

注：此处的抽象并不是指抽象类或接口这种高层概念，实现也不是指继承或接口实现。抽象与实现其实指的是两种独立变化的维度。其中，抽象包含实现，因此，一个抽象类的变化可能涉及多种维度的变化。

17.1.2 桥接模式的应用场景

在生活场景中，桥接模式随处可见，比如连接起两个空间维度的桥、连接虚拟网络与真实网络的链接。

当一个类内部具备两种或多种变化维度时，使用桥接模式可以解耦这些变化的维度，使高层代码架构稳定。桥接模式适用于以下几种业务场景。

（1）在抽象和具体实现之间需要增加更多灵活性的场景。

（2）一个类存在两个（或多个）独立变化的维度，而这两个（或多个）维度都需要独立进行扩展。

（3）不希望使用继承，或因为多层继承导致系统类的个数剧增。

注：桥接模式的一个常见使用场景就是替换继承。我们知道，继承拥有很多优点，比如，抽象、封装、多态等，父类封装共性，子类实现特性。继承可以很好地实现代码复用（封装）的功能，但这也是继承的一大缺点。因为父类拥有的方法，子类也会继承得到，无论子类需不需要，这说明继承具备强侵入性（父类代码侵入子类），同时会导致子类臃肿。因此，在设计模式中，有一个原则为优先使用组合/聚合，而不是继承。

很多时候，我们分不清该使用继承还是组合/聚合或其他方式等，其实可以从现实语义进行思考。因为软件最终还是提供给现实生活中的人使用的，是服务于人类社会的，软件是具备现实场景的。当我们从纯代码角度无法看清问题时，现实角度可能会提供更加开阔的思路。

17.1.3 桥接模式的 UML 类图

桥接模式的 UML 类图如下。

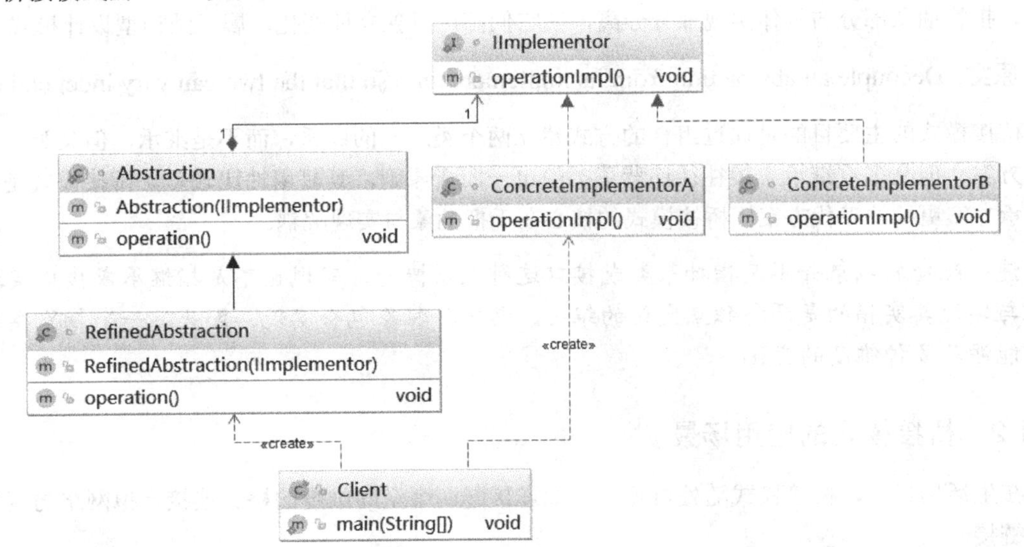

由上图可以看到，桥接模式主要包含 4 个角色。

（1）抽象（Abstraction）：该类持有一个对实现角色的引用，抽象角色中的方法需要实现角色来实现。抽象角色一般为抽象类（构造函数规定子类要传入一个实现对象）。

（2）修正抽象（RefinedAbstraction）：Abstraction 的具体实现，对 Abstraction 的方法进行完善和扩展。

（3）实现（IImplementor）：确定实现维度的基本操作，提供给 Abstraction 使用。该类一般为接口或抽象类。

（4）具体实现（ConcreteImplementor）：Implementor 的具体实现。

17.1.4　桥接模式的通用写法

下面是桥接模式的通用写法。

```
public class Client {
    public static void main(String[] args) {
        //创建一个具体角色
        IImplementor imp = new ConcreteImplementorA();
        //创建一个抽象角色，聚合实现
        Abstraction abs = new RefinedAbstraction(imp);
        //执行操作
        abs.operation();
    }

    //抽象
    static abstract class Abstraction {

        protected IImplementor implementor;

        public Abstraction(IImplementor implementor) {
            this.implementor = implementor;
        }

        public void operation() {
            this.implementor.operationImpl();
        }
    }

    //修正抽象
    static class RefinedAbstraction extends Abstraction {
```

```
        public RefinedAbstraction(IImplementor implementor) {
            super(implementor);
        }

        @Override
        public void operation() {
            super.operation();
            System.out.println("refined operation");
        }
    }

    //抽象实现
    interface IImplementor {
        void operationImpl();
    }

    //具体实现A
    static class ConcreteImplementorA implements IImplementor {

        public void operationImpl() {
            System.out.println("I'm ConcreteImplementor A");
        }
    }

    //具体实现B
    static class ConcreteImplementorB implements IImplementor {

        public void operationImpl() {
            System.out.println("I'm ConcreteImplementor B");
        }
    }
}
```

17.2　使用桥接模式设计复杂消息系统

举个例子，我们在平时办公的时候经常通过邮件消息、短信消息或者系统内消息与同事进行沟通。尤其在走一些审批流程的时候，我们需要记录这些过程以备查。根据类型来划分，消息可以分为邮件消息、短信消息和系统内消息。但是，根据紧急程度来划分，消息可以分为普通消息、加急消息和特急消息。显然，整个消息系统可以划分为两个维度，如下图所示。

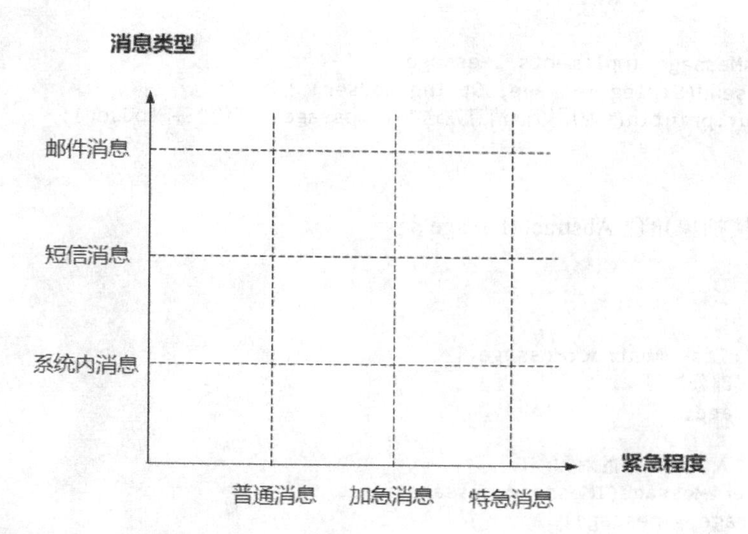

如果我们用继承，则情况就复杂了，而且也不利于扩展。邮件消息可以是普通的，也可以是加急的；短信消息可以是普通的，也可以是加急的。下面我们用桥接模式来解决这个问题。

首先创建一个 IMessage 接口担任桥接的角色。

```
/**
 * 实现消息发送的统一接口
 */
public interface IMessage {
    //要发送的消息的内容和接收人
    void send(String message, String toUser);
}
```

创建邮件消息实现 EmailMessage 类。

```
/**
 * 邮件消息的实现类
 */
public class EmailMessage implements IMessage {
    public void send(String message, String toUser) {
        System.out.println("使用邮件消息发送" + message + "给" + toUser);
    }
}
```

创建短信消息实现 SmsMessage 类。

```
/**
 * 短信消息的实现类
 * SMS(Short IMessage Service)短信消息服务
```

```
 */
public class SmsMessage implements IMessage {
    public void send(String message, String toUser) {
        System.out.println("使用短信消息发送" + message + "给" + toUser);
    }
}
```

然后创建桥接抽象角色 AbstractMessage 类。

```
/**
 * 抽象消息类
 */
public abstract class AbstractMessage {
    //持有一个实现部分的对象
    IMessage message;

    //构造方法，传入实现部分的对象
    public AbstractMessage(IMessage message) {
        this.message = message;
    }

    //发送消息，委派给实现部分的方法
    public void sendMessage(String message, String toUser) {
        this.message.send(message, toUser);
    }
}
```

创建具体实现普通消息 NomalMessage 类。

```
/**
 * 普通消息类
 */
public class NomalMessage extends AbstractMessage {

    //构造方法，传入实现部分的对象
    public NomalMessage(IMessage message) {
        super(message);
    }

    @Override
    public void sendMessage(String message, String toUser) {
        //对于普通消息，直接调用父类方法发送消息即可
        super.sendMessage(message, toUser);
    }
}
```

创建具体实现加急消息 UrgencyMessage 类。

```
/**
 * 加急消息类
```

```
 */
public class UrgencyMessage extends AbstractMessage {

    //构造方法
    public UrgencyMessage(IMessage message) {
        super(message);
    }

    @Override
    public void sendMessage(String message, String toUser) {
        message = "加急：" + message;
        super.sendMessage(message, toUser);
    }

    //扩展它功能，监控某个消息的处理状态
    public Object watch(String messageId) {
        //根据给出的消息编码（messageId）查询消息的处理状态
        //组织成监控的处理状态，然后返回
        return null;
    }
}
```

最后编写客户端测试代码。

```
public static void main(String[] args) {
    IMessage message = new SmsMessage();
    AbstractMessage abstractMessage = new NomalMessage(message);
    abstractMessage.sendMessage("加班申请速批", "王总");

    message = new EmailMessage();
    abstractMessage = new UrgencyMessage(message);
    abstractMessage.sendMessage("加班申请速批", "王总");
}
```

运行结果如下图所示。

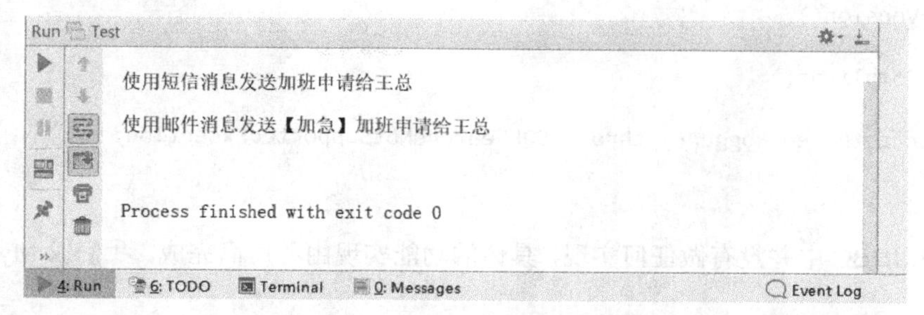

在上面的案例中，我们采用桥接模式解耦了“消息类型”和“消息紧急程度”这两个独立变化的维度。后续如果有更多的消息类型，比如微信、钉钉等，则直接新建一个类继承 IMessage 即

可；如果紧急程度需要新增，则同样只需新建一个类实现 AbstractMessage 类即可。

17.3 桥接模式在 JDK 源码中的应用

大家非常熟悉的 JDBC API，其中有一个 Driver 类就是桥接对象。我们都知道，在使用的时候通过 Class.forName()方法可以动态加载各个数据库厂商实现的 Driver 类。以 MySQL 的实现为例，具体客户端应用代码如下。

```
 //1.加载驱动
  Class.forName("com.mysql.jdbc.Driver");  //反射机制加载驱动类
 //2.获取连接 Connection
   //主机:端口号/数据库名
 Connection
conn=DriverManager.getConnection("jdbc:mysql://localhost:3306/test","root","root");
 //3.得到执行 SQL 语句的对象 Statement
Statement stmt = conn.createStatement();
 //4.执行 SQL 语句，并返回结果
 ResultSet rs=stmt.executeQuery("select *from table");
```

首先来看 Driver 接口的定义。

```
public interface Driver {

Connection connect(String url, java.util.Properties info) throws SQLException;

boolean acceptsURL(String url) throws SQLException;

DriverPropertyInfo[]  getPropertyInfo(String  url,  java.util.Properties  info)  throws
SQLException;

int getMajorVersion();

int getMinorVersion();

boolean jdbcCompliant();

public Logger getParentLogger() throws SQLFeatureNotSupportedException;

}
```

Driver 在 JDBC 中并没有做任何实现，具体的功能实现由各厂商完成，我们以 MySQL 的实现为例。

```
public class Driver extends NonRegisteringDriver implements java.sql.Driver {
  public Driver() throws SQLException {
```

```
    }

    static {
        try {
            DriverManager.registerDriver(new Driver());
        } catch (SQLException var1) {
            throw new RuntimeException("Can't register driver!");
        }
    }
}
```

当我们执行 Class.forName("com.mysql.jdbc.Driver")方法的时候，就会执行 com.mysql.jdbc.Driver 类的静态块中的代码。而静态块中的代码只是调用了一下 DriverManager 的 registerDriver()方法，然后将 Driver 对象注册到 DriverManager 中。继续跟进 DriverManager 类，来看相关的代码。

```
public class DriverManager {

    private final static CopyOnWriteArrayList<DriverInfo> registeredDrivers = new
CopyOnWriteArrayList<>();
    ...
    private DriverManager(){}
    static {
        loadInitialDrivers();
        println("JDBC DriverManager initialized");
    }

   ...
public static synchronized void registerDriver(java.sql.Driver driver)
        throws SQLException {

        registerDriver(driver, null);
    }

    public static synchronized void registerDriver(java.sql.Driver driver,
            DriverAction da)
        throws SQLException {
        if(driver != null) {
            registeredDrivers.addIfAbsent(new DriverInfo(driver, da));
        } else {
            throw new NullPointerException();
        }

        println("registerDriver: " + driver);

}
...
}
```

在注册之前，将传过来的 Driver 对象封装成一个 DriverInfo 对象。接下来继续执行客户端代码的第二步，调用 DriverManager 的 getConnection()方法获取连接对象，跟进源码。

```
public class DriverManager {

  ...

   public static Connection getConnection(String url,
      java.util.Properties info) throws SQLException {

      return (getConnection(url, info, Reflection.getCallerClass()));
   }


   public static Connection getConnection(String url,
      String user, String password) throws SQLException {
      java.util.Properties info = new java.util.Properties();

      if (user != null) {
         info.put("user", user);
      }
      if (password != null) {
         info.put("password", password);
      }

      return (getConnection(url, info, Reflection.getCallerClass()));
   }


   public static Connection getConnection(String url)
      throws SQLException {

      java.util.Properties info = new java.util.Properties();
      return (getConnection(url, info, Reflection.getCallerClass()));
}

  ...


   private static Connection getConnection(String url, java.util.Properties info, Class<?>
caller) throws SQLException {

      ClassLoader callerCL = caller != null ? caller.getClassLoader() : null;
```

```
        synchronized(DriverManager.class) {
            if (callerCL == null) {
                callerCL = Thread.currentThread().getContextClassLoader();
            }
        }

        if(url == null) {
            throw new SQLException("The url cannot be null", "08001");
        }

        println("DriverManager.getConnection(\"" + url + "\")");

        SQLException reason = null;

        for(DriverInfo aDriver : registeredDrivers) {
            if(isDriverAllowed(aDriver.driver, callerCL)) {
                try {
                    println("    trying " + aDriver.driver.getClass().getName());
                    Connection con = aDriver.driver.connect(url, info);
                    if (con != null) {
                        println("getConnection returning " + aDriver.driver.getClass().getName());
                        return (con);
                    }
                } catch (SQLException ex) {
                    if (reason == null) {
                        reason = ex;
                    }
                }

            } else {
                println("    skipping: " + aDriver.getClass().getName());
            }

        }
        if (reason != null)    {
            println("getConnection failed: " + reason);
            throw reason;
        }

        println("getConnection: no suitable driver found for "+ url);
        throw new SQLException("No suitable driver found for "+ url, "08001");
    }
    ...
}
```

在 getConnection()中，又会调用各自厂商实现的 Driver 的 connect()方法获得连接对象。这样就巧妙地避开了使用继承，为不同的数据库提供了相同的接口。JDBC API 中的 DriverManager 就是桥，如下图所示。

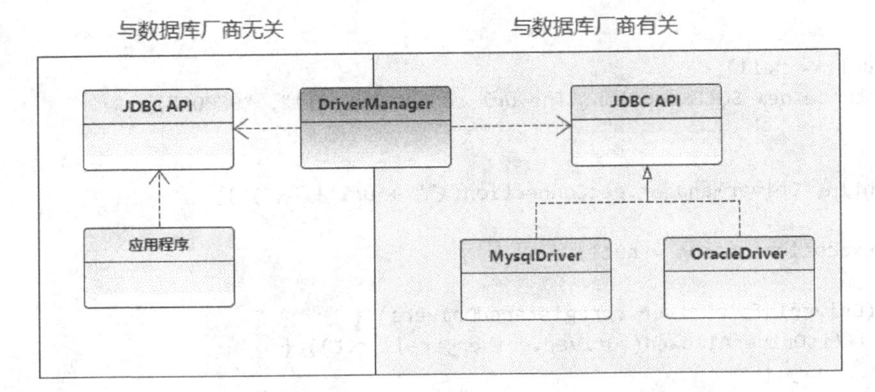

17.4 桥接模式扩展

通过上面的例子，我们能很好地感知到桥接模式遵循了里氏替换原则和依赖倒置原则，最终实现了开闭原则，对修改关闭，对扩展开放。这里将桥接模式的优缺点总结如下。

17.4.1 桥接模式的优点

（1）分离抽象部分及其具体实现部分。

（2）提高了系统的扩展性。

（3）符合开闭原则。

（4）符合合成复用原则。

17.4.2 桥接模式的缺点

（1）增加了系统的理解与设计难度。

（2）需要正确地识别系统中两个独立变化的维度。

4

第 4 篇

行为型设计模式

18

第 18 章 委派模式

18.1 委派模式概述

18.1.1 委派模式的定义

委派模式（Delegate Pattern）又叫作委托模式，是一种面向对象的设计模式，允许对象组合实现与继承相同的代码重用。它的基本作用就是负责任务的调用和分配，是一种特殊的静态代理模式，可以理解为全权代理模式，但是代理模式注重过程，而委派模式注重结果。委派模式属于行为型设计模式，不属于 GoF 的 23 种设计模式。

18.1.2 委派模式的应用场景

现实生活中，常有委派模式的场景发生，例如，老板给项目经理下达任务，项目经理会根据实际情况给每个员工都分配工作任务，待员工把工作任务完成之后，再由项目经理向老板汇报工作进度和结果。再比如，我们经常写授权委托书，授权他人代办事务。

委派模式适用于以下应用场景。

（1）需要实现表现层和业务层之间的松耦合。

（2）需要编排多个服务之间的调用。

（3）需要封装一层服务查找和调用。

18.1.3　委派模式的 UML 类图

委派模式的 UML 类图如下。

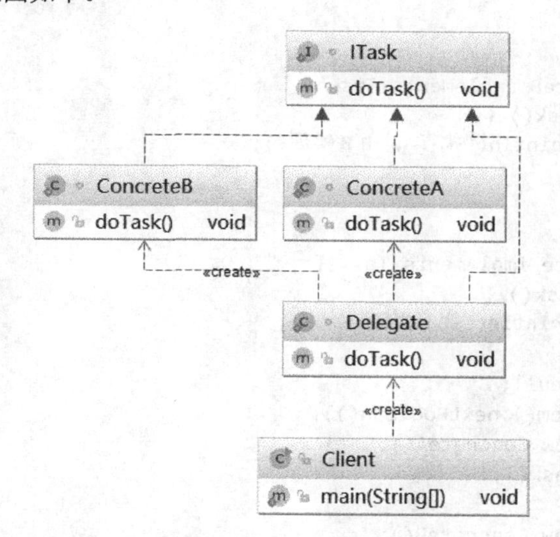

由上图可以看到，委派模式有 3 个参与角色。

（1）抽象任务角色（ITask）：定义一个抽象接口，它有若干实现类。

（2）委派者角色（Delegate）：负责在各个具体角色实例之间做出决策，判断并调用具体实现的方法。

（3）具体任务角色（Concrete）：真正执行任务的角色。

18.1.4　委派模式的通用写法

以下是委派模式的通用写法。

```
public class Client {
    public static void main(String[] args) {
```

```
        new Delegate().doTask();
    }

    //抽象任务角色
    interface Task {
        void doTask();
    }
    //具体任务角色 A
    static class ConcreteA implements Task {
        public void doTask() {
            System.out.println("执行 , 由 A 实现");
        }
    }
    //具体任务角色 B
    static class ConcreteB implements Task {
        public void doTask() {
            System.out.println("执行 , 由 B 实现");
        }
    }
    //委派者角色
    static class Delegate implements Task {
        public void doTask() {
            System.out.println("代理执行开始。);

            Task task = null;
            if (new Random().nextBoolean()){
                task = new ConcreteA();
                task.doTask();
            }else{
                task = new ConcreteB();
                task.doTask();
            }

            System.out.println("代理执行完毕。");
        }
    }
}
```

18.2 使用委派模式模拟任务分配场景

我们用代码来模拟老板给员工分配任务的业务场景。

首先创建 IEmployee 员工接口。

```
public interface IEmployee {
    void doing(String task);
}
```

创建员工 EmployeeA 类。

```
public class EmployeeA implements IEmployee {
    protected String goodAt = "编程";
    public void doing(String task) {
        System.out.println("我是员工 A，我擅长" + goodAt + ",现在开始做" +task + "工作");
    }
}
```

创建员工 EmployeeB 类。

```
public class EmployeeB implements IEmployee {
    protected String goodAt = "平面设计";
    public void doing(String task) {
        System.out.println("我是员工 B，我擅长" + goodAt + ",现在开始做" +task + "工作");
    }
}
```

创建项目经理 Leader 类。

```
public class Leader implements IEmployee {

    private Map<String,IEmployee> employee = new HashMap<String,IEmployee>();

    public Leader(){
        employee.put("爬虫",new EmployeeA());
        employee.put("海报图",new EmployeeB());
    }

    public void doing(String task) {
        if(!employee.containsKey(task)){
            System.out.println("这个任务" +task + "超出我的能力范围");
            return;
        }
        employee.get(task).doing(task);
    }
}
```

然后创建 Boss 类下达命令。

```
public class Boss {
    public void command(String task,Leader leader){
        leader.doing(task);
    }
}
```

最后编写客户端测试代码。

```
public class Test {
    public static void main(String[] args) {
        new Boss().command("海报图",new Leader());
        new Boss().command("爬虫",new Leader());
        new Boss().command("卖手机",new Leader());
    }
}
```

通过上面代码，我们生动地还原了老板分配任务的业务场景，这也是委派模式的生动体现。其类图如下。

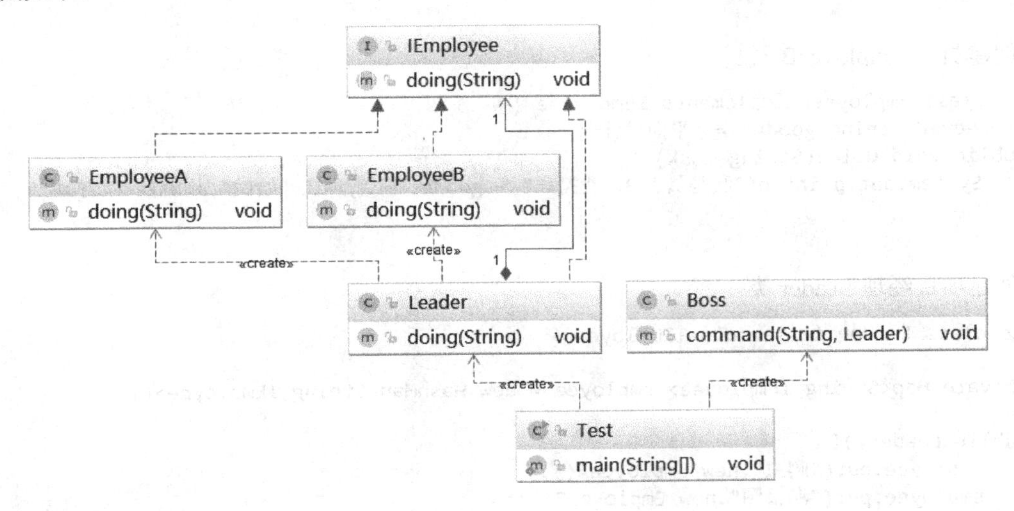

18.3 委派模式在框架源码中的应用

18.3.1 委派模式在 JDK 源码中的应用

JDK 中有一个典型的委派，众所周知，JVM 在加载类时用双亲委派模型，这又是什么呢？一个类加载器在加载类时，先把这个请求委派给自己的父类加载器去执行。如果父类加载器还存在父类加载器，则继续向上委派，直到顶层的启动类加载器；如果父类加载器能够完成类加载，则成功返回；如果父类加载器无法完成加载，则子加载器尝试自己去加载。从定义中可以看到，当双亲委派加载的一个类加载器加载类时，首先不是自己加载，而是委派给父类加载器。下面来看 loadClass()方法的源码，此方法在 ClassLoader 中。在这个类里定义了一个双亲，用于下面的类加载。

```
public abstract class ClassLoader {
    ...
    private final ClassLoader parent;
```

```
    ...
    protected Class<?> loadClass(String name, boolean resolve)
        throws ClassNotFoundException
    {
        synchronized (getClassLoadingLock(name)) {
            Class<?> c = findLoadedClass(name);
            if (c == null) {
                long t0 = System.nanoTime();
                try {
                    if (parent != null) {
                        c = parent.loadClass(name, false);
                    } else {
                        c = findBootstrapClassOrNull(name);
                    }
                } catch (ClassNotFoundException e) {
                }

                if (c == null) {
                    long t1 = System.nanoTime();
                    c = findClass(name);

                    sun.misc.PerfCounter.getParentDelegationTime().addTime(t1 - t0);
                    sun.misc.PerfCounter.getFindClassTime().addElapsedTimeFrom(t1);
                    sun.misc.PerfCounter.getFindClasses().increment();
                }
            }
            if (resolve) {
                resolveClass(c);
            }
            return c;
        }
    }
    ...
}
```

同样，在 Method 类里，常用的代理执行方法 invoke()也存在类似机制，代码如下。

```
public Object invoke(Object obj, Object... args)
        throws IllegalAccessException, IllegalArgumentException,
           InvocationTargetException
    {
        if (!override) {
            if (!Reflection.quickCheckMemberAccess(clazz, modifiers)) {
                Class<?> caller = Reflection.getCallerClass();
                checkAccess(caller, clazz, obj, modifiers);
            }
        }
```

```
        MethodAccessor ma = methodAccessor;             // read volatile
        if (ma == null) {
            ma = acquireMethodAccessor();
        }
        return ma.invoke(obj, args);
    }
```

看完代码，相信小伙伴们把委派模式和代理模式的区别弄清楚了吧。

18.3.2 委派模式在 Spring 源码中的应用

下面来看委派模式在Spring中的应用，Spring IoC模块中的DefaultBeanDefinitionDocumentReader类，当调用 doRegisterBeanDefinitions()方法时，即在 BeanDefinition 进行注册的过程中，会设置 BeanDefinitionParserDelegate 类型的 Delegate 对象传给 this.delegate，并将这个对象作为一个参数传入 parseBeanDefinitions(root, this.delegate)。主要的解析工作就是由 delegate 作为主要角色来完成的，代码如下。

```
protected void parseBeanDefinitions(Element root, BeanDefinitionParserDelegate delegate) {
   //判断节点是否属于同一命名空间，如果是，则执行后续的解析
   if (delegate.isDefaultNamespace(root)) {
      NodeList nl = root.getChildNodes();
      for (int i = 0; i < nl.getLength(); i++) {
         Node node = nl.item(i);
         if (node instanceof Element) {
            Element ele = (Element) node;
            if (delegate.isDefaultNamespace(ele)) {
               parseDefaultElement(ele, delegate);
            }
            else {
               //注解定义的 Context 的 nameSpace 进入这个分支
               delegate.parseCustomElement(ele);
            }
         }
      }
   }
   else {
      delegate.parseCustomElement(root);
   }
}
```

上面代码中的 parseDefaultElement(ele, delegate)方法，主要功能是针对不同的节点类型，完成 Bean 的注册操作，而在这个过程中，delegate 会调用 element 的 parseBeanDefinitionElement()方法，从而得到一个 BeanDefinitionHolder 类型的对象，之后通过这个对象完成注册。

再来还原一下 Spring MVC 的 DispatcherServlet 是如何实现委派模式的。创建业务类 MemberController。

```
public class MemberController {

    public void getMemberById(String mid){

    }

}
```

创建 OrderController 类。

```
public class OrderController {

    public void getOrderById(String mid){

    }

}
```

创建 SystemController 类。

```
public class SystemController {

    public void logout(){

    }
}
```

创建 DispatcherServlet 类。

```
public class DispatcherServlet extends HttpServlet {

    private Map<String,Method> handlerMapping = new HashMap<String,Method>();

    @Override
    protected void service(HttpServletRequest req, HttpServletResponse resp) throws
ServletException, IOException {
        doDispatch(req,resp);
    }

    private void doDispatch(HttpServletRequest req, HttpServletResponse resp) {
        String url = req.getRequestURI();
        Method method = handlerMapping.get(url);
       //此处省略反射调用方法的代码
           ...
```

```
    }

    @Override
    public void init() throws ServletException {
        try {
            handlerMapping.put("/web/getMemeberById.json",
MemberController.class.getMethod("getMemberById", new Class[]{String.class}));
        }catch (Exception e){
            e.printStackTrace();
        }
    }
}
```

配置 web.xml 文件。

```
<?xml version="1.0" encoding="UTF-8"?>
<web-app xmlns:xsi="http://www.w3.org/2001/XMLSchema-instance"
   xmlns="http://java.sun.com/xml/ns/j2ee" xmlns:javaee="http://java.sun.com/xml/ns/javaee"
   xmlns:web="http://java.sun.com/xml/ns/javaee/web-app_2_5.xsd"
   xsi:schemaLocation="http://java.sun.com/xml/ns/j2ee
http://java.sun.com/xml/ns/j2ee/web-app_2_4.xsd"
   version="2.4">
   <display-name>Gupao Web Application</display-name>


   <servlet>
      <servlet-name>delegateServlet</servlet-name>

<servlet-class>com.gupaoedu.vip.pattern.delegate.mvc.DispatcherServlet</servlet-class>
      <load-on-startup>1</load-on-startup>
   </servlet>

   <servlet-mapping>
      <servlet-name>delegateServlet</servlet-name>
      <url-pattern>/*</url-pattern>
   </servlet-mapping>

</web-app>
```

这样，一个完整的委派模式就实现了。当然，在 Spring 中运用委派模式的情况还有很多，大家通过命名就可以识别。在 Spring 源码中，只要以 Delegate 结尾的都实现了委派模式。例如，BeanDefinitionParserDelegate 根据不同的类型委派不同的逻辑来解析 BeanDefinition。

18.4　委派模式扩展

18.4.1　委派模式的优点

通过任务委派能够将一个大型任务细化，然后通过统一管理这些子任务的完成情况实现任务的跟进，加快任务执行的效率。

18.4.2　委派模式的缺点

任务委派方式需要根据任务的复杂程度进行不同的改变，在任务比较复杂的情况下，可能需要进行多重委派，容易造成紊乱。

19

第19章 模板方法模式

19.1 模板方法模式概述

19.1.1 模板方法模式的定义

模板方法模式（Template Method Pattern）又叫作模板模式，指定义一个操作中的算法的框架，而将一些步骤延迟到子类中，使得子类可以不改变一个算法的结构即可重定义该算法的某些特定步骤，属于行为型设计模式。

原文： Define the skeleton of an algorithm in an operation, deferring some steps to subclasses. Template Method lets subclasses redefine certain steps of an algorithm without changing the algorithm's structure.

模板方法模式实际上封装了一个固定流程，该流程由几个步骤组成，具体步骤可以由子类进行不同的实现，从而让固定的流程产生不同的结果。它非常简单，其实就是类的继承机制，但它却是一个应用非常广泛的模式。模板方法模式的本质是抽象封装流程，具体进行实现。

19.1.2 模板方法模式的应用场景

当完成一个操作具有固定的流程时，由抽象固定流程步骤，具体步骤交给子类进行具体实现

（固定的流程，不同的实现）。

在日常生活中，模板方法模式很常见。比如入职流程：填写入职登记表 → 打印简历 → 复印学历证书 → 复印身份证 → 签订劳动合同 → 建立花名册 → 办理工牌 → 安排工位等。再比如，笔者平时在家炒菜的流程：洗锅 → 点火 → 热锅 → 上油 → 下原料 → 翻炒 → 放调料 → 出锅。再比如赵本山、宋丹丹的小品《钟点工》中有这么一段台词。赵本山问："要把大象装冰箱，总共分几步?"宋丹丹答："三步。第一步，把冰箱门打开；第二步，把大象装进去；第三步，把冰箱门带上。"其流程如下图所示。此后又有人做脑筋急转弯延伸，问："怎么把长颈鹿放进冰箱?"有人回答："第一步，打开冰箱门；第二步，把大象拿出来；第三步，把长颈鹿塞进去；第四步，关闭冰箱门。"虽说小品中的情景只是一个笑话，但从现实中来看，这些都是模板方法模式的体现。

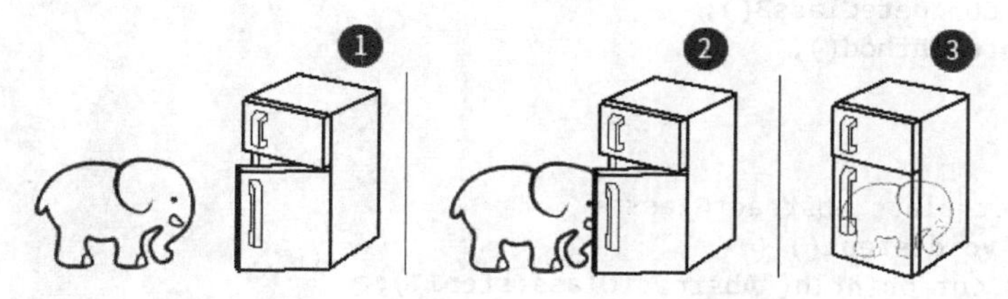

模板方法模式适用于以下应用场景。

（1）一次性实现一个算法的不变的部分，并将可变的行为留给子类来实现。

（2）各子类中公共的行为被提取出来，集中到一个公共的父类中，从而避免代码重复。

19.1.3　模板方法模式的 UML 类图

模板方法模式的 UML 类图如下。

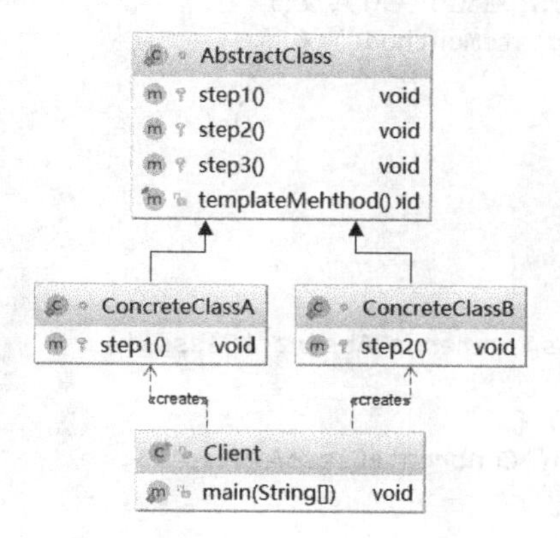

由上图可以看到，模板方法模式主要包含 2 个角色。

（1）抽象模板（AbstractClass）：抽象模板类，定义了一套算法框架/流程。

（2）具体实现（ConcreteClass）：具体实现类，对算法框架/流程的某些步骤进行了实现。

19.1.4 模板方法模式的通用写法

下面是模板方法模式的通用写法。

```
public class Client {
    public static void main(String[] args) {
        AbstractClass abc = new ConcreteClassA();
        abc.templateMehthod();
        abc = new ConcreteClassB();
        abc.templateMehthod();
    }

    //抽象模板类
    static abstract class AbstractClass {
        protected void step1() {
            System.out.println("AbstractClass:step1");
        }

        protected void step2() {
            System.out.println("AbstractClass:step2");
        }

        protected void step3() {
            System.out.println("AbstractClass:step3");
        }

        //声明为 final 类型的方法，避免子类对其覆写
        public final void templateMehthod() {
            this.step1();
            this.step2();
            this.step3();
        }
    }

    //具体实现类 A
    static class ConcreteClassA extends AbstractClass {
        @Override
        protected void step1() {
            System.out.println("ConcreateClassA:step1");
        }
```

```
    }

    //具体实现类 B
    static class ConcreteClassB extends AbstractClass {
        @Override
        protected void step2() {
            System.out.println("ConcreateClassB:step2");
        }
    }
}
```

通常把抽象模板类 AbstractClass 的模板方法 templateMethod()定义成 final 类型的方法，避免子类对其覆写，并定义算法结构/流程的语义。

19.2 使用模板方法模式解决实际问题

19.2.1 模板方法模式中的钩子方法

我们以咕泡学院的课程创建流程为例：发布预习资料 → 制作课件 PPT → 在线直播 → 提交课堂笔记 → 提交源码 → 布置作业 → 检查作业。首先创建 AbastractCourse 抽象类。

```
public abstract class AbastractCourse {

    public final void createCourse(){
        //1.发布预习资料
        postPreResoucse();

        //2.制作课件 PPT
        createPPT();

        //3.在线直播
        liveVideo();

        //4.上传课后资料
        postResource();

        //5.布置作业
        postHomework();

        if(needCheckHomework()){
            checkHomework();
        }
    }
```

```
    protected abstract void checkHomework();

    //钩子方法
    protected boolean needCheckHomework(){return  false;}

    protected void postHomework(){
        System.out.println("布置作业");
    }

    protected void postResource(){
        System.out.println("上传课后资料");
    }

    protected void liveVideo(){
        System.out.println("直播授课");
    }

    protected void createPPT(){
        System.out.println("制作课件");
    }

    protected void postPreResoucse(){
        System.out.println("发布预习资料");
    }

}
```

上面代码中有个钩子方法，可能有些小伙伴不是太理解，在此笔者稍做解释。设计钩子方法的主要目的是干预执行流程，使得控制行为流程更加灵活，更符合实际业务的需求。钩子方法的返回值一般为适合条件分支语句的返回值（如 boolean、int 等）。小伙伴们可以根据自己的业务场景决定是否需要使用钩子方法。

然后创建 JavaCourse 类。

```
public class JavaCourse extends AbastractCourse {
    private boolean needCheckHomework = false;

    public void setNeedCheckHomework(boolean needCheckHomework) {
        this.needCheckHomework = needCheckHomework;
    }

    @Override
    protected boolean needCheckHomework() {
        return this.needCheckHomework;
    }
```

```
    protected void checkHomework() {
        System.out.println("检查 Java 作业");
    }
}
```

创建 PythonCourse 类。

```
public class PythonCourse extends AbastractCourse {
    protected void checkHomework() {
        System.out.println("检查 Python 作业");
    }
}
```

最后编写客户端测试代码。

```
    public static void main(String[] args) {
        System.out.println("=========架构师课程=========");
        JavaCourse java = new JavaCourse();
        java.setNeedCheckHomework(false);
        java.createCourse();


        System.out.println("=========Python 课程=========");
        PythonCourse python = new PythonCourse();
        python.createCourse();
    }
```

通过这样一个案例，相信小伙伴们对模板方法模式有了一个基本的印象。为了加深理解，我们结合一个常见的业务场景进行介绍。

19.2.2 使用模板方法模式重构 JDBC 业务操作

创建一个模板类 JdbcTemplate，封装所有的 JDBC 操作。以查询为例，每次查询的表都不同，返回的数据结构也就都不一样。我们针对不同的数据，都要封装成不同的实体对象。而每个实体封装的逻辑都是不一样的，但封装前和封装后的处理流程是不变的，因此，可以使用模板方法模式设计这样的业务场景。首先创建约束 ORM 逻辑的接口 RowMapper。

```
/**
 * ORM 映射定制化的接口
 */
public interface RowMapper<T> {
    T mapRow(ResultSet rs,int rowNum) throws Exception;
}
```

然后创建封装了所有处理流程的抽象类 JdbcTemplate。

```
public abstract class JdbcTemplate {
    private DataSource dataSource;

    public JdbcTemplate(DataSource dataSource) {
        this.dataSource = dataSource;
    }

    public final List<?> executeQuery(String sql,RowMapper<?> rowMapper,Object[] values){
        try {
            //1.获取连接
            Connection conn = this.getConnection();
            //2.创建语句集
            PreparedStatement pstm = this.createPrepareStatement(conn,sql);
            //3.执行语句集
            ResultSet rs = this.executeQuery(pstm,values);
            //4.处理结果集
            List<?> result = this.parseResultSet(rs,rowMapper);
            //5.关闭结果集
            rs.close();
            //6.关闭语句集
            pstm.close();
            //7.关闭连接
            conn.close();
            return result;
        }catch (Exception e){
            e.printStackTrace();
        }
        return null;
    }

    private List<?> parseResultSet(ResultSet rs, RowMapper<?> rowMapper) throws Exception {
        List<Object> result = new ArrayList<Object>();
        int rowNum = 0;
        while (rs.next()){
            result.add(rowMapper.mapRow(rs,rowNum++));
        }
        return result;
    }


    private ResultSet executeQuery(PreparedStatement pstm, Object[] values) throws
SQLException {
        for (int i = 0; i < values.length; i++) {
            pstm.setObject(i,values[i]);
        }
        return pstm.executeQuery();
    }
```

```
    private PreparedStatement createPrepareStatement(Connection conn, String sql) throws
SQLException {
        return conn.prepareStatement(sql);
    }

    private Connection getConnection() throws SQLException {
        return this.dataSource.getConnection();
    }
}
```

创建实体对象 Member 类。

```
public class Member {

    private String username;
    private String password;
    private String nickname;
    private int age;
    private String addr;

    public String getUsername() {
        return username;
    }

    public void setUsername(String username) {
        this.username = username;
    }

    public String getPassword() {
        return password;
    }

    public void setPassword(String password) {
        this.password = password;
    }

    public String getNickname() {
        return nickname;
    }

    public void setNickname(String nickname) {
        this.nickname = nickname;
    }

    public int getAge() {
        return age;
```

```
    }

    public void setAge(int age) {
        this.age = age;
    }

    public String getAddr() {
        return addr;
    }

    public void setAddr(String addr) {
        this.addr = addr;
    }
}
```

创建数据库操作类 MemberDao。

```
public class MemberDao extends JdbcTemplate {
    public MemberDao(DataSource dataSource) {
        super(dataSource);
    }

    public List<?> selectAll(){
        String sql = "select * from t_member";
        return super.executeQuery(sql, new RowMapper<Member>() {
            public Member mapRow(ResultSet rs, int rowNum) throws Exception {
                Member member = new Member();
                //字段过多，原型模式
                member.setUsername(rs.getString("username"));
                member.setPassword(rs.getString("password"));
                member.setAge(rs.getInt("age"));
                member.setAddr(rs.getString("addr"));
                return member;
            }
        },null);
    }
}
```

最后编写客户端测试代码。

```
public static void main(String[] args) {
        MemberDao memberDao = new MemberDao(null);
        List<?> result = memberDao.selectAll();
        System.out.println(result);
    }
```

希望通过这两个案例的业务场景分析，小伙伴们能够对模板方法模式有更深的理解。

19.3 模板方法模式在框架源码中的应用

19.3.1 模板方法模式在 JDK 源码中的应用

先来看 JDK 中的 AbstractList，代码如下。

```
package java.util;

public abstract class AbstractList<E> extends AbstractCollection<E> implements List<E> {
    ...
    abstract public E get(int index);
    ...
}
```

我们看到，get()是一个抽象方法，它的逻辑交给子类来实现，大家所熟知的 ArrayList 就是 AbstractList 的子类。同理，有 AbstractList 就有 AbstractSet 和 AbstractMap，有兴趣的小伙伴可以去看它们的源码实现。还有一个每天都在用的 HttpServlet，有 service()、doGet()和 doPost() 3 种方法，这些都是模板方法的抽象实现。

19.3.2 模板方法模式在 MyBatis 源码中的应用

在 MyBatis 源码中，模板方法模式有很多经典的应用场景。不妨来看一个 BaseExecutor 类，它是一个基础的 SQL 执行类，实现了大部分 SQL 执行逻辑，然后把几个方法交给子类定制化完成，源码如下。

```
...
public abstract class BaseExecutor implements Executor {
    ...
    protected abstract int doUpdate(MappedStatement var1, Object var2) throws SQLException;

    protected abstract List<BatchResult> doFlushStatements(boolean var1) throws SQLException;

    protected abstract <E> List<E> doQuery(MappedStatement var1, Object var2, RowBounds var3,
ResultHandler var4, BoundSql var5) throws SQLException;

    protected abstract <E> Cursor<E> doQueryCursor(MappedStatement var1, Object var2, RowBounds
var3, BoundSql var4) throws SQLException;
...
}
```

例如，doUpdate、doFlushStatements、doQuery 和 doQueryCursor 这几个方法就是交由子类来实现的，BaseExecutor 的子类有 ReuseExecutor、SimpleExecutor、BatchExecutor 和 ClosedExecutor，其类图如下。

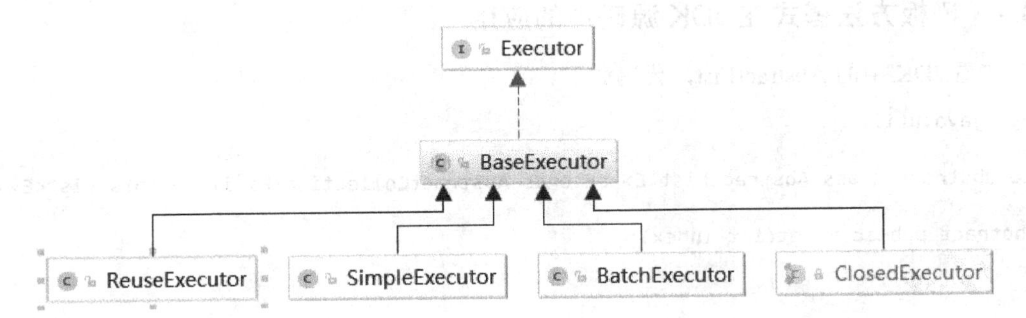

下面是 SimpleExecutor 的 doUpdate()方法实现。

```
public int doUpdate(MappedStatement ms, Object parameter) throws SQLException {
    Statement stmt = null;

    int var6;
    try {
        Configuration configuration = ms.getConfiguration();
        StatementHandler handler = configuration.newStatementHandler(this, ms, parameter,
RowBounds.DEFAULT, (ResultHandler)null, (BoundSql)null);
        stmt = this.prepareStatement(handler, ms.getStatementLog());
        var6 = handler.update(stmt);
    } finally {
        this.closeStatement(stmt);
    }

    return var6;
}
```

再来对比一下 BatchExecutor 的 doUpate()方法实现。

```
public int doUpdate(MappedStatement ms, Object parameterObject) throws SQLException {
    Configuration configuration = ms.getConfiguration();
    StatementHandler handler = configuration.newStatementHandler(this, ms, parameterObject,
RowBounds.DEFAULT, (ResultHandler)null, (BoundSql)null);
    BoundSql boundSql = handler.getBoundSql();
    String sql = boundSql.getSql();
    Statement stmt;
    if(sql.equals(this.currentSql) && ms.equals(this.currentStatement)) {
        int last = this.statementList.size() - 1;
        stmt = (Statement)this.statementList.get(last);
        this.applyTransactionTimeout(stmt);
```

```
        handler.parameterize(stmt);
        BatchResult batchResult = (BatchResult)this.batchResultList.get(last);
        batchResult.addParameterObject(parameterObject);
    } else {
        Connection connection = this.getConnection(ms.getStatementLog());
        stmt = handler.prepare(connection, this.transaction.getTimeout());
        handler.parameterize(stmt);
        this.currentSql = sql;
        this.currentStatement = ms;
        this.statementList.add(stmt);
        this.batchResultList.add(new BatchResult(ms, sql, parameterObject));
    }

    handler.batch(stmt);
    return -2147482646;
}
```

细心的小伙伴一定看出了差异，BatchExecutor 的处理逻辑比 SimpleExecutor 更为复杂。调用的核心 API 也有区别，SimpleExecutor 调用的核心方法是 handler.update()方法，BatchExecutor 调用的核心方法是 handler.batch()方法。这里暂时不对 MyBatis 源码进行深入分析，感兴趣的小伙伴可以自行继续深入研究。

19.4　模板方法模式扩展

19.4.1　模板方法模式的优点

（1）利用模板方法将相同处理逻辑的代码放到抽象父类中，可以提高代码的复用性。

（2）将不同的算法逻辑分离到不同的子类中，通过对子类的扩展增加新的行为，提高代码的可扩展性。

（3）把不变的行为写在父类上，去除子类的重复代码，提供了一个很好的代码复用平台，符合开闭原则。

19.4.2　模板方法模式的缺点

（1）每一个抽象类都需要一个子类来实现，这样导致类数量增加。

（2）类数量的增加，间接地增加了系统实现的复杂度。

（3）由于继承关系自身的缺点，如果父类添加新的抽象方法，则所有子类都要改一遍。

第 20 章 策略模式

20.1 策略模式概述

20.1.1 策略模式的定义

策略模式（Strategy Pattern）又叫作政策模式（Policy Pattern），它将定义的算法家族分别封装起来，让它们之间可以互相替换，从而让算法的变化不会影响到使用算法的用户，属于行为型设计模式。

原文：Define a family of algorithms, encapsulate each one, and make them interchangeable.

策略模式使用的就是面向对象的继承和多态机制，从而实现同一行为在不同的场景下具备不同的实现。

20.1.2 策略模式的应用场景

策略模式在生活场景中的应用非常多。比如，一个人的纳税比率与他的工资有关，不同的工资水平对应不同的税率。再比如，在互联网移动支付的大背景下，我们每次下单后付款前，都需要选择支付方式，如下图所示。

阶梯个税

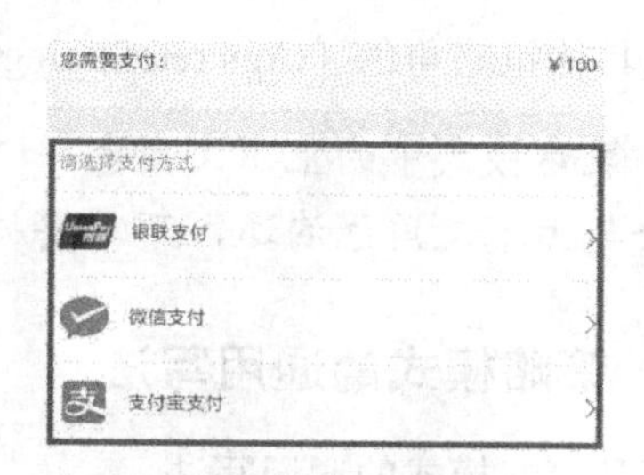

支付方式选择

策略模式可以解决在有多种相似算法的情况下使用 if...else 或 switch...case 所带来的复杂性和臃肿性问题。在日常业务开发中，策略模式适用于以下应用场景。

（1）针对同一类型问题，有多种处理方式，每一种都能独立解决问题。

（2）需要自由切换算法的场景。

（3）需要屏蔽算法规则的场景。

20.1.3　策略模式的 UML 类图

策略模式的 UML 类图如下。

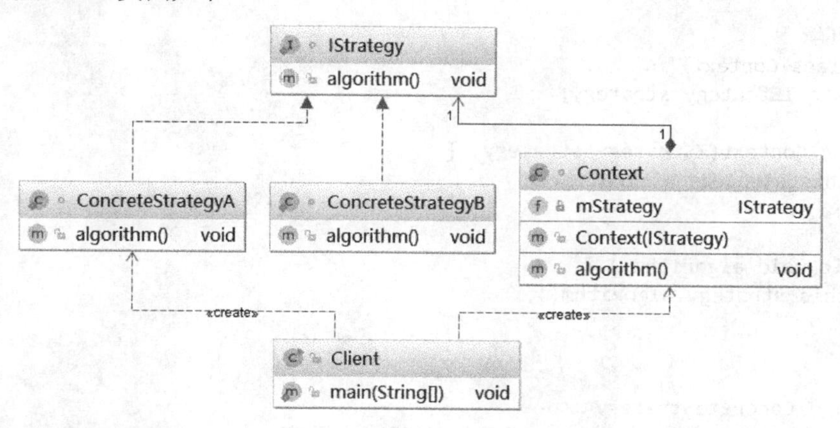

由上图可以看到，策略模式主要包含 3 个角色。

（1）上下文角色（Context）：用来操作策略的上下文环境，屏蔽高层模块（客户端）对策略、算法的直接访问，封装可能存在的变化。

（2）抽象策略角色（IStrategy）：规定策略或算法的行为。

（3）具体策略角色（ConcreteStrategy）：具体的策略或算法实现。

注：策略模式中的上下文角色（Context），其职责是隔离客户端与策略类的耦合，让客户端完全与上下文角色沟通，不需要关心具体策略。

20.1.4 策略模式的通用写法

以下是策略模式的通用写法。

```
public class Client {
    public static void main(String[] args) {
        //选择一个具体策略
        IStrategy strategy = new ConcreteStrategyA();
        //创建一个上下文环境
        Context context = new Context(strategy);
        //客户端直接让上下文角色执行算法
        context.algorithm();
    }

    //抽象策略类 IStrategy
    interface IStrategy {
        void algorithm();
    }

    //上下文环境
    static class Context {
        private IStrategy strategy;

        public Context(IStrategy strategy) {
            this.strategy = strategy;
        }

        public void algorithm() {
            this.strategy.algorithm();
        }
    }

    //具体策略类 ConcreteStrategyA
    static class ConcreteStrategyA implements IStrategy {
        public void algorithm() {
            System.out.println("Strategy A");
        }
    }

    //具体策略类 ConcreteStrategyB
    static class ConcreteStrategyB implements IStrategy {
```

```
        public void algorithm() {
            System.out.println("Strategy B");
        }
    }
}
```

20.2 使用策略模式解决实际问题

20.2.1 使用策略模式实现促销优惠方案选择

大家都知道，咕泡学院的架构师课程经常会有优惠活动，优惠策略有很多种可能，如领取优惠券抵扣、返现促销、拼团优惠等。下面用代码来模拟，首先创建一个促销策略的抽象 PromotionStrategy。

```
/**
 * 促销策略抽象
 */
public interface IPromotionStrategy {
    void doPromotion();
}
```

然后分别创建优惠券抵扣策略 CouponStrategy 类、返现促销策略 CashbackStrategy 类、拼团优惠策略 GroupbuyStrategy 类和无优惠策略 EmptyStrategy 类。

CouponStrategy 类的代码如下。

```
public class CouponStrategy implements IPromotionStrategy {
    public void doPromotion() {
        System.out.println("使用优惠券抵扣");
    }
}
```

CashbackStrategy 类的代码如下。

```
public class CashbackStrategy implements IPromotionStrategy {
    public void doPromotion() {
        System.out.println("返现，直接打款到支付宝账号");
    }
}
```

GroupbuyStrategy 类的代码如下。

```
public class GroupbuyStrategy implements IPromotionStrategy {
    public void doPromotion() {
        System.out.println("5人成团，可以优惠");
```

```
    }
}
```

EmptyStrategy 类的代码如下。

```
public class EmptyStrategy implements IPromotionStrategy {
    public void doPromotion() {
        System.out.println("无优惠");
    }
}
```

接着创建促销活动方案 PromotionActivity 类。

```
public class PromotionActivity {
    private IPromotionStrategy strategy;

    public PromotionActivity(IPromotionStrategy strategy) {
        this.strategy = strategy;
    }

    public void execute(){
        strategy.doPromotion();
    }
}
```

最后编写客户端测试代码。

```
public static void main(String[] args) {
    PromotionActivity activity618 = new PromotionActivity(new CouponStrategy());
    PromotionActivity activity1111 = new PromotionActivity(new CashbackStrategy());

    activity618.execute();
    activity1111.execute();
}
```

此时，小伙伴们会发现，如果把上面这段客户端测试代码放到实际的业务场景中，其实并不实用。因为我们做活动的时候往往要根据不同的需求对促销策略进行动态选择，并不会一次性执行多种优惠。所以代码通常会这样写。

```
public static void main(String[] args) {
    PromotionActivity promotionActivity = null;

    String promotionKey = "COUPON";

    if(StringUtils.equals(promotionKey,"COUPON")){
        promotionActivity = new PromotionActivity(new CouponStrategy());
    }else if(StringUtils.equals(promotionKey,"CASHBACK")){
        promotionActivity = new PromotionActivity(new CashbackStrategy());
```

```
    }
    promotionActivity.execute();
}
```

这样改造之后，代码满足了业务需求，客户可根据自己的需求选择不同的优惠策略。但是，经过一段时间的业务积累，促销活动会越来越多。于是，程序员就开始经常加班，每次上活动之前都要通宵改代码，而且要做重复测试，判断逻辑可能也会变得越来越复杂。此时，我们要思考代码是否需要重构。回顾之前学过的设计模式，我们应该如何来优化这段代码呢？其实，可以结合单例模式和简单工厂模式，创建 PromotionStrategyFactory 类。

```
public class PromotionStrategyFacory {

    private static Map<String,IPromotionStrategy> PROMOTIONS = new HashMap<String,
IPromotionStrategy>();

    static {
        PROMOTIONS.put(PromotionKey.COUPON,new CouponStrategy());
        PROMOTIONS.put(PromotionKey.CASHBACK,new CashbackStrategy());
        PROMOTIONS.put(PromotionKey.GROUPBUY,new GroupbuyStrategy());
    }

    private static final IPromotionStrategy EMPTY = new EmptyStrategy();

    private PromotionStrategyFacory(){}

    public static IPromotionStrategy getPromotionStrategy(String promotionKey){
        IPromotionStrategy strategy = PROMOTIONS.get(promotionKey);
        return strategy == null ? EMPTY : strategy;
    }
    private interface PromotionKey{
        String COUPON = "COUPON";
        String CASHBACK = "CASHBACK";
        String GROUPBUY = "GROUPBUY";
    }

    public static  Set<String> getPromotionKeys(){
        return PROMOTIONS.keySet();
    }
}
```

这时候，客户端测试代码如下。

```
public static void main(String[] args) {
      PromotionStrategyFacory.getPromotionKeys();
      String promotionKey = "COUPON";
```

```
        IPromotionStrategy promotionStrategy = PromotionStrategyFacory.getPromotionStrategy
(promotionKey);
        promotionStrategy.doPromotion();
}
```

代码优化之后，程序员的维护工作也变得轻松了。每次上新活动都不影响原来的代码逻辑。

20.2.2 使用策略模式重构支付方式选择场景

为了加深对策略模式的理解，我们再举一个案例。相信小伙伴们都用过支付宝、微信支付、银联支付及京东白条，一个常见的应用场景就是大家在下单支付时会提示选择支付方式，如果用户未选，系统也会默认好推荐的支付方式进行结算。来看如下图所示的类图，我们用策略模式来模拟此业务场景。

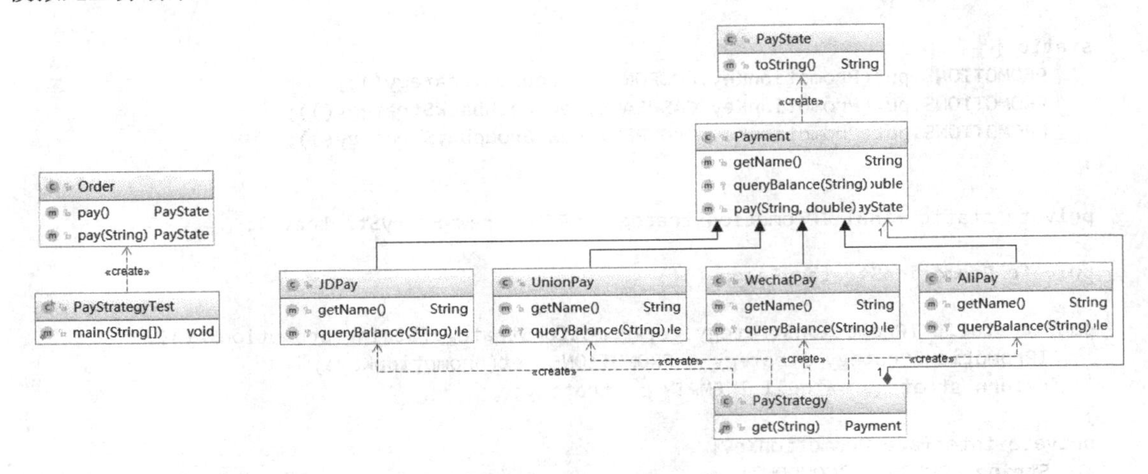

首先创建 Payment 抽象类，定义支付规范和支付逻辑，代码如下。

```
import com.gupaoedu.vip.pattern.strategy.pay.PayState;

/**
 * 支付渠道
 */
public abstract class Payment {

    public abstract String getName();

    //通用逻辑被放到抽象类里实现
    public MsgResult pay(String uid, double amount){
        //余额是否足够
```

```
        if(queryBalance(uid) < amount){
            return new MsgResult(500,"支付失败","余额不足");
        }
        return new MsgResult(200,"支付成功","支付金额" + amount);
    }

    protected abstract double queryBalance(String uid);
}
```

然后分别创建具体的支付方式，支付宝 AliPay 类的代码如下。

```
public class AliPay extends Payment {
    public String getName() {
        return "支付宝";
    }

    protected double queryBalance(String uid) {
        return 900;
    }
}
```

京东白条 JDPay 类的代码如下。

```
public class JDPay extends Payment {
    public String getName() {
        return "京东白条";
    }

    protected double queryBalance(String uid) {
        return 500;
    }
}
```

微信支付 WechatPay 类的代码如下。

```
public class WechatPay extends Payment {
    public String getName() {
        return "微信支付";
    }

    protected double queryBalance(String uid) {
        return 263;
    }
}
```

银联支付 UnionPay 类的代码如下。

```
public class UnionPay extends Payment {
    public String getName() {
```

```
        return "银联支付";
    }

    protected double queryBalance(String uid) {
        return 120;
    }
}
```

接着创建支付状态的包装类 MsgResult。

```
/**
 * 支付完成以后的状态
 */
public class MsgResult {
    private int code;
    private Object data;
    private String msg;

    public MsgResult(int code, String msg, Object data) {
        this.code = code;
        this.data = data;
        this.msg = msg;
    }

    @Override
    public String toString() {
        return "MsgResult{" +
                "code=" + code +
                ", data=" + data +
                ", msg='" + msg + '\'' +
                '}';
    }
}
```

创建支付策略管理类。

```
import java.util.HashMap;
import java.util.Map;

/**
 * 支付策略管理
 */
public class PayStrategy {
    public static  final String ALI_PAY = "AliPay";
    public static  final String JD_PAY = "JdPay";
    public static  final String WECHAT_PAY = "WechatPay";
    public static  final String UNION_PAY = "UnionPay";
    public static  final String DEFAULT_PAY = ALI_PAY;
```

```
    private static Map<String,Payment> strategy = new HashMap<String,Payment>();

    static {
        strategy.put(ALI_PAY,new AliPay());
        strategy.put(JD_PAY,new JDPay());
        strategy.put(WECHAT_PAY,new WechatPay());
        strategy.put(UNION_PAY,new UnionPay());
    }

    public static Payment get(String payKey){
        if(!strategy.containsKey(payKey)){
            return strategy.get(DEFAULT_PAY);
        }
        return strategy.get(payKey);
    }
}
```

创建订单 Order 类。

```
import com.gupaoedu.vip.pattern.strategy.pay.payport.PayStrategy;
import com.gupaoedu.vip.pattern.strategy.pay.payport.Payment;

public class Order {
    private String uid;
    private String orderId;
    private double amount;

    public Order(String uid, String orderId, double amount) {
        this.uid = uid;
        this.orderId = orderId;
        this.amount = amount;
    }

    public MsgResult pay(){
        return pay(PayStrategy.DEFAULT_PAY);
    }

    public MsgResult pay(String payKey){
        Payment payment = PayStrategy.get(payKey);
        System.out.println("欢迎使用" + payment.getName());
        System.out.println("本次交易金额为" + amount + "，开始扣款");
        return payment.pay(uid,amount);
    }
}
```

最后编写客户端测试代码。

```
public static void main(String[] args) {
        Order order = new Order("1","2020031401000323",324.5);
        System.out.println(order.pay(PayStrategy.ALI_PAY));
    }
```

运行结果如下图所示。

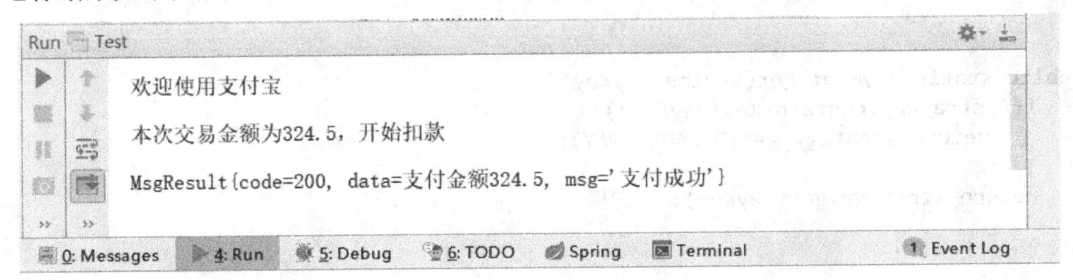

通过常见的业务场景举例，希望小伙伴们能够更深刻地理解策略模式。

20.2.3 策略模式和委派模式结合使用

上面代码中列举了常见的几个业务场景，相信小伙伴们对委派模式和策略模式有了非常深刻的理解。再来回顾一下 DispatcherServlet 的委派逻辑，代码如下。

```
package com.gupaoedu.vip.pattern.delegate.mvc;

...

public class DispatcherServlet extends HttpServlet{

    private void doDispatch(HttpServletRequest request, HttpServletResponse response) throws
Exception{

        String uri = request.getRequestURI();
        String mid = request.getParameter("mid");

        if("getMemberById".equals(uri)){
            new MemberController().getMemberById(mid);
        }else if("getOrderById".equals(uri)){
            new OrderController().getOrderById(mid);
        }else if("logout".equals(uri)){
            new SystemController().logout();
        }else {
            response.getWriter().write("404 Not Found!!");
        }
```

```
    }
...
}
```

这样的代码扩展性不太优雅，也不现实，因为实际项目中一定不止这几个 Controller，往往有成千上万个 Controller，显然，我们不能写成千上万个 if...else 。那么如何改造呢？小伙伴们一定先想到了策略模式，来看一下笔者是怎么优化的。

```
package com.gupaoedu.vip.pattern.delegate.mvc;


/**
 * 相当于项目经理的角色
 */
public class DispatcherServlet extends HttpServlet{

    private List<Handler> handlerMapping = new ArrayList<Handler>();

    public void init() throws ServletException {
        try {
            Class<?> memberControllerClass = MemberController.class;
            handlerMapping.add(new Handler()
                    .setController(memberControllerClass.newInstance())
                    .setMethod(memberControllerClass.getMethod("getMemberById", new Class[]{String.class}))
                    .setUrl("/web/getMemberById.json"));
        }catch(Exception e){

        }
    }

    private void doDispatch(HttpServletRequest request, HttpServletResponse response){

        //1.获取用户请求的 URL
        //如果按照 J2EE 的标准，则每个 URL 都对应一个 Serlvet，URL 由浏览器输入
       String uri = request.getRequestURI();

        //2.Servlet 获得 URL 以后，要做权衡（要做判断、选择）
        //根据用户请求的 URL，找到这个 URL 对应的某一个 Java 类的方法

        //3.通过获得的 URL 在 handlerMapping 中找到对应的 Handler
        Handler handle = null;
        for (Handler h: handlerMapping) {
            if(uri.equals(h.getUrl())){
                handle = h;
                break;
```

```
            }
        }

        //4.将具体的任务分发给Method（通过反射去调用其对应的方法）
        Object object = null;
        try {
            object = handle.getMethod().invoke(handle.getController(),request.getParameter("mid"));
        } catch (IllegalAccessException e) {
            e.printStackTrace();
        } catch (InvocationTargetException e) {
            e.printStackTrace();
        }

        //5.获取到Method执行的结果，通过Response返回出去
//      response.getWriter().write();
    }


    protected void service(HttpServletRequest req, HttpServletResponse resp) throws
ServletException, IOException {
        try {
            doDispatch(req,resp);
        } catch (Exception e) {
            e.printStackTrace();
        }
    }

    class Handler{

        private Object controller;
        private Method method;
        private String url;

        public Object getController() {
            return controller;
        }

        public Handler setController(Object controller) {
            this.controller = controller;
            return this;
        }

        public Method getMethod() {
            return method;
        }

        public Handler setMethod(Method method) {
```

```
            this.method = method;
            return this;
        }

        public String getUrl() {
            return url;
        }

        public Handler setUrl(String url) {
            this.url = url;
            return this;
        }
    }
}
```

在上面代码中，笔者结合了策略模式、简单工厂模式、单例模式。当然，这个优化方案不一定是最完美的，仅代表个人观点。感兴趣的小伙伴可以继续思考，让这段代码变得更优雅。

20.3 策略模式在框架源码中的应用

20.3.1 策略模式在 JDK 源码中的应用

首先来看 JDK 中一个比较常用的比较器 Comparator 接口，大家常用的 compare()方法就是一个策略抽象实现。

```
public interface Comparator<T> {
    int compare(T o1, T o2);
    ...
}
```

Comparator 抽象下有非常多实现类，我们经常会把 Comparator 作为参数传入当做排序策略，例如 Arrays 类的 parallelSort 方法等。

```
public class Arrays {
    ...
    public static <T> void parallelSort(T[] a, int fromIndex, int toIndex,
                                        Comparator<? super T> cmp) {
    ...
    }
    ...
}
```

还有 TreeMap 的构造方法。

```
public class TreeMap<K,V>
```

```
    extends AbstractMap<K,V>
    implements NavigableMap<K,V>, Cloneable, java.io.Serializable {
   ...
    public TreeMap(Comparator<? super K> comparator) {
        this.comparator = comparator;
    }
    ...
}
```

这就是 Comparator 在 JDK 源码中的应用。

20.3.2 策略模式在 Spring 源码中的应用

再来看策略模式在 Spring 源码中的应用，来看 Resource 类的代码。

```
package org.springframework.core.io;

import java.io.File;
import java.io.IOException;
import java.net.URI;
import java.net.URL;
import java.nio.channels.Channels;
import java.nio.channels.ReadableByteChannel;
import org.springframework.lang.Nullable;

public interface Resource extends InputStreamSource {
    boolean exists();

    default boolean isReadable() {
        return true;
    }

    default boolean isOpen() {
        return false;
    }

    default boolean isFile() {
        return false;
    }

    URL getURL() throws IOException;

    URI getURI() throws IOException;

    File getFile() throws IOException;

    default ReadableByteChannel readableChannel() throws IOException {
```

```
        return Channels.newChannel(this.getInputStream());
    }

    long contentLength() throws IOException;

    long lastModified() throws IOException;

    Resource createRelative(String var1) throws IOException;

    @Nullable
    String getFilename();

    String getDescription();
}
```

虽然没有直接使用 Resource 类，但是经常使用它的子类，如下图所示。

```
 * @see WritableResource
 * @see ContextResource
 * @see UrlResource
 * @see ClassPathResource
 * @see FileSystemResource
 * @see PathResource
 * @see ByteArrayResource
 * @see InputStreamResource
 */
public interface Resource extends InputStreamSource {
```

还有一个非常典型的场景，Spring 的初始化也采用了策略模式，不同类型的类采用不同的初始化策略。首先有一个 InstantiationStrategy 接口，源码如下。

```
package org.springframework.beans.factory.support;

import java.lang.reflect.Constructor;
import java.lang.reflect.Method;
import org.springframework.beans.BeansException;
import org.springframework.beans.factory.BeanFactory;
import org.springframework.lang.Nullable;

public interface InstantiationStrategy {
    Object instantiate(RootBeanDefinition var1, @Nullable String var2, BeanFactory var3) throws
BeansException;

    Object instantiate(RootBeanDefinition var1, @Nullable String var2, BeanFactory var3,
Constructor<?> var4, @Nullable Object... var5) throws BeansException;
```

```
    Object instantiate(RootBeanDefinition var1, @Nullable String var2, BeanFactory var3,
@Nullable Object var4, Method var5, @Nullable Object... var6) throws BeansException;
}
```

顶层的策略抽象非常简单，但是它下面有 SimpleInstantiationStrategy 和 CglibSubclassingInstantiationStrategy 两种策略，其类图如下。

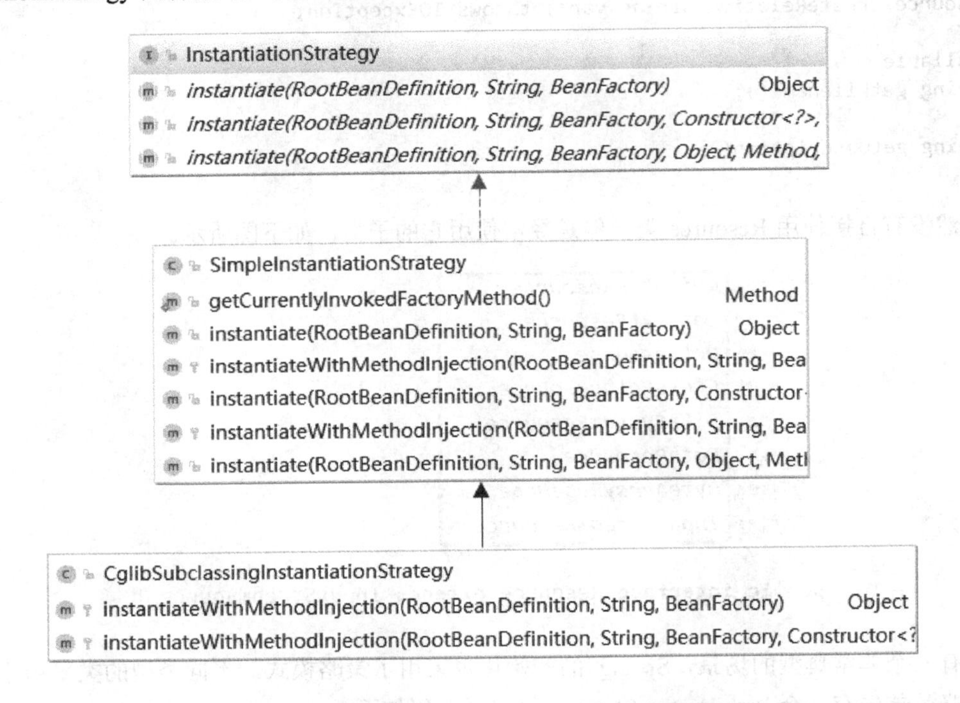

打开类图，我们发现 CglibSubclassingInstantiationStrategy 策略类还继承了 SimpleInstantiationStrategy 类，说明在实际应用中，多种策略之间可以继承使用。小伙伴们可以把它作为一个参考，在实际业务场景中根据需要来设计。

20.4 策略模式扩展

20.4.1 策略模式的优点

（1）策略模式符合开闭原则。

（2）避免使用多重条件转移语句，如 if...else 语句、switch...case 语句

（3）使用策略模式可以提高算法的保密性和安全性。

20.4.2　策略模式的缺点

（1）客户端必须知道所有的策略，并且自行决定使用哪一个策略类。

（2）代码中会产生非常多策略类，增加维护难度。

21

第 21 章 责任链模式

21.1 责任链模式概述

21.1.1 责任链模式的定义

责任链模式（Chain of Responsibility Pattern）将链中每一个节点都看作一个对象，每个节点处理的请求均不同，且内部自动维护下一个节点对象。当一个请求从链式的首端发出时，会沿着责任链预设的路径依次传递到每一个节点对象，直至被链中的某个对象处理为止，属于行为型设计模式。

> **原文**：Avoid coupling the sender of a request to its receiver by giving more than one object a chance to handle the request. Chain the receiving objects and pass the request along the chain until an object handles it.

21.1.2 责任链模式的应用场景

在日常生活中，责任链模式是比较常见的。我们平时处理工作中的一些事务，往往是各部门协同合作来完成某一个任务的。而每个部门都有各自的职责，因此，很多时候事情完成一半，便会转交到下一个部门，直到所有部门都审批通过，事情才能完成。还有我们平时说的“过五关，斩六将”其实就是闯关，也是责任链模式的一种应用场景。

责任链模式主要解耦了请求与处理，客户只需将请求发送到链上即可，不需要关心请求的具体内容和处理细节，请求会自动进行传递，直至有节点对象进行处理。责任链模式主要适用于以下应用场景。

（1）多个对象可以处理同一请求，但具体由哪个对象处理则在运行时动态决定。

（2）在不明确指定接收者的情况下，向多个对象中的一个提交请求。

（3）可动态指定一组对象处理请求。

21.1.3　责任链模式的 UML 类图

责任链模式的 UML 类图如下。

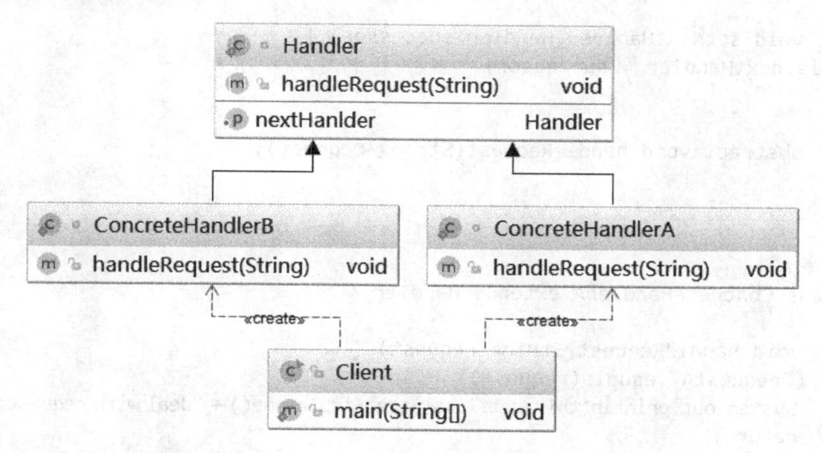

由上图可以看到，责任链模式主要包含 2 个角色。

（1）抽象处理者（Handler）：定义一个请求处理的方法，并维护一个下一处理节点 Handler 对象的引用。

（2）具体处理者（ConcreteHandler）：对请求进行处理，如果不感兴趣，则进行转发。

责任链模式的本质是解耦请求与处理，让请求在处理链中能进行传递与被处理；理解责任链模式应当理解的是其模式（道）而不是其具体实现（术），责任链模式的独到之处是其将节点处理者组合成了链式结构，并允许节点自身决定是否进行请求处理或转发，相当于让请求流动起来。

21.1.4 责任链模式的通用写法

以下是责任链模式的通用写法。

```
public class Client {
    public static void main(String[] args) {
        Handler handlerA = new ConcreteHandlerA();
        Handler handlerB = new ConcreteHandlerB();
        handlerA.setNextHanlder(handlerB);
        handlerA.handleRequest("requestB");
    }

    //抽象处理者
    static abstract class Handler {

        protected Handler nextHandler;

        public void setNextHanlder(Handler successor) {
            this.nextHandler = successor;
        }

        public abstract void handleRequest(String request);

    }

    //具体处理者 A
    static class ConcreteHandlerA extends Handler {

        public void handleRequest(String request) {
            if ("requestA".equals(request)) {
                System.out.println(this.getClass().getSimpleName() + "deal with request: " + request);
                return;
            }
            if (this.nextHandler != null) {
                this.nextHandler.handleRequest(request);
            }
        }
    }

    //具体处理者 B
    static class ConcreteHandlerB extends Handler {

        public void handleRequest(String request) {
            if ("requestB".equals(request)) {
                System.out.println(this.getClass().getSimpleName() + "deal with request: " + request);
                return;
```

```
            }
            if (this.nextHandler != null) {
                this.nextHandler.handleRequest(request);
            }
        }
    }
}
```

在上面代码中，我们把消息硬编码为 String 类型，而在真实业务中，消息是具备多样性的，可以是 int、String 或者自定义类型。因此，在上面代码的基础上，可以对消息类型进行抽象 Request，增强了消息的兼容性。

21.2　使用责任链模式解决实际问题

21.2.1　使用责任链模式设计热插拔权限控制

首先创建一个实体类 Member。

```
public class Member {
    private String loginName;
    private String loginPass;
    private String roleName;

    public Member(String loginName, String loginPass) {
        this.loginName = loginName;
        this.loginPass = loginPass;
    }

    public String getLoginName() {
        return loginName;
    }

    public String getLoginPass() {
        return loginPass;
    }

    public String getRoleName() {
        return roleName;
    }

    public void setRoleName(String roleName) {
        this.roleName = roleName;
    }
```

```
    @Override
    public String toString() {
        return "Member{" +
                "loginName='" + loginName + '\'' +
                ", loginPass='" + loginPass + '\'' +
                '}';
    }
}
```

然后来看一段我们经常写的代码。

```
public class MemberService {

    public void login(String loginName,String loginPass){
        if(StringUtils.isEmpty(loginName) ||
                StringUtils.isEmpty(loginPass)){
            System.out.println("用户名和密码校验成功，可以往下执行");
            return;
        }
        System.out.println("用户名和密码不为空，可以往下执行");

        Member member = checkExists(loginName,loginPass);
        if(null == member){
            System.out.println("用户不存在");
            return;
        }
        System.out.println("登录成功！");

        if(!"管理员".equals(member.getRoleName())){
            System.out.println("您不是管理员，没有操作权限");
            return;
        }
        System.out.println("允许操作");

    }

    private Member checkExists(String loginName,String loginPass){
        Member member = new Member(loginName,loginPass);
        member.setRoleName("管理员");
        return member;
    }

    public static void main(String[] args) {
        MemberService service = new MemberService();
        service.login("tom","666");
    }

}
```

在上面代码中，主要做了登录前的数据验证。其判断逻辑是有先后顺序的。首先做非空判断，然后检查账号是否有效，最终获得用户角色。根据用户角色所拥有的权限匹配是否有操作权限。那么这样的检验性代码一般都是必不可少的，但是写在具体的业务代码中又显得非常臃肿，因此可以用责任链模式，将这些检查步骤串联起来，而且不影响代码美观，可以使我们在编码时更加专注于某一个具体的业务逻辑处理。

下面用责任链模式来优化代码，首先创建一个 Handler 类。

```
public abstract class Handler {
    protected Handler chain;

    public void next(Handler handler){
        this.chain = handler;
    }

    public abstract void doHandle(Member member);

}
```

然后分别创建非空校验 ValidateHandler 类、登录校验 LoginHandler 类和权限校验 AuthHandler 类。ValidateHandler 类的代码如下。

```
public class ValidateHandler extends Handler {
    public void doHandle(Member member) {
        if(StringUtils.isEmpty(member.getLoginName()) ||
        StringUtils.isEmpty(member.getLoginPass())){
            System.out.println("用户名或者密码为空");
            return;
        }
        System.out.println("用户名和密码校验成功，可以往下执行");
        chain.doHandle(member);
    }
}
```

LoginHandler 类的代码如下。

```
public class LoginHandler extends Handler {

    public void doHandle(Member member) {
        System.out.println("登录成功！");
        member.setRoleName("管理员");
        chain.doHandle(member);
    }
}
```

AuthHandler 类的代码如下。

```
public class AuthHandler extends Handler {
```

```
    public void doHandle(Member member) {
        if(!"管理员".equals(member.getRoleName())){
            System.out.println("您不是管理员，没有操作权限");
            return;
        }
        System.out.println("您是管理员，允许操作");
    }
}
```

接着修改 MemberService 中的代码，其实只需要将前面定义好的几个 Handler 根据业务需求串联起来，形成一条链即可。

```
public class MemberService {
    public void login(String loginName,String loginPass){
        Handler validateHandler = new ValidateHandler();
        Handler loginHandler = new LoginHandler();
        Handler authHandler = new AuthHandler();

        validateHandler.next(loginHandler);
        loginHandler.next(authHandler);

        validateHandler.doHandle(new Member(loginName,loginPass));
    }
}
```

最后编写客户端调用代码。

```
public class Test {
    public static void main(String[] args) {
        MemberService service = new MemberService();
        service.login("tom","666");
    }
}
```

其运行结果如下图所示。

其实我们平时使用的很多权限校验框架都是运用这个原理的，将各个维度的权限处理解耦之后再串联起来，只处理各自相关的职责。如果职责与自己不相关，则抛给链上的下一个 Handler，俗称“踢皮球”。

21.2.2 责任链模式和建造者模式结合使用

因为责任链模式具备链式结构，而在上面代码中，负责组装链式结构的角色是 MemberService。当链式结构较长时，MemberService 的工作会非常烦琐，并且 MemberService 的代码相对臃肿，且后续更改处理者或消息类型时，都必须在 MemberService 中进行修改，不符合开闭原则。产生这些问题的原因就是链式结构的组装过于复杂，而对于复杂结构的创建，我们很自然地就会想到建造者模式。使用建造者模式，完全可以对 MemberService 指定的处理节点对象进行自动链式组装，客户只需指定处理节点对象，其他任何事情都不用关心，并且客户指定的处理节点对象的顺序不同，构造出来的链式结构也随之不同。我们来改造一下，首先修改 Handler 的代码。

```
public abstract class Handler<T> {
    protected Handler chain;

    public void next(Handler handler){
        this.chain = handler;
    }

    public abstract void doHandle(Member member);

    public static class Builder<T> {
        private Handler<T> head;
        private Handler<T> tail;

        public Builder<T> addHandler(Handler<T> handler) {
                if (this.head == null) {
                    this.head = this.tail = handler;
                    return this;
                }
                this.tail.next(handler);
                this.tail = handler;

            return this;
        }

        public Handler<T> build() {
            return this.head;
        }
    }

}
```

然后修改 MemberService 的代码。

```
public class MemberService {
    public void login(String loginName,String loginPass){

        Handler.Builder builder = new Handler.Builder();
        builder.addHandler(new ValidateHandler())
               .addHandler(new LoginHandler())
               .addHandler(new AuthHandler());

        builder.build().doHandle(new Member(loginName,loginPass));

    }
}
```

因为建造者模式要构建的是节点处理者，所以我们把 Builder 作为 Handler 的静态内部类，并且因为客户端不需要进行链式组装，所以还可以把链式组装方法 next()方法设置为 private，使 Handler 更加高聚合，代码如下。

```
public abstract class Handler<T> {
    protected Handler chain;

    private void next(Handler handler){
        this.chain = handler;
    }
    ...
}
```

通过这个案例，小伙伴们应该已经感受到责任链模式和建造者模式结合的精髓了。

21.3 责任链模式在框架源码中的应用

21.3.1 责任链模式在 JDK 源码中的应用

首先来看责任链模式在 JDK 中的应用，来看一个 J2EE 标准中常见的 Filter 类。

```
public interface Filter {

    public void init(FilterConfig filterConfig) throws ServletException;

    public void doFilter(ServletRequest request, ServletResponse response, FilterChain chain)
throws IOException, ServletException;
```

```
    public void destroy();
}
```

这个 Filter 接口的定义非常简单，相当于责任链模式中的 Handler 抽象角色，那么它是如何形成一条责任链的呢？来看另外一个类，其实由 doFilter()方法的最后一个参数可以看到其类型是 FilterChain 类，源码如下。

```
public interface FilterChain {
    public void doFilter(ServletRequest request, ServletResponse response) throws IOException,
ServletException;
}
```

FilterChain 类中只定义了一个 doFilter()方法，那么它们是怎么串联成一个责任链的呢？实际上，J2EE 只定义了一个规范，具体处理逻辑是由使用者自己来实现的。我们来看一个 Spring 的实现 MockFilterChain 类。

```
public class MockFilterChain implements FilterChain {

    @Nullable
    private ServletRequest request;

    @Nullable
    private ServletResponse response;

    private final List<Filter> filters;

    @Nullable
    private Iterator<Filter> iterator;

    ...

public void doFilter(ServletRequest request, ServletResponse response) throws IOException,
ServletException {
            Assert.notNull(request, "Request must not be null");
            Assert.notNull(response, "Response must not be null");
            Assert.state(this.request == null, "This FilterChain has already been called!");

            if (this.iterator == null) {
                    this.iterator = this.filters.iterator();
            }

            if (this.iterator.hasNext()) {
                    Filter nextFilter = this.iterator.next();
                    nextFilter.doFilter(request, response, this);
            }
```

```
            this.request = request;
            this.response = response;
        }
        ...
}
```

它把链条中的所有 Filter 都放到 List 中，然后在调用 doFilter()方法时循环迭代 List，也就是说 List 中的 Filter 会按顺序执行。

21.3.2 责任链模式在 Netty 源码中的应用

再来看一个例子，在 Netty 中非常经典的串行化处理 Pipeline 就采用了责任链模式。它底层采用双向链表的数据结构，将链上的各个处理器串联起来。客户端每一个请求的到来，Netty 都认为 Pipeline 中的所有处理器都有机会处理它。因此，对于入栈的请求全部从头节点开始往后传播，一直传播到尾节点才会把消息释放掉。来看一个 Netty 的责任处理器接口 ChannelHandler。

```
public interface ChannelHandler {

    //当 handler 被添加到真实的上下文中，并且在准备处理事件时被调用
    //handler 被添加进去的回调
    void handlerAdded(ChannelHandlerContext var1) throws Exception;

    //handler 被移出后的回调
    void handlerRemoved(ChannelHandlerContext var1) throws Exception;

    /** @deprecated */
    @Deprecated
    void exceptionCaught(ChannelHandlerContext var1, Throwable var2) throws Exception;

    @Inherited
    @Documented
    @Target({ElementType.TYPE})
    @Retention(RetentionPolicy.RUNTIME)
    public @interface Sharable {
    }
    ...
}
```

Netty 对责任处理接口做了更细粒度的划分，处理器被分成了两种，一种是入栈处理器 ChannelInboundHandler，另一种是出栈处理器 ChannelOutboundHandler，这两个接口都继承自 ChannelHandler。而所有处理器最终都被添加在 Pipeline 上。所以，添加删除责任处理器的接口的行为在 Netty 的 ChannelPipeline 中进行了规定。

```
public interface ChannelPipeline
        extends ChannelInboundInvoker, ChannelOutboundInvoker, Iterable<Entry<String,
        ChannelHandler>> {

    ChannelPipeline addFirst(String name, ChannelHandler handler);

    ChannelPipeline addFirst(EventExecutorGroup group, String name, ChannelHandler handler);

    ChannelPipeline addLast(String name, ChannelHandler handler);

    ChannelPipeline addLast(EventExecutorGroup group, String name, ChannelHandler handler);

    ChannelPipeline addBefore(String baseName, String name, ChannelHandler handler);
    ...
}
```

在默认实现类中，将所有 Handler 都串成了一个链表。

```
public class DefaultChannelPipeline implements ChannelPipeline {

    static final InternalLogger logger = InternalLoggerFactory.getInstance(DefaultChannel
    Pipeline.class);
    ...
    final AbstractChannelHandlerContext head;
    final AbstractChannelHandlerContext tail;
    ...
}
```

在 Pipeline 中的任意一个节点，只要我们不手动地往下传播，这个事件就会在当前节点终止传播。对于入栈数据，默认会传递到尾节点进行回收。如果不进行下一步传播，则事件就会终止在当前节点。对于出栈数据，把数据写回客户端也意味着事件的终止。

当然，在很多安全框架中也会大量使用责任链模式，比如 Spring Security、Apache Shiro 都会用到责任链模式，感兴趣的小伙伴可以尝试研究一下。

在大部分框架中，无论怎么实现，所有实现都是大同小异的。其实如果我们站在设计者的角度看源码，对学习源码和提升编码内功是非常有益的。因为这样，我们可以站在更高的角度来学习优秀的思想，而不是钻到某一个代码细节里。我们需要对所有设计都有一个宏观概念，这样学习起来才更加轻松。

21.4 责任链模式扩展

21.4.1 责任链模式的优点

（1）将请求与处理解耦。

（2）请求处理者（节点对象）只需关注自己感兴趣的请求进行处理即可，对于不感兴趣的请求，直接转发给下一个节点对象。

（3）具备链式传递处理请求功能，请求发送者不需要知晓链路结构，只需等待请求处理结果即可。

（4）链路结构灵活，可以通过改变链路结构动态地新增或删减责任。

（5）易于扩展新的请求处理类（节点），符合开闭原则。

21.4.2 责任链模式的缺点

（1）责任链太长或者处理时间过长，会影响整体性能。

（2）如果节点对象存在循环引用，则会造成死循环，导致系统崩溃。

22

第 22 章 迭代器模式

22.1 迭代器模式概述

22.1.1 迭代器模式的定义

迭代器模式（Iterator Pattern）又叫作游标模式（Cursor Pattern），它提供一种按顺序访问集合/容器对象元素的方法，而又无须暴露集合内部表示。迭代器模式可以为不同的容器提供一致的遍历行为，而不用关心容器内元素的组成结构，属于行为型设计模式。

原文：Provide a way to access the elements of an aggregate object sequentially without exposing its underlying representation.

迭代器模式的本质是把集合对象的迭代行为抽离到迭代器中，提供一致的访问接口。

22.1.2 迭代器模式的应用场景

迭代器模式在生活中应用得比较广泛。比如，物流系统中的传送带，不管传送的是什么物品，都被打包成一个个箱子，并且有一个统一的二维码。这样我们不需要关心箱子里是什么，在分发时只需要一个个检查发送的目的地即可。再比如，我们平时乘坐交通工具，都是统一刷卡或者刷脸进站，而不需要关心是男性还是女性、是残疾人还是正常人等信息。

我们把多个对象聚在一起形成的总体称为集合（Aggregate），集合对象是能够包容一组对象的容器对象。不同的集合其内部元素的聚合结构可能不同，而迭代器模式屏蔽了内部元素的获取细节，为外部提供一致的元素访问行为，解耦了元素迭代与集合对象间的耦合，并且通过提供不同的迭代器，可以为同一个集合对象提供不同顺序的元素访问行为，扩展了集合对象元素迭代功能，符合开闭原则。迭代器模式适用于以下应用场景。

（1）访问一个集合对象的内容而无须暴露它的内部表示。

（2）为遍历不同的集合结构提供一个统一的访问接口。

22.1.3 迭代器模式的 UML 类图

迭代器模式的 UML 类图如下。

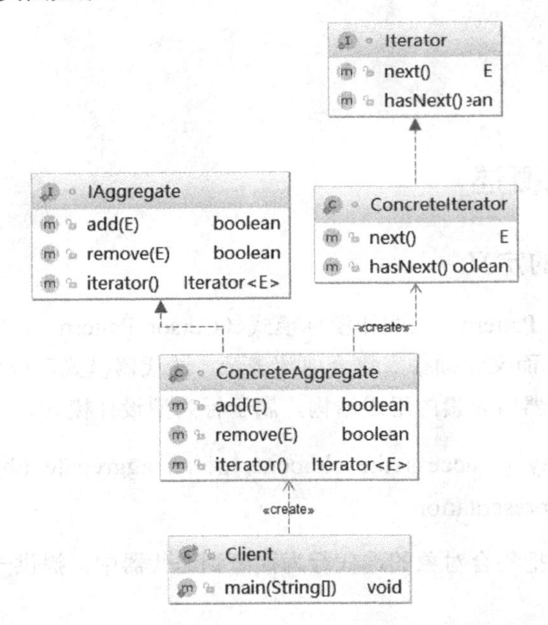

由上图可以看到，迭代器模式主要包含 4 个角色。

（1）抽象迭代器（Iterator）：抽象迭代器负责定义访问和遍历元素的接口。

（2）具体迭代器（ConcreteIterator）：提供具体的元素遍历行为。

（3）抽象容器（IAggregate）：负责定义提供具体迭代器的接口。

（4）具体容器（ConcreteAggregate）：创建具体迭代器。

22.1.4　迭代器模式的通用写法

以下是迭代器模式的通用写法。

```
public class Client {
    public static void main(String[] args) {
        //创建一个容器对象
        IAggregate<String> aggregate = new ConcreteAggregate<String>();
        //添加元素
        aggregate.add("one");
        aggregate.add("two");
        aggregate.add("three");
        //获取容器对象迭代器
        Iterator<String> iterator = aggregate.iterator();
        //遍历
        while (iterator.hasNext()) {
            String element = iterator.next();
            System.out.println(element);
        }

    }

    //抽象迭代器
    interface Iterator<E> {

        E next();

        boolean hasNext();
    }

    //抽象容器
    interface IAggregate<E> {
        boolean add(E element);

        boolean remove(E element);

        Iterator<E> iterator();
    }

    //具体迭代器
    static class ConcreteIterator<E> implements Iterator<E> {
        private List<E> list;
```

```
        private int cursor = 0;

        public ConcreteIterator(List<E> list) {
            this.list = list;
        }


        public E next() {
            return this.list.get(this.cursor ++);
        }


        public boolean hasNext() {
            return this.cursor < this.list.size();
        }
    }

    //具体容器
    static class ConcreteAggregate<E> implements IAggregate<E> {
        private List<E> list = new ArrayList<E>();

        public boolean add(E element) {
            return this.list.add(element);
        }

        public boolean remove(E element) {
            return this.list.remove(element);
        }

        public Iterator<E> iterator() {
            return new ConcreteIterator<E>(this.list);
        }
    }
}
```

通过上面代码，我们可以更清楚地了解到迭代器模式的作用是将集合对象的迭代行为抽取到迭代器中，解耦了元素迭代与集合对象之间的耦合，使得不同的集合对象对外都能提供一致的迭代行为。

22.2 手写自定义的集合迭代器

总体来说，迭代器模式是非常简单的。还是以课程为例，我们创建一个课程集合，集合中的每一个元素都是课程对象，然后手写一个迭代器，将每一个课程对象的信息都读出来。首先创建集合元素课程 Course 类。

```
public class Course {
    private String name;

    public Course(String name) {
        this.name = name;
    }

    public String getName() {
        return name;
    }
}
```

然后创建自定义迭代器 Iterator 接口。

```
public interface Iterator<E> {

    E next();

    boolean hasNext();

}
```

创建自定义的课程集合 CourseAggregate 接口。

```
public interface CourseAggregate {

    void add(Course course);

    void remove(Course course);

    Iterator<Course> iterator();
}
```

接着分别实现迭代器接口和集合接口，创建 IteratorImpl 实现类。

```
public class IteratorImpl<E> implements Iterator<E> {

    private List<E> list;
    private int cursor;
    private E element;
    public IteratorImpl(List list){
        this.list = list;
    }
```

```
    public E next() {
        System.out.print("当前位置" + cursor + ": ");
        element = list.get(cursor);
        cursor ++;
        return element;
    }

    public boolean hasNext(){
        if(cursor > list.size() - 1){
            return false;
        }
        return true;
    }
}
```

创建课程集合 CourseAggregateImpl 实现类。

```
public class CourseAggregateImpl implements CourseAggregate {

    private List courseList;

    public CourseAggregateImpl() {
        this.courseList = new ArrayList();
    }

    public void add(Course course) {
        courseList.add(course);
    }

    public void remove(Course course) {
        courseList.remove(course);
    }

    public Iterator<Course> iterator() {
        return new IteratorImpl(courseList);
    }
}
```

最后编写客户端测试代码。

```
public static void main(String[] args) {
        Course java = new Course("Java 架构");
        Course javaBase = new Course("Java 入门");
        Course design = new Course("Java 设计模式精讲");
```

```
        Course ai = new Course("人工智能");


        CourseAggregate courseAggregate = new CourseAggregateImpl();

        courseAggregate.add(java);
        courseAggregate.add(javaBase);
        courseAggregate.add(design);
        courseAggregate.add(ai);

        System.out.println("-----课程列表-----");
        printCourses(courseAggregate);

        courseAggregate.remove(ai);

        System.out.println("-----删除操作之后的课程列表-----");
        printCourses(courseAggregate);
    }


    public static void printCourses(CourseAggregate courseAggregate){
        Iterator<Course> iterator = courseAggregate.iterator();
        while(!iterator.hasNext()){
            Course course = iterator.next();
            System.out.println("《" + course.getName() + "》");
        }
    }
```

运行结果如下图所示。

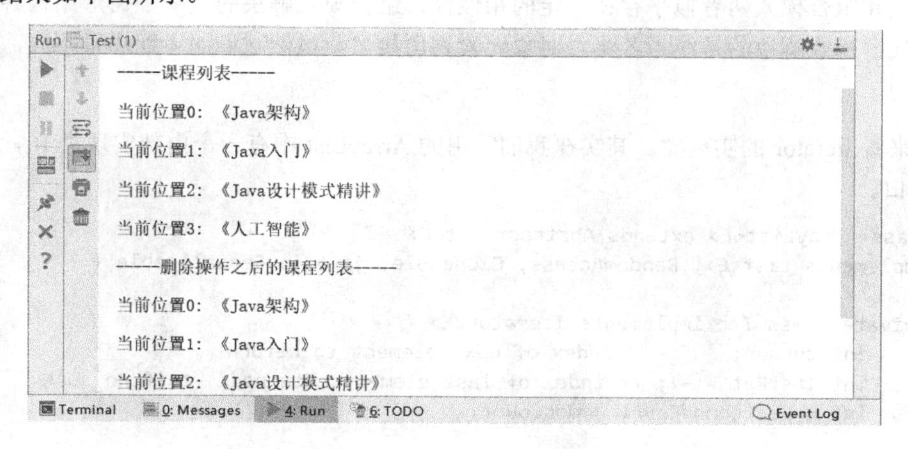

看到这里，小伙伴们肯定有一种似曾相识的感觉，让人不禁想起每天都在用的 JDK 自带的集合迭代器。下面就来看源码中是如何运用迭代器模式的。

22.3 迭代器模式在框架源码中的应用

22.3.1 迭代器模式在 JDK 源码中的应用

先来看 JDK 中大家非常熟悉的 Iterator 源码。

```
public interface Iterator<E> {
    boolean hasNext();

    E next();

    default void remove() {
        throw new UnsupportedOperationException("remove");
    }
    default void forEachRemaining(Consumer<? super E> action) {
        Objects.requireNonNull(action);
        while (hasNext())
            action.accept(next());
    }
}
```

从上面代码中，我们看到定义了两个主要方法 hasNext()方法和 next()方法，和我们自己写的完全一致。

另外，从上面代码中，我们看到 remove()方法实现似曾相识。其实是在组合模式中见过的。迭代器模式和组合模式两者似乎存在一定的相似性，组合模式解决的是统一树形结构各层次访问接口，迭代器模式解决的是统一各集合对象元素遍历接口。虽然它们的适配场景不同，但核心理念是相通的。

接着来看 Iterator 的实现类，其实在我们常用的 ArrayList 中有一个内部实现类 Itr，它实现了 Iterator 接口。

```
public class ArrayList<E> extends AbstractList<E>
        implements List<E>, RandomAccess, Cloneable, java.io.Serializable {
        ...
        private class Itr implements Iterator<E> {
            int cursor;       // index of next element to return
            int lastRet = -1; // index of last element returned; -1 if no such
            int expectedModCount = modCount;
```

```
        public boolean hasNext() {
            return cursor != size;
        }

        @SuppressWarnings("unchecked")
        public E next() {
            checkForComodification();
            int i = cursor;
            if (i >= size)
                throw new NoSuchElementException();
            Object[] elementData = ArrayList.this.elementData;
            if (i >= elementData.length)
                throw new ConcurrentModificationException();
            cursor = i + 1;
            return (E) elementData[lastRet = i];
        }
        ...
    }
    ...
}
```

其中，hasNext()方法和 next()方法的实现也非常简单，继续往下看，在 ArrayList 内部还有几个迭代器对 Itr 进行了进一步扩展，首先看 ListItr。

```
private class ListItr extends Itr implements ListIterator<E> {
        ListItr(int index) {
        super();
        cursor = index;
    }

    public boolean hasPrevious() {
        return cursor != 0;
    }

    public int nextIndex() {
        return cursor;
    }

    public int previousIndex() {
        return cursor - 1;
    }
  ...
}
```

它增加了 hasPrevious()方法，主要用于判断是否还有上一个元素。另外，还有 SubList 对子集合的迭代处理。

22.3.2 迭代器模式在 MyBatis 源码中的应用

当然，迭代器模式在 MyBatis 中也是必不可少的，来看一个 DefaultCursor 类。

```
public class DefaultCursor<T> implements Cursor<T> {

    ...

    private final CursorIterator cursorIterator = new CursorIterator();
    ...
}
```

它实现了 Cursor 接口，而且定义了一个成员变量 cursorIterator，其定义的类型为 CursorIterator。继续查看 CursorIterator 类的源码实现，它是 DefaultCursor 的一个内部类，并且实现了 JDK 中的 Iterator 接口。

22.4 迭代器模式扩展

22.4.1 迭代器模式的优点

（1）多态迭代：为不同的聚合结构提供一致的遍历接口，即一个迭代接口可以访问不同的集合对象。

（2）简化集合对象接口：迭代器模式将集合对象本身应该提供的元素迭代接口抽取到迭代器中，使集合对象无须关心具体迭代行为。

（3）元素迭代功能多样化：每个集合对象都可以提供一个或多个不同的迭代器，使得同种元素的聚合结构可以有不同的迭代行为。

（4）解耦迭代与集合：迭代器模式封装了具体的迭代算法，迭代算法的变化不会影响到集合对象的架构。

22.4.2 迭代器模式的缺点

对于比较简单的遍历（如数组或者有序列表），使用迭代器模式遍历较为烦琐。

在日常开发中，我们几乎不会自己写迭代器。除非需要定制一个自己实现的数据结构对应的迭代器，否则，开源框架提供的 API 完全够用。

23

第 23 章 命令模式

23.1 命令模式概述

23.1.1 命令模式的定义

命令模式（Command Pattern）是对命令的封装，每一个命令都是一个操作：请求方发出请求要求执行一个操作；接收方收到请求，并执行操作。命令模式解耦了请求方和接收方，请求方只需请求执行命令，不用关心命令怎样被接收、怎样被操作及是否被执行等。命令模式属于行为型设计模式。

> **原文**: Encapsulate a request as an object, thereby letting you parameterize clients with different requests, queue or log requests, and support undoable operations.

在软件系统中，行为请求者与行为实现者通常是一种紧耦合关系，因为这样的实现简单明了。但紧耦合关系缺乏扩展性，在某些场合中，当需要对行为进行记录、撤销或重做等处理时，只能修改源码。而命令模式通过在请求与实现间引入一个抽象命令接口，解耦了请求与实现，并且中间件是抽象的，它由不同的子类实现，因此具备扩展性。所以，命令模式的本质是解耦命令请求与处理。

23.1.2 命令模式的应用场景

在日常生活中，命令模式是很常见的。比如，经历过黑白电视机年代的小伙伴应该都有过这样的经历。那个年代在看电视的时候，想要换个频道简直不容易。我们得走到电视机前扭动换台的旋钮，一顿“咔咔咔”折腾才能完成频道的切换。如今，遥控器的发明简直就是“解放战争”，我们躺在沙发上只需要轻轻一按遥控器即可完成频道的切换。这就是命令模式，将换台请求和换台处理完全解耦了。

再比如，我们去餐厅吃饭。餐厅菜单不是等到客人来了才定制的，而是已经预先配置好的。这样，客人来了就只需要点菜，而不是任由客人临时定制。在小品《不差钱》中就生动地描绘了这一场景，客人来到一家高档酒店点餐，其中有一段台词，客人问服务员说：“孩子，来一个小野鸡炖蘑菇。”服务员说：“没有。”客人示意说：“这个可以有！”服务员说：“这个真没有！”因此，餐厅提供的菜单就相当于把请求和处理进行了解耦，这就是命令模式的体现。

当系统的某项操作具备命令语义，且命令实现不稳定（变化）时，可以通过命令模式解耦请求与实现，使用抽象命令接口使请求方的代码架构稳定，封装接收方具体命令的实现细节。接收方与抽象命令接口呈现弱耦合（内部方法无须一致），具备良好的扩展性。命令模式主要适用于以下应用场景。

（1）现实语义中具备“命令”的操作（如命令菜单、Shell 命令等）。

（2）请求的调用者和接收者需要解耦，使得调用者和接收者不直接交互。

（3）需要抽象出等待执行的行为，比如撤销（Undo）操作和恢复（Redo）等操作。

（4）需要支持命令宏（即命令组合操作）。

23.1.3 命令模式的 UML 类图

命令模式的 UML 类图如下。

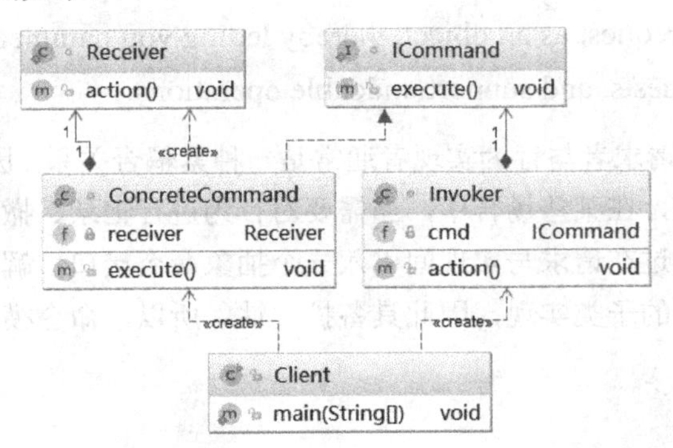

由上图可以看到，命令模式主要包含 4 个角色。

（1）接收者角色（Receiver）：该类负责具体实施或执行一个请求。

（2）命令角色（ICommand）：定义需要执行的所有命令行为。

（3）具体命令角色（ConcreteCommand）：该类内部维护一个 Receiver，在其 execute()方法中调用 Receiver 的相关方法。

（4）请求者角色（Invoker）：接收客户端的命令，并执行命令。

从命令模式的 UML 类图中，其实可以很清晰地看出：ICommand 的出现就是作为 Receiver 和 Invoker 的中间件，解耦了彼此。而之所以引入 ICommand 中间件，是以下两方面原因。

（1）解耦请求与实现：即解耦了 Invoker 和 Receiver，因为在 UML 类图中，Invoker 是一个具体的实现，等待接收客户端传入的命令（即 Invoker 与客户端耦合），Invoker 处于业务逻辑区域，应当是一个稳定的结构。而 Receiver 属于业务功能模块，是经常变动的；如果没有 ICommand，则 Invoker 紧耦合 Receiver，一个稳定的结构依赖了一个不稳定的结构，就会导致整个结构都不稳定。这就是引入 ICommand 的原因：不仅仅是解耦请求与实现，同时稳定（Invoker）依赖稳定（Command），结构还是稳定的。

（2）增强扩展性：扩展性又体现在以下两个方面。

① Receiver 属于底层细节，可以通过更换不同的 Receiver 达到不同的细节实现。

② ICommand 接口本身就是抽象的，具备扩展性；而且由于命令对象本身具备抽象性，如果结合装饰器模式，则功能扩展简直如鱼得水。

注：在一个系统中，不同的命令对应不同的请求，也就是说无法把请求抽象化，因此命令模式中的 Receiver 是具体实现；但是如果在某一个模块中，可以对 Receiver 进行抽象，其实这就变相使用到了桥接模式（ICommand 类具备两个变化的维度：ICommand 和 Receiver），这样扩展性会更加优秀。

23.1.4　命令模式的通用写法

以下是命令模式的通用写法。

```
public class Client {
    public static void main(String[] args) {
        ICommand cmd = new ConcreteCommand();
        Invoker invoker = new Invoker(cmd);
```

```
        invoker.action();
    }

    //接收者
    static class Receiver {
        public void action() {
            System.out.println("执行具体操作");
        }
    }

    //抽象命令接口
    interface ICommand {
        void execute();
    }

    //具体命令
    static class ConcreteCommand implements ICommand {
        //直接创建接收者，不暴露给客户端
        private Receiver receiver = new Receiver();

        public void execute() {
            this.receiver.action();
        }
    }

    //请求者
    static class Invoker {
        private ICommand cmd;

        public Invoker(ICommand cmd) {
            this.cmd = cmd;
        }

        public void action() {
            this.cmd.execute();
        }
    }
}
```

23.2 使用命令模式重构播放器控制条

假如我们开发一个播放器，播放器有播放功能、拖动进度条功能、停止播放功能、暂停功能，我们在操作播放器的时候并不是直接调用播放器的方法，而是通过一个控制条去传达指令给播放器内核，具体传达什么指令，会被封装为一个个按钮。那么每个按钮就相当于对一条命令的封装。用控制条实现了用户发送指令与播放器内核接收指令的解耦。下面来看代码，首先创建播放器内核 GPlayer 类。

```
public class GPlayer {
    public void play(){
        System.out.println("正常播放");
    }

    public void speed(){
        System.out.println("拖动进度条");
    }

    public void stop(){
        System.out.println("停止播放");
    }

    public void pause(){
        System.out.println("暂停播放");
    }
}
```

创建命令接口 IAction 类。

```
public interface IAction {
    void execute();
}
```

然后分别创建操作播放器可以接收的指令，播放指令 PlayAction 类的代码如下。

```
public class PlayAction implements IAction {
    private GPlayer gplayer;

    public PlayAction(GPlayer gplayer) {
        this.gplayer = gplayer;
    }

    public void execute() {
        gplayer.play();
    }
}
```

暂停指令 PauseAction 类的代码如下。

```
public class PauseAction implements IAction {
    private GPlayer gplayer;

    public PauseAction(GPlayer gplayer) {
```

```
        this.gplayer = gplayer;
    }

    public void execute() {
        gplayer.pause();
    }
}
```

拖动进度条指令 SpeedAction 类的代码如下。

```
public class SpeedAction implements IAction {
    private GPlayer gplayer;

    public SpeedAction(GPlayer gplayer) {
        this.gplayer = gplayer;
    }

    public void execute() {
        gplayer.speed();
    }
}
```

停止播放指令 StopAction 类的代码如下。

```
public class StopAction implements IAction {
    private GPlayer gplayer;

    public StopAction(GPlayer gplayer) {
        this.gplayer = gplayer;
    }

    public void execute() {
        gplayer.stop();
    }
}
```

最后创建控制条 Controller 类。

```
public class Controller {
    private List<IAction> actions = new ArrayList<IAction>();
    public void addAction(IAction action){
        actions.add(action);
    }

    public void execute(IAction action){
        action.execute();
    }

    public void executes(){
```

```
        for(IAction action : actions){
            action.execute();
        }
        actions.clear();
    }
}
```

从上面代码来看，控制条可以执行单条命令，也可以批量执行多条命令。下面来看客户端测试代码。

```
    public static void main(String[] args) {

        GPlayer player = new GPlayer();
        Controller controller = new Controller();
        controller.execute(new PlayAction(player));

        controller.addAction(new PauseAction(player));
        controller.addAction(new PlayAction(player));
        controller.addAction(new StopAction(player));
        controller.addAction(new SpeedAction(player));
        controller.executes();
    }
```

由于控制条已经与播放器内核解耦了，以后如果想扩展新命令，只需增加命令即可，控制条的结构无须改动。

23.3　命令模式在框架源码中的应用

23.3.1　命令模式在 JDK 源码中的应用

首先来看 JDK 中的 Runnable 接口，Runnable 相当于命令的抽象，只要是实现了 Runnable 接口的类都被认为是一个线程。

```
public interface Runnable {
    public abstract void run();
}
```

实际上调用线程的 start()方法之后，就有资格去抢 CPU 资源，而不需要编写获得 CPU 资源的逻辑。而线程抢到 CPU 资源后，就会执行 run()方法中的内容，用 Runnable 接口把用户请求和 CPU 执行进行解耦。

23.3.2 命令模式在 JUnit 源码中的应用

再来看一个大家非常熟悉的 junit.framework.Test 接口。

```
package junit.framework;

public interface Test {
    public abstract int countTestCases();

    public abstract void run(TestResult result);
}
```

Test 接口中有两个方法，第一个是 countTestCases()方法，用来统计当前需要执行的测试用例总数。第二个是 run()方法，用来执行具体的测试逻辑，其参数 TestResult 是用来返回测试结果的。实际上，我们在平时编写测试用例的时候，只需要实现 Test 接口就被认为是一个测试用例，那么在执行的时候就会被自动识别。通常做法都是继承 TestCase 类，不妨来看一下 TestCase 的源码。

```
public abstract class TestCase extends Assert implements Test {
        ...
        public void run(TestResult result) {
        result.run(this);
    }
    ...
}
```

实际上，TestCase 类也实现了 Test 接口。我们继承 TestCase 类，相当于也实现了 Test 接口，自然就会被扫描成为一个测试用例。

23.4 命令模式扩展

23.4.1 命令模式的优点

（1）通过引入中间件（抽象接口），解耦了命令请求与实现。

（2）扩展性良好，可以很容易地增加新命令。

（3）支持组合命令，支持命令队列。

（4）可以在现有命令的基础上，增加额外功能。比如日志记录，结合装饰器模式会更加灵活。

23.4.2　命令模式的缺点

（1）具体命令类可能过多。

（2）命令模式的结果其实就是接收方的执行结果，但是为了以命令的形式进行架构、解耦请求与实现，引入了额外类型结构（引入了请求方与抽象命令接口），增加了理解上的困难。不过这也是设计模式的通病，抽象必然会额外增加类的数量；代码抽离肯定比代码聚合更难理解。

24

第 24 章 状态模式

24.1 状态模式概述

24.1.1 状态模式的定义

状态模式（State Pattern）也叫作状态机模式（State Machine Pattern），允许对象在内部状态发生改变时改变它的行为，对象看起来好像修改了它的类，属于行为型设计模式。

原文：Allow an object to alter its behavior when its internal state changes. The object will appear to change its class.

状态模式中类的行为是由状态决定的，在不同的状态下有不同的行为。其意图是让一个对象在其内部改变的时候，行为也随之改变。状态模式的核心是状态与行为绑定，不同的状态对应不同的行为。

24.1.2 状态模式的应用场景

状态模式在生活场景中比较常见。例如，我们平时网购的订单状态变化。另外，我们平时坐电梯时电梯的状态变化。

在软件开发过程中，对于某一项操作，可能存在不同的情况。通常处理多情况的问题最直接的方式就是使用 if...else 或 switch...case 条件语句进行枚举。但是这种做法对于复杂状态的判断存在天然弊端：条件判断语句过于臃肿，可读性差，且不具备扩展性，维护难度也大。而如果转换思维，将这些不同状态独立起来，用各个不同的类进行表示，系统处于哪种情况，直接使用相应的状态类对象进行处理，消除了 if...else、switch...case 等冗余语句，代码更有层次性，并且具备良好的扩展力。

状态模式主要解决的就是控制一个对象状态的条件表达式过于复杂时的情况。通过把状态的判断逻辑转移到表示不同状态的一系列类中，可以把复杂的判断逻辑简化。对象的行为依赖它的状态（属性），并且行为会随着它的状态改变而改变。状态模式主要适用于以下应用场景。

（1）行为随状态改变而改变的场景。

（2）一个操作中含有庞大的多分支结构，并且这些分支取决于对象的状态。

24.1.3　状态模式的 UML 类图

状态模式的 UML 类图如下。

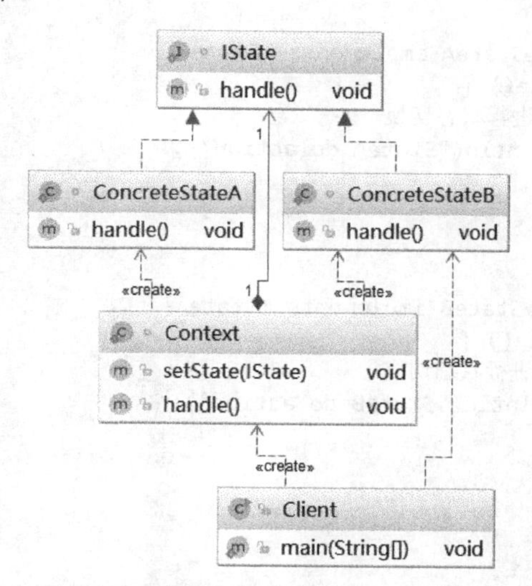

由上图可以看到，状态模式主要包含 3 个角色。

（1）环境类角色（Context）：定义客户端需要的接口，内部维护一个当前状态实例，并负责具体状态的切换。

（2）抽象状态角色（IState）：定义该状态下的行为，可以有一个或多个行为。

（3）具体状态角色（ConcreteState）：具体实现该状态对应的行为，并且在需要的情况下进行状态切换。

24.1.4 状态模式的通用写法

以下是状态模式的通用写法。

```
public class Client {
    public static void main(String[] args) {
        Context context = new Context();
        context.setState(new ConcreteStateB());
        context.handle();
    }

    //抽象状态：State
    interface IState {
        void handle();
    }

    //具体状态类
    static class ConcreteStateA implements IState {
        public void handle() {
            //必要时刻需要进行状态切换
            System.out.println("StateA do action");
        }
    }

    //具体状态类
    static class ConcreteStateB implements IState {
        public void handle() {
            //必要时刻需要进行状态切换
            System.out.println("StateB do action");
        }
    }

    //环境类
    static class Context {
        private static final IState STATE_A = new ConcreteStateA();
        private static final IState STATE_B = new ConcreteStateB();
        //默认状态A
        private IState currentState = STATE_A;

        public void setState(IState state) {
            this.currentState = state;
```

```
        }

        public void handle() {
            this.currentState.handle();
        }
    }
}
```

上面代码很好地展现了状态模式下状态分离的好处，代码清晰。不过上面代码还未能完全展示状态模式的全貌，因为不同的状态之间可能存在自动切换的场景（比如手机处于开机状态后就会立即切换到屏幕点亮状态），但是上面代码未体现出场景切换的功能。我们可以对上面代码进行修改，使其满足状态切换的功能，具体代码如下。

```
public class Client {
    public static void main(String[] args) {
        Context context = new Context();
        context.setState(new ConcreteStateA());
        context.handle();
    }

    //抽象状态类 State
    static abstract class State {
        protected Context context;

        public void setContext(Context context) {
            this.context = context;
        }

        public abstract void handle();
    }

    //具体状态类 A
    static class ConcreteStateA extends State {
        @Override
        public void handle() {
            System.out.println("StateA do action");
            //A 状态完成后自动切换到 B 状态
            this.context.setState(Context.STATE_B);
            this.context.getState().handle();
        }
    }

    //具体状态类 B
    static class ConcreteStateB extends State {
        @Override
        public void handle() {
            System.out.println("StateB do action");
```

```
        }
    }

    //环境类
    static class Context {
        public static final State STATE_A = new ConcreteStateA();
        public static final State STATE_B = new ConcreteStateB();
        //默认状态 A
        private State currentState = STATE_A;
        {
            STATE_A.setContext(this);
            STATE_B.setContext(this);
        }

        public void setState(State state) {
            this.currentState = state;
            this.currentState.setContext(this);
        }

        public State getState() {
            return this.currentState;
        }

        public void handle() {
            this.currentState.handle();
        }
    }
}
```

在上面代码中，客户端访问的是 A 状态，但是程序运行后会自动切换到 B 状态。主要是将抽象状态类 State 由接口改成抽象类，增加对环境类 Context 的维护，让具体状态 ConcreteState 可以不必耦合其他具体状态类，而是借由 Context 间接进行功能切换。

24.2 使用状态模式解决实际问题

24.2.1 使用状态模式实现登录状态自由切换

当我们在 GPer 社区阅读文章时，如果觉得文章写得很好，我们就会评论、收藏两连发。如果处于登录情况下，则可以直接做评论、收藏这些行为。否则，跳转到登录界面，登录后再继续执行先前的动作。这里涉及的状态有两种：登录与未登录；行为有两种：评论和收藏。下面使用状态模式来实现这个逻辑，代码如下。

首先创建抽象状态角色 UserState 类。

```
public abstract class UserState {
    protected AppContext context;

    public void setContext(AppContext context) {
        this.context = context;
    }

    public abstract void favorite();

    public abstract void comment(String comment);
}
```

然后创建登录状态 LogInState 类。

```
public class LoginInState extends UserState {
    @Override
    public void favorite() {
        System.out.println("收藏成功！");
    }

    @Override
    public void comment(String comment) {
        System.out.println(comment);
    }
}
```

创建未登录状态 UnloginState 类。

```
public class UnLoginState extends UserState {
    @Override
    public void favorite() {
        this.switch2Login();
        this.context.getState().favorite();
    }

    @Override
    public void comment(String comment) {
        this.switch2Login();
        this.context.getState().comment(comment);
    }

    private void switch2Login() {
        System.out.println("跳转到登录页面!");
        this.context.setState(this.context.STATE_LOGIN);
    }
}
```

创建上下文角色 AppContext 类。

```
public class AppContext {
    public static final UserState STATE_LOGIN = new LoginInState();
    public static final UserState STATE_UNLOGIN = new UnLoginState();
    private UserState currentState = STATE_UNLOGIN;
    {
        STATE_LOGIN.setContext(this);
        STATE_UNLOGIN.setContext(this);
    }

    public void setState(UserState state) {
        this.currentState = state;
        this.currentState.setContext(this);
    }

    public UserState getState() {
        return this.currentState;
    }

    public void favorite() {
        this.currentState.favorite();
    }

    public void comment(String comment) {
        this.currentState.comment(comment);
    }
}
```

最后编写客户端测试代码。

```
public static void main(String[] args) {
      AppContext context = new AppContext();
      context.favorite();
      context.comment("评论：好文章，360个赞!");
   }
```

运行结果如下图所示。

24.2.2　使用状态机实现订单状态流转控制

状态机是状态模式的一种应用，相当于上下文角色的一个升级版。在工作流或游戏等各种系统中有大量使用，如各种工作流引擎，它几乎是状态机的子集和实现，封装状态的变化规则。Spring也提供了一个很好的解决方案。Spring 中的组件名称就叫作状态机（StateMachine）。状态机帮助开发者简化状态控制的开发过程，让状态机结构更加层次化。下面用 Spring 状态机模拟订单状态流转的过程。

（1）添加依赖。

```
<dependency>
    <groupId>org.springframework.statemachine</groupId>
    <artifactId>spring-statemachine-core</artifactId>
    <version>2.0.1.RELEASE</version>
</dependency>
```

（2）创建订单实体 Order 类。

```
public class Order {
    private int id;
    private OrderStatus status;
    public void setStatus(OrderStatus status) {
        this.status = status;
    }

    public OrderStatus getStatus() {
        return status;
    }

    public void setId(int id) {
        this.id = id;
    }

    public int getId() {
        return id;
    }

    @Override
    public String toString() {
        return "订单号：" + id + ", 订单状态：" + status;
    }
}
```

（3）创建订单状态枚举类和状态转换枚举类。

```
/**
 * 订单状态
 */
public enum OrderStatus {
    //待支付，待发货，待收货，订单结束
    WAIT_PAYMENT, WAIT_DELIVER, WAIT_RECEIVE, FINISH;
}

/**
 * 订单状态改变事件
 */
public enum OrderStatusChangeEvent {
    //支付，发货，确认收货
    PAYED, DELIVERY, RECEIVED;
}
```

（4）添加状态流转配置。

```
/**
 * 订单状态机配置
 */
@Configuration
@EnableStateMachine(name = "orderStateMachine")
public class OrderStateMachineConfig extends StateMachineConfigurerAdapter<OrderStatus,
OrderStatusChangeEvent> {

    /**
     * 配置状态
     * @param states
     * @throws Exception
     */
    public void configure(StateMachineStateConfigurer<OrderStatus, OrderStatusChangeEvent>
    states) throws Exception {
        states
                .withStates()
                .initial(OrderStatus.WAIT_PAYMENT)
                .states(EnumSet.allOf(OrderStatus.class));
    }

    /**
     * 配置状态转换事件关系
     * @param transitions
     * @throws Exception
     */
    public void configure(StateMachineTransitionConfigurer<OrderStatus, OrderStatusChangeEvent>
```

```
    transitions) throws Exception {
        transitions
                .withExternal().source(OrderStatus.WAIT_PAYMENT).target(OrderStatus.WAIT_DELIVER)
                .event(OrderStatusChangeEvent.PAYED)
                .and()
                .withExternal().source(OrderStatus.WAIT_DELIVER).target(OrderStatus.WAIT_RECEIVE)
                .event(OrderStatusChangeEvent.DELIVERY)
                .and()
                .withExternal().source(OrderStatus.WAIT_RECEIVE).target(OrderStatus.FINISH)
                .event(OrderStatusChangeEvent.RECEIVED);
    }

    /**
     * 持久化配置
     * 在实际使用中，可以配合 Redis 等进行持久化操作
     * @return
     */
    @Bean
    public DefaultStateMachinePersister persister(){
        return new DefaultStateMachinePersister<>(new StateMachinePersist<Object, Object,
         Order>() {
            @Override
            public void write(StateMachineContext<Object, Object> context, Order order) throws
             Exception {
                //此处并没有进行持久化操作
            }

            @Override
            public StateMachineContext<Object, Object> read(Order order) throws Exception {
                //此处直接获取 Order 中的状态，其实并没有进行持久化读取操作
                return new DefaultStateMachineContext(order.getStatus(), null, null, null);
            }
        });
    }
}
```

（5）添加订单状态监听器。

```
@Component("orderStateListener")
@WithStateMachine(name = "orderStateMachine")
public class OrderStateListenerImpl{

    @OnTransition(source = "WAIT_PAYMENT", target = "WAIT_DELIVER")
    public boolean payTransition(Message<OrderStatusChangeEvent> message) {
        Order order = (Order) message.getHeaders().get("order");
        order.setStatus(OrderStatus.WAIT_DELIVER);
        System.out.println("支付，状态机反馈信息：" + message.getHeaders().toString());
        return true;
```

```
    }

    @OnTransition(source = "WAIT_DELIVER", target = "WAIT_RECEIVE")
    public boolean deliverTransition(Message<OrderStatusChangeEvent> message) {
        Order order = (Order) message.getHeaders().get("order");
        order.setStatus(OrderStatus.WAIT_RECEIVE);
        System.out.println("发货，状态机反馈信息：" + message.getHeaders().toString());
        return true;
    }

    @OnTransition(source = "WAIT_RECEIVE", target = "FINISH")
    public boolean receiveTransition(Message<OrderStatusChangeEvent> message){
        Order order = (Order) message.getHeaders().get("order");
        order.setStatus(OrderStatus.FINISH);
        System.out.println("收货，状态机反馈信息：" + message.getHeaders().toString());
        return true;
    }
}
```

（6）创建 IOrderService 接口。

```
public interface IOrderService {
    //创建新订单
    Order create();
    //发起支付
    Order pay(int id);
    //订单发货
    Order deliver(int id);
    //订单收货
    Order receive(int id);
    //获取所有订单信息
    Map<Integer, Order> getOrders();
}
```

（7）在 Service 业务逻辑中应用。

```
@Service("orderService")
public class OrderServiceImpl implements IOrderService {

    @Autowired
    private StateMachine<OrderStatus, OrderStatusChangeEvent> orderStateMachine;

    @Autowired
    private StateMachinePersister<OrderStatus, OrderStatusChangeEvent, Order> persister;

    private int id = 1;
    private Map<Integer, Order> orders = new HashMap<>();

    public Order create() {
```

```
        Order order = new Order();
        order.setStatus(OrderStatus.WAIT_PAYMENT);
        order.setId(id++);
        orders.put(order.getId(), order);
        return order;
    }

    public Order pay(int id) {
        Order order = orders.get(id);
        System.out.println("线程名称：" + Thread.currentThread().getName() + " 尝试支付，订单
号：" + id);
        Message message = MessageBuilder.withPayload(OrderStatusChangeEvent.PAYED).
setHeader("order", order).build();
        if (!sendEvent(message, order)) {
            System.out.println("线程名称：" + Thread.currentThread().getName() + " 支付失败，状
态异常，订单号：" + id);
        }
        return orders.get(id);
    }

    public Order deliver(int id) {
        Order order = orders.get(id);
        System.out.println("线程名称：" + Thread.currentThread().getName() + " 尝试发货，订单
号：" + id);
        if (!sendEvent(MessageBuilder.withPayload(OrderStatusChangeEvent.DELIVERY)
.setHeader("order", order).build(), orders.get(id))) {
            System.out.println("线程名称：" + Thread.currentThread().getName() + " 发货失败，
状态异常，订单号：" + id);
        }
        return orders.get(id);
    }

    public Order receive(int id) {
        Order order = orders.get(id);
        System.out.println("线程名称：" + Thread.currentThread().getName() + " 尝试收货，订单
号：" + id);
        if (!sendEvent(MessageBuilder.withPayload(OrderStatusChangeEvent.RECEIVED)
.setHeader("order", order).build(), orders.get(id))) {
            System.out.println("线程名称：" + Thread.currentThread().getName() + " 收货失败，
状态异常，订单号：" + id);
        }
        return orders.get(id);
    }


    public Map<Integer, Order> getOrders() {
        return orders;
```

```
    }

    /**
     * 发送订单状态转换事件
     *
     * @param message
     * @param order
     * @return
     */
    private synchronized boolean sendEvent(Message<OrderStatusChangeEvent> message, Order
order) {
        boolean result = false;
        try {
            orderStateMachine.start();
            //尝试恢复状态机状态
            persister.restore(orderStateMachine, order);
            //添加延迟用于线程安全测试
            Thread.sleep(1000);
            result = orderStateMachine.sendEvent(message);
            //持久化状态机状态
            persister.persist(orderStateMachine, order);
        } catch (Exception e) {
            e.printStackTrace();
        } finally {
            orderStateMachine.stop();
        }
        return result;
    }
}
```

（8）编写客户端测试代码。

```
@SpringBootApplication
public class Test {
    public static void main(String[] args) {

        Thread.currentThread().setName("主线程");

        ConfigurableApplicationContext context = SpringApplication.run(Test.class,args);

        IOrderService orderService = (IOrderService)context.getBean("orderService");

        orderService.create();
        orderService.create();

        orderService.pay(1);
```

```
        new Thread("客户线程"){
            @Override
            public void run() {
                orderService.deliver(1);
                orderService.receive(1);
            }
        }.start();

        orderService.pay(2);
        orderService.deliver(2);
        orderService.receive(2);

        System.out.println("全部订单状态： " + orderService.getOrders());

    }
}
```

通过这个真实的业务案例，相信小伙伴们已经对状态模式有了一个非常深刻的理解。

24.3　状态模式在 JSF 源码中的应用

状态模式的具体应用在源码中非常少见，源码中一般只提供一种通用的解决方案。如果一定要找，当然也是能找到的。经历千辛万苦，持续烧脑，下面来看一个在 JSF 源码中的 Lifecycle 类。JSF 算是一个比较经典的前端框架，没用过的小伙伴也没关系，这里只是分析一下其设计思想。在 JSF 中，所有页面的处理分为 6 个阶段，被定义在 PhaseId 类中，用不同的常量来表示生命周期阶段，源码如下。

```
public class PhaseId implements Comparable {
   ...
    private static final PhaseId[] values =
    {
        ANY_PHASE,                   //任意一个生命周期阶段
        RESTORE_VIEW,                //恢复视图阶段
        APPLY_REQUEST_VALUES,        //应用请求值阶段
        PROCESS_VALIDATIONS,         //处理输入校验阶段
        UPDATE_MODEL_VALUES,         //更新模型的值阶段
        INVOKE_APPLICATION,          //调用应用阶段
        RENDER_RESPONSE              //显示响应阶段
    };
    ...
}
```

这些状态的切换都在 Lifecycle 的 execute()方法中进行，其中会传入一个参数 FacesContext 对象，最终所有状态都被 FacesContext 保存。在此不做深入分析。

24.4 状态模式扩展

24.4.1 状态模式与责任链模式的区别

状态模式和责任链模式都能消除 if...else 分支过多的问题。但在某些情况下，状态模式中的状态可以理解为责任，那么在这种情况下，两种模式都可以使用。

从定义来看，状态模式强调的是一个对象内在状态的改变，而责任链模式强调的是外部节点对象间的改变。

从代码实现上来看，两者最大的区别就是状态模式的各个状态对象知道自己要进入的下一个状态对象，而责任链模式并不清楚其下一个节点处理对象，因为链式组装由客户端负责。

24.4.2 状态模式与策略模式的区别

状态模式和策略模式的 UML 类图架构几乎完全一样，但两者的应用场景是不一样的。策略模式的多种算法行为择其一都能满足，彼此之间是独立的，用户可自行更换策略算法；而状态模式的各个状态间存在相互关系，彼此之间在一定条件下存在自动切换状态的效果，并且用户无法指定状态，只能设置初始状态。

24.4.3 状态模式的优点

（1）结构清晰：将状态独立为类，消除了冗余的 if...else 或 switch...case 语句，使代码更加简洁，提高了系统的可维护性。

（2）将状态转换显示化：通常对象内部都是使用数值类型来定义状态的，状态的切换通过赋值进行表现，不够直观；而使用状态类，当切换状态时，是以不同的类进行表示的，转换目的更加明确。

（3）状态类职责明确且具备扩展性。

24.4.4　状态模式的缺点

（1）类膨胀：如果一个事物具备很多状态，则会造成状态类太多。

（2）状态模式的结构与实现都较为复杂，如果使用不当，将导致程序结构和代码的混乱。

（3）状态模式对开闭原则的支持并不太好，对于可以切换状态的状态模式，增加新的状态类需要修改那些负责状态转换的源码，否则无法切换到新增状态，而且修改某个状态类的行为也需要修改对应类的源码。

25

第 25 章 备忘录模式

25.1 备忘录模式概述

25.1.1 备忘录模式的定义

备忘录模式（Memento Pattern）又叫作快照模式（Snapshot Pattern）或令牌模式（Token Pattern），指在不破坏封装的前提下，捕获一个对象的内部状态，并在对象之外保存这个状态。这样以后就可将该对象恢复到原先保存的状态，属于行为型设计模式。

原文： Without violating encapsulation, capture and externalize an object's internal state so that the object can be restored to this state later.

在软件系统中，备忘录模式可以提供一种“后悔药”的机制，它通过存储系统各个历史状态的快照，使得在任一时刻都可以将系统回滚到某一个历史状态。

备忘录模式的本质是从发起人实体类（Originator）隔离存储功能，降低实体类的职责。同时由于存储信息（Memento）独立，且存储信息的实体交由管理类（Caretaker）管理，则可以通过为管理类扩展额外的功能对存储信息进行扩展操作（比如增加历史快照功能）。

25.1.2　备忘录模式的应用场景

对于程序员来说，可能天天都在使用备忘录模式，比如我们每天使用的 Git、SVN 都可以提供一种代码版本撤回的功能。还有一个比较贴切的现实场景就是游戏的存档功能，通过将游戏当前进度存储到本地文件系统或数据库中，使得下次继续游戏时，玩家可以从之前的位置继续进行。

备忘录模式主要适用于以下应用场景。

（1）需要保存历史快照的场景。

（2）希望在对象之外保存状态，且除了自己，其他类对象无法访问状态保存的具体内容。

25.1.3　备忘录模式的 UML 类图

备忘录模式的 UML 类图如下。

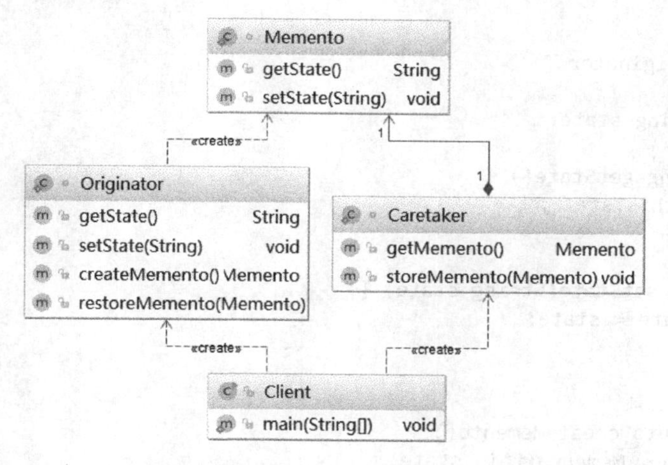

由上图可以看到，备忘录模式主要包含 3 个角色。

（1）发起人角色（Originator）：负责创建一个备忘录，记录自身需要保存的状态；具备状态回滚功能。

（2）备忘录角色（Memento）：用于存储 Originator 的内部状态，且可以防止 Originator 以外的对象进行访问。

（3）备忘录管理员角色（Caretaker）：负责存储、提供管理 Memento，无法对 Memento 的内容进行操作和访问。

25.1.4 备忘录模式的通用写法

以下是备忘录模式的通用写法。

```
public class Client {
    public static void main(String[] args) {
        //创建一个发起人角色
        Originator originator = new Originator();
        //创建一个备忘录管理员角色
        Caretaker caretaker = new Caretaker();
        //管理员存储发起人的备忘录
        caretaker.storeMemento(originator.createMemento());
        //发起人从管理员获取备忘录进行回滚
        originator.restoreMemento(caretaker.getMemento());

    }

    //发起人角色
    static class Originator {
        //内部状态
        private String state;

        public String getState() {
            return this.state;
        }

        public void setState(String state) {
            this.state = state;
        }

        //创建一个备忘录
        public Memento createMemento() {
            return new Memento(this.state);
        }

        //从备忘录恢复
        public void restoreMemento(Memento memento) {
            this.setState(memento.getState());
        }
    }

    //备忘录角色
    static class Memento {
        private String state;

        public Memento(String state){
```

```
            this.state = state;
        }

        public String getState() {
            return this.state;
        }

        public void setState(String state) {
            this.state = state;
        }
    }

    //备忘录管理者角色
    static class Caretaker {
        //备忘录对象
        private Memento memento;

        public Memento getMemento() {
            return this.memento;
        }

        public void storeMemento(Memento memento) {
            this.memento = memento;
        }
    }
}
```

备忘录模式要求备忘录（Memento）的内容只对发起人（Originator）可见，对其他对象（Caretaker）不可见；但是在上面代码中，备忘录管理员（Caretaker）其实可以通过备忘录（Memento）提供的相关方法 getState()、setState()获取和修改内部状态，这违背了备忘录模式的要求，可能造成备忘录内部状态无意间被其他对象修改，导致发起人状态恢复错误，系统稳定性下降。

```
interface IMemento {
}

static class Caretaker {
    //备忘录对象
    private IMemento memento;

    public IMemento getMemento() {
        return this.memento;
    }

    public void storeMemento(IMemento memento) {
        this.memento = memento;
    }
```

```
}

static class Originator {
    //内部状态
    private String state;
    ...
    ...
    //从备忘录恢复
    public void restoreMemento(IMemento memento) {
        this.state = ((Memento) memento).state;
    }

    p'rstatic class Memento implements IMemento {
        private String state;

        private Memento(String state) {
            this.state = state;
        }
    }
}
```

在上面代码中，我们把 Memento 作为 Originator 的私有静态内部类，这样就满足了 Originator 对 Memento 具有宽访问权限，但是直接这样做，其他类（Caretaker）是无法访问 Memento 的。因此，在最上面通过定义一个空接口 IMemento，然后让 Memento 实现 IMemento，变相将 Memento 扩展为公有 IMemento，这样就实现了其他类（Caretaker）对 Memento 具备窄访问权限。这种形式才算严格满足备忘录模式的设计要求。

25.2 使用备忘录模式实现草稿箱功能

大家肯定都用过网页中的富文本编辑器，编辑器通常都会附带草稿箱、撤销等操作。下面用一段代码来实现这样的功能。假设我们在 GPer 社区中发布一篇文章，文章编辑的过程需要花很长时间，中间也会不停地撤销、修改，甚至可能要花好几天才能写出一篇精品文章，因此可能会将已经编辑好的内容实时保存到草稿箱。

首先创建发起人角色编辑器 Editor 类。

```
public class Editor {

    private String title;
    private String content;
    private String imgs;
```

```
public Editor(String title, String content, String imgs) {
    this.title = title;
    this.content = content;
    this.imgs = imgs;
}

public String getTitle() {
    return title;
}

public void setTitle(String title) {
    this.title = title;
}

public String getContent() {
    return content;
}

public void setContent(String content) {
    this.content = content;
}

public String getImgs() {
    return imgs;
}

public void setImgs(String imgs) {
    this.imgs = imgs;
}

public ArticleMemento saveToMemento() {
    ArticleMemento articleMemento = new ArticleMemento(this.title,this.content,this.imgs);
    return articleMemento;
}

public void undoFromMemento(ArticleMemento articleMemento) {

    this.title = articleMemento.getTitle();
    this.content = articleMemento.getContent();
    this.imgs = articleMemento.getImgs();
```

```
    }

    @Override
    public String toString() {
        return "Editor{" +
                "title='" + title + '\'' +
                ", content='" + content + '\'' +
                ", imgs='" + imgs + '\'' +
                '}';
    }
}
```

然后创建备忘录角色 ArticleMemento 类。

```
public class ArticleMemento {
    private String title;
    private String content;
    private String imgs;

    public ArticleMemento(String title, String content, String imgs) {
        this.title = title;
        this.content = content;
        this.imgs = imgs;
    }

    public String getTitle() {
        return title;
    }

    public String getContent() {
        return content;
    }

    public String getImgs() {
        return imgs;
    }

    @Override
    public String toString() {
        return "ArticleMemento{" +
                "title='" + title + '\'' +
                ", content='" + content + '\'' +
                ", imgs='" + imgs + '\'' +
```

```
            '}';
    }
}
```

接着创建备忘录管理角色草稿箱 DraftsBox 类。

```
public class DraftsBox {

    private final Stack<ArticleMemento> STACK = new Stack<ArticleMemento>();

    public ArticleMemento getMemento() {
        ArticleMemento articleMemento= STACK.pop();
        return articleMemento;
    }

    public void addMemento(ArticleMemento articleMemento) {
        STACK.push(articleMemento);
    }

}
```

草稿箱中定义的 Stack 类是 Vector 的一个子类，它实现了一个标准的后进先出的栈。如下表所示，主要定义了以下方法。

方法定义	方法描述
boolean empty()	测试堆栈是否为空
Object peek( )	查看堆栈顶部的对象，但不从堆栈中移除它
Object pop( )	移除堆栈顶部的对象，并作为此函数的值返回该对象
Object push(Object element)	把对象压入堆栈顶部
int search(Object element)	返回对象在堆栈中的位置，以1为基数

最后编写客户端测试代码。

```
    public static void main(String[] args) {
        DraftsBox draftsBox = new DraftsBox();

        Editor editor = new Editor("我是这样手写 Spring 的，麻雀虽小五脏俱全",
                "本文节选自《Spring5 核心原理与 30 个类手写实战》一书，Tom 著，电子工业出版社出版。
",
                "35576a9ef6fc407aa088eb8280fb1d9d.png");

        ArticleMemento articleMemento = editor.saveToMemento();
        draftsBox.addMemento(articleMemento);
```

```
        System.out.println("标题：" + editor.getTitle() + "\n" +
                           "内容：" + editor.getContent() + "\n" +
                           "插图：" + editor.getImgs() + "\n 暂存成功");

        System.out.println("完整的信息" + editor);


        System.out.println("==========首次修改文章===========");
        editor.setTitle("【Tom 原创】我是这样手写 Spring 的，麻雀虽小五脏俱全");
        editor.setContent("本文节选自《Spring5 核心原理与 30 个类手写实战》一书，Tom 著");

        System.out.println("==========首次修改文章完成===========");

        System.out.println("完整的信息" + editor);

        articleMemento = editor.saveToMemento();

        draftsBox.addMemento(articleMemento);

        System.out.println("==========保存到草稿箱===========");


        System.out.println("==========第 2 次修改文章===========");
        editor.setTitle("手写 Spring");
        editor.setContent("本文节选自《Spring5 核心原理与 30 个类手写实战》一书，Tom 著");
        System.out.println("完整的信息" + editor);
        System.out.println("==========第 2 次修改文章完成===========");

        System.out.println("==========第 1 次撤销===========");
        articleMemento = draftsBox.getMemento();
        editor.undoFromMemento(articleMemento);
        System.out.println("完整的信息" + editor);
        System.out.println("==========第 1 次撤销完成===========");


        System.out.println("==========第 2 次撤销===========");
        articleMemento = draftsBox.getMemento();
        editor.undoFromMemento(articleMemento);
        System.out.println("完整的信息" + editor);
        System.out.println("==========第 2 次撤销完成===========");

    }
```

运行结果如下图所示。

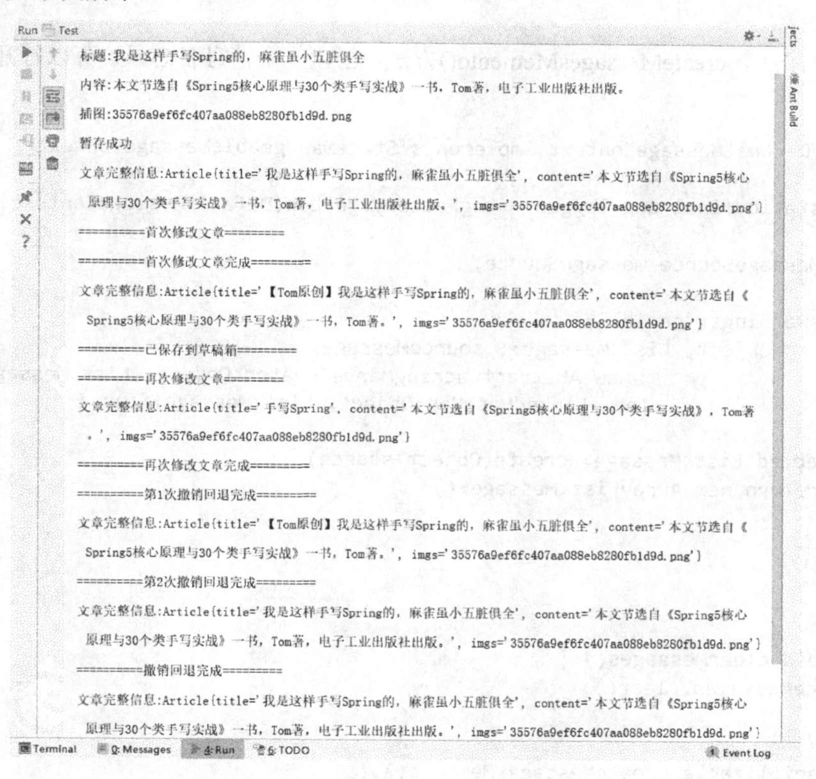

25.3 备忘录模式在 Spring 源码中的应用

备忘录模式在框架源码中的应用是比较少的，主要还是结合具体的应用场景来使用。笔者在 JDK 源码里一顿找，目前为止还是没找到具体的应用，包括在 MyBatis 中也没有找到对应的源码。在 Spring 的 Webflow 源码中找到一个 StateManageableMessageContext 接口，源码如下。

```
public interface StateManageableMessageContext extends MessageContext {

    public Serializable createMessagesMemento();

    public void restoreMessages(Serializable messagesMemento);

```

```
    public void setMessageSource(MessageSource messageSource);
}
```

我们看到有一个 createMessagesMemento()方法，创建一个消息备忘录。可以打开它的实现类，源码如下。

```
public class DefaultMessageContext implements StateManageableMessageContext {

    private static final Log logger = LogFactory.getLog(DefaultMessageContext.class);

    private MessageSource messageSource;

    @SuppressWarnings("serial")
    private Map<Object, List<Message>> sourceMessages =
                           new AbstractCachingMapDecorator<Object, List<Message>>(
                          new LinkedHashMap<Object, List<Message>>()) {

        protected List<Message> create(Object source) {
            return new ArrayList<Message>();
        }
    };

    ...

    public void clearMessages() {
        sourceMessages.clear();
    }

    public Serializable createMessagesMemento() {
        return new LinkedHashMap<Object, List<Message>>(sourceMessages);
    }

    @SuppressWarnings("unchecked")
    public void restoreMessages(Serializable messagesMemento) {
        sourceMessages.putAll((Map<Object, List<Message>>) messagesMemento);
    }

    public void setMessageSource(MessageSource messageSource) {
        if (messageSource == null) {
            messageSource = new DefaultTextFallbackMessageSource();
        }
        this.messageSource = messageSource;
    }

    ...

}
```

我们看到其主要逻辑就相当于给 Message 留一个备份，以备恢复之用。

25.4　备忘录模式扩展

25.4.1　备忘录模式的优点

（1）简化发起人实体类（Originator）的职责，隔离状态存储与获取，实现了信息的封装，客户端无须关心状态的保存细节。

（2）提供状态回滚功能。

25.4.2　备忘录模式的缺点

备忘录模式的缺点主要是消耗资源。如果需要保存的状态过多，则每一次保存都会消耗很多内存。

26

第 26 章 中介者模式

26.1 中介者模式概述

26.1.1 中介者模式的定义

中介者模式（Mediator Pattern）又叫作调解者模式或调停者模式。用一个中介对象封装一系列对象交互，中介者使各对象不需要显式地相互作用，从而使其耦合松散，而且可以独立地改变它们之间的交互，属于行为型设计模式。

原文： Define an object that encapsulates how a set of objects interact. Mediator promotes loose coupling by keeping objects from referring to each other explicitly, and it lets you vary their interaction independently.

中介者模式包装了一系列对象相互作用的方式，使得这些对象不必相互明显作用，从而使它们可以松散耦合。当某些对象之间的作用发生改变时，不会立即影响其他一些对象之间的作用。保证这些作用可以彼此独立地变化。其核心思想是，通过中介者解耦系统各层次对象的直接耦合，层次对象的对外依赖通信全部交由中介者转发。

26.1.2　中介者模式的应用场景

在现实生活中，中介者的存在是不可缺少的，如果没有了中介者，我们就不能与远方的朋友进行交流。各个同事对象将会相互进行引用，如果每个对象都与多个对象进行交互，则会形成如下图所示的网状结构。

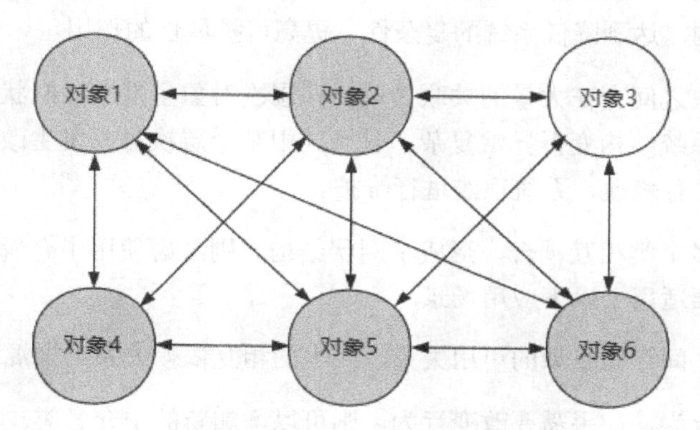

从上图可以看到，每个对象之间都过度耦合，这样既不利于信息的复用也不利于扩展。如果引入中介者模式，则对象之间的关系将变成星形结构，如下图所示。

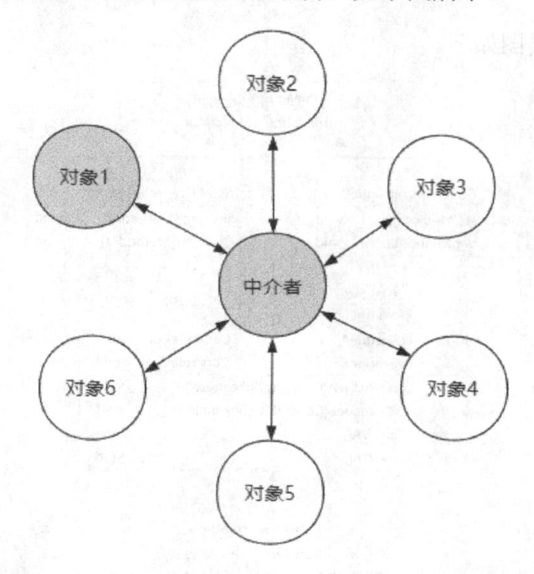

从上图可以看到，使用中介者模式后，任何一个类的变化，只会影响中介者和类本身，不像之前的设计，任何一个类的变化都会引起其关联的所有类的变化。这样的设计大大减少了系统的耦合度。

其实日常生活中我们每天都在刷的朋友圈，就是一个中介者。还有我们所见的信息交易平台，也是中介者模式的体现。

中介者模式是用来降低多个对象和类之间的通信复杂性的。这种模式通过提供一个中介类，将系统各层次对象间的多对多关系变成一对多关系，中介者对象可以将复杂的网状结构变成以中介者为中心的星形结构，达到降低系统的复杂性、提高可扩展性的作用。

若系统各层次对象之间存在大量的关联关系，即层次对象呈复杂的网状结构，如果直接让它们紧耦合通信，会使系统结构变得异常复杂，且当其中某个层次对象发生改变时，则与其紧耦合的相应层次对象也需进行修改，系统很难进行维护。

简单地说，如果多个类相互耦合，形成了网状结构，则考虑使用中介者模式进行优化。总结一下，中介者模式主要适用于以下应用场景。

（1）系统中对象之间存在复杂的引用关系，产生的相互依赖关系结构混乱且难以理解。

（2）交互的公共行为，如果需要改变行为，则可以增加新的中介者类。

26.1.3 中介者模式的 UML 类图

中介者模式的 UML 类图如下。

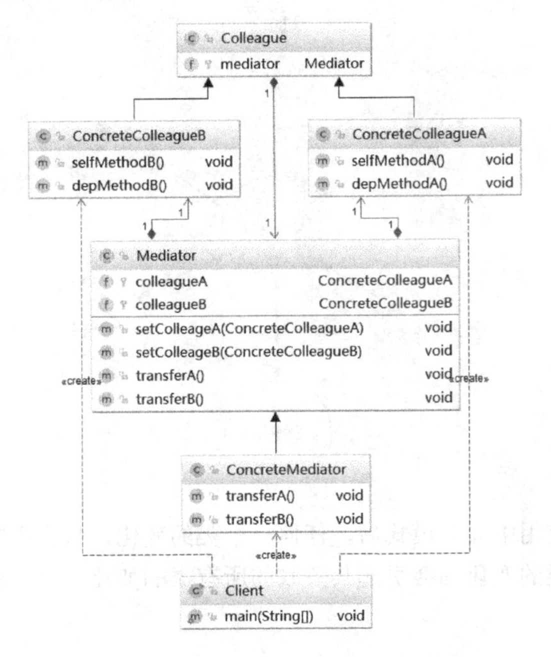

由上图可以看到，中介者模式主要包含 4 个角色。

（1）抽象中介者（Mediator）：定义统一的接口，用于各个同事角色之间的通信。

（2）具体中介者（ConcreteMediator）：从具体的同事对象接收消息，向具体同事对象发出命令，协调各同事间的协作。

（3）抽象同事类（Colleague）：每一个同事对象均需要依赖中介者角色，与其他同事间通信时，交由中介者进行转发协作。

（4）具体同事类（ConcreteColleague）：负责实现自发行为（Self-Method），转发依赖方法（Dep-Method）交由中介者进行协调。

26.1.4 中介者模式的通用写法

以下是中介者模式的通用写法。

```
public class Client {
    public static void main(String[] args) {
        Mediator mediator = new ConcreteMediator();
        ConcreteColleagueA colleagueA = new ConcreteColleagueA(mediator);
        ConcreteColleagueB colleagueB = new ConcreteColleagueB(mediator);
        colleagueA.depMethodA();
        System.out.println("-------------------------");
        colleagueB.depMethodB();
    }



    //抽象同事类
    static abstract class Colleague {
        protected Mediator mediator;

        public Colleague(Mediator mediator) {
            this.mediator = mediator;
        }
    }

    //具体同事类 A
    static class ConcreteColleagueA extends Colleague {
        public ConcreteColleagueA(Mediator mediator) {
            super(mediator);
            this.mediator.setColleageA(this);
        }
```

```
    //自发行为：Self-Method
    public void selfMethodA() {
        //处理自己的逻辑
        System.out.println(this.getClass().getSimpleName() + ", self-Method");
    }

    //依赖方法：Dep-Method
    public void depMethodA() {
        //处理自己的逻辑
        System.out.println(this.getClass().getSimpleName() + ":depMethod: delegate to Mediator");
        //无法处理的业务逻辑委托给中介者处理
        this.mediator.transferA();
    }
}

//具体同事类 B
static class ConcreteColleagueB extends Colleague {
    public ConcreteColleagueB(Mediator mediator) {
        super(mediator);
        this.mediator.setColleageB(this);
    }

    //自发行为：Self-Method
    public void selfMethodB() {
        // 处理自己的逻辑
        System.out.println(this.getClass().getSimpleName() + ", self-Method");
    }

    //依赖方法：Dep-Method
    public void depMethodB() {
        //处理自己的逻辑
        System.out.println(this.getClass().getSimpleName() + ":depMethod: delegate to Mediator");
        //无法处理的业务逻辑委托给中介者处理
        this.mediator.transferB();
    }
}

//抽象中介者
static abstract class Mediator {
    protected ConcreteColleagueA colleagueA;
    protected ConcreteColleagueB colleagueB;

    public void setColleageA(ConcreteColleagueA colleague) {
        this.colleagueA = colleague;
    }

    public void setColleageB(ConcreteColleagueB colleague) {
```

```
            this.colleagueB = colleague;
        }

        //中介者业务逻辑
        public abstract void transferA();

        public abstract void transferB();
    }

    //具体中介者
    static class ConcreteMediator extends Mediator {
        @Override
        public void transferA() {
            //协调行为：A 转发到 B
            this.colleagueB.selfMethodB();
        }

        @Override
        public void transferB() {
            //协调行为：B 转发到 A
            this.colleagueA.selfMethodA();
        }
    }
}
```

从上面代码可以看到，其实 Colleague 的作用只是硬性规定让同事类都依赖 Mediator，而且 Mediator 内部直接引用的是 ConcreteColleague，这是因为在真实业务中，ConcreteColleagueA 和 ConcreteColleagueB 这些同事类分别代表不同的具体业务类，因此它们之间无法进行共性抽象，也就是说 Colleague 在平常的编码中可以忽略，直接在 Mediator 中引用真实业务类即可。

26.2　使用中介者模式设计群聊场景

假设我们要构建一个聊天室系统，用户可以向聊天室发送消息，聊天室会向所有用户显示消息。实际上就是用户发信息与聊天室显示的通信过程，不过用户无法直接将信息发给聊天室，而需要将信息先发到服务器上，然后服务器再将该消息发给聊天室进行显示，具体代码如下。首先创建 User 类。

```
public class User {
    private String name;
    private ChatRoom chatRoom;

    public User(String name, ChatRoom chatRoom) {
```

```
        this.name = name;
        this.chatRoom = chatRoom;
    }

    public void sendMessage(String msg) {
        this.chatRoom.showMsg(this, msg);
    }

    public String getName() {
        return name;
    }
}
```

然后创建 ChatRoom 类。

```
public class ChatRoom {
    public void showMsg(User user, String msg) {
        System.out.println("[" + user.getName() + "] :" + msg);
    }
}
```

最后编写客户端测试代码。

```
    public static void main(String[] args) {
        ChatRoom room = new ChatRoom();

        User tom = new User("Tom",room);
        User jerry = new User("Jerry",room);
        tom.sendMessage("Hi! I am Tom.");
        jerry.sendMessage("Hello! My name is Jerry.");
    }
```

运行结果如下图所示。

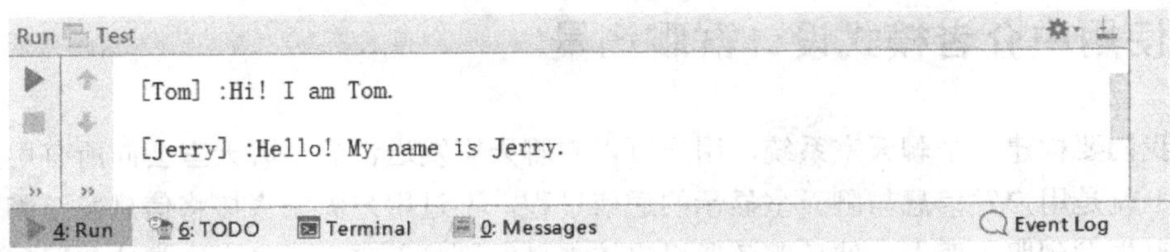

26.3 中介者模式在 JDK 源码中的应用

首先来看 JDK 中的 Timer 类。打开 Timer 的结构，我们发现 Timer 类中有很多 schedule()重载方法，如下图所示。

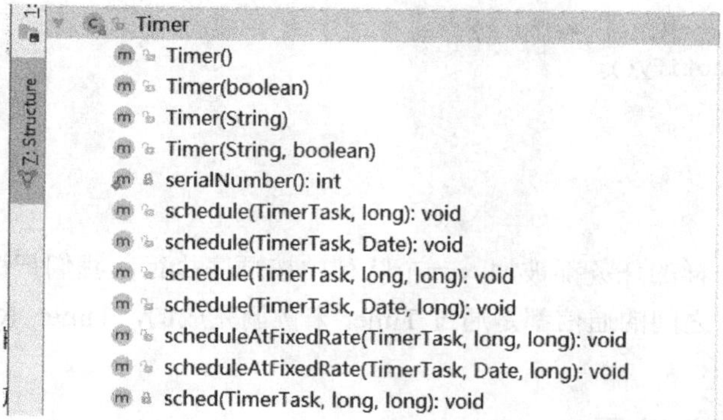

7: Structure

Timer
Timer()
Timer(boolean)
Timer(String)
Timer(String, boolean)
serialNumber(): int
schedule(TimerTask, long): void
schedule(TimerTask, Date): void
schedule(TimerTask, long, long): void
schedule(TimerTask, Date, long): void
scheduleAtFixedRate(TimerTask, long, long): void
scheduleAtFixedRate(TimerTask, Date, long): void
sched(TimerTask, long, long): void

任意点开其中一个方法，我们发现所有方法最终都调用了私有的 schedule()方法，源码如下。

```
public class Timer {
    ...
    public void schedule(TimerTask task, long delay) {
        if (delay < 0)
            throw new IllegalArgumentException("Negative delay.");
        sched(task, System.currentTimeMillis()+delay, 0);
    }
    ...
    private void sched(TimerTask task, long time, long period) {
        if (time < 0)
            throw new IllegalArgumentException("Illegal execution time.");

        if (Math.abs(period) > (Long.MAX_VALUE >> 1))
            period >>= 1;

        synchronized(queue) {
            if (!thread.newTasksMayBeScheduled)
                throw new IllegalStateException("Timer already cancelled.");

            synchronized(task.lock) {
                if (task.state != TimerTask.VIRGIN)
                    throw new IllegalStateException(
                        "Task already scheduled or cancelled");
                task.nextExecutionTime = time;
                task.period = period;
                task.state = TimerTask.SCHEDULED;
            }

            queue.add(task);
```

```
            if (queue.getMin() == task)
                queue.notify();
        }
    }
    ...
}
```

而且，不管是什么样的任务都被加入一个队列中按顺序执行。我们把这个队列中的所有对象都称为“同事”。同事之间的通信都是通过 Timer 来协调完成的，Timer 承担了中介者的角色。

26.4 中介者模式扩展

26.4.1 中介者模式的优点

（1）减少类间依赖，将多对多依赖转化成一对多，降低了类间耦合。

（2）类间各司其职，符合迪米特法则。

26.4.2 中介者模式的缺点

中介者模式将原本多个对象直接的相互依赖变成了中介者和多个同事类的依赖关系。当同事类越多时，中介者就会越臃肿，变得复杂且难以维护。

27

第 27 章 解释器模式

27.1 解释器模式概述

27.1.1 解释器模式的定义

解释器模式（Interpreter Pattern）指给定一门语言，定义它的文法的一种表示，并定义一个解释器，该解释器使用该表示来解释语言中的句子。解释器模式是一种按照规定的文法（语法）进行解析的模式，属于行为型设计模式。

原文： Given a language, define a representation for its grammar along with an interpreter that uses the representation to interpret sentences in the language.

就比如编译器可以将源码编译解释为机器码，让 CPU 能进行识别并运行。解释器模式的作用其实与编译器一样，都是将一些固定的文法（语法）进行解释，构建出一个解释句子的解释器。简单理解，解释器是一个简单文法分析工具，它可以识别句子语义，分离终结符号和非终结符号，提取出需要的信息，让我们能针对不同的信息做出相应的处理。其核心思想是识别文法，构建解释。

27.1.2 解释器模式的应用场景

其实，我们每天都生活在解释器模式中，我们平时听到的音乐都可以通过简谱记录下来；还有战争年代发明的摩尔斯电码（Morse code，又译为摩斯密码），其实也是一种解释器，如下图所示。

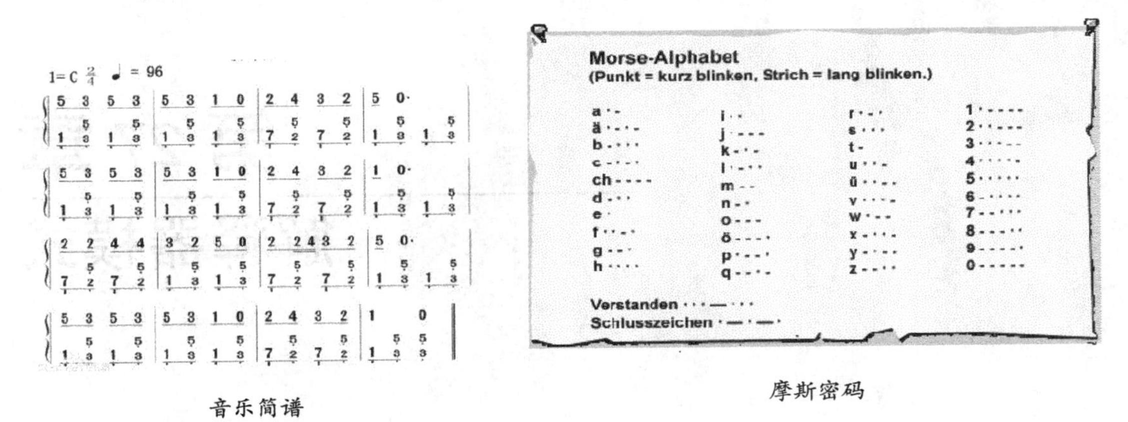

音乐简谱　　摩斯密码

在程序中，如果存在一种特定类型的问题，该类型问题涉及多个不同实例，但是具备固定文法描述，则可以使用解释器模式对该类型问题进行解释，分离出需要的信息，根据获取的信息做出相应的处理。简而言之，对于一些固定文法，构建一个特定的解释器。解释器模式主要适用于以下应用场景。

（1）一些重复出现的问题可以用一种简单的语言进行表达。

（2）一个简单语法需要解释的场景。

27.1.3 解释器模式的 UML 类图

解释器模式的 UML 类图如下。

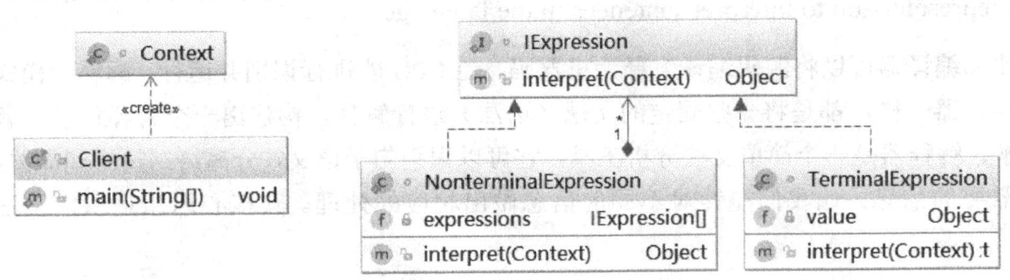

由上图可以看到，解释器模式主要包含 4 个角色。

（1）抽象表达式（IExpression）：负责定义一个解释方法 interpret，交由具体子类进行具体解释。

（2）终结符表达式（TerminalExpression）：实现文法中与终结符有关的解释操作。文法中的每一个终结符都有一个具体终结表达式与之相对应，比如公式 R=R1+R2，R1 和 R2 就是终结符，对应的解析 R1 和 R2 的解释器就是终结符表达式。通常一个解释器模式中只有一个终结符表达式，但有多个实例，对应不同的终结符（如 R1、R2）。

（3）非终结符表达式（NonterminalExpression）：实现文法中与非终结符有关的解释操作。文法中的每条规则都对应一个非终结符表达式。非终结符表达式一般是文法中的运算符或者其他关键字，比如在公式 R=R1+R2 中，“+”就是非终结符，解析“+”的解释器就是一个非终结符表达式。非终结符表达式根据逻辑的复杂程度而增加，原则上每个文法规则都对应一个非终结符表达式。

（4）上下文环境类（Context）：包含解释器之外的全局信息。它一般用来存放文法中各个终结符所对应的具体值，比如 R=R1+R2，给 R1 赋值 100，给 R2 赋值 200，这些信息需要被存放到环境中。

27.1.4　解释器模式的通用写法

以下是解释器模式的通用写法。

```
public class Client {
    public static void main(String[] args) {
        try {
            Context context = new Context();
            //定义一个语法容器，用于存储一个具体表达式
            Stack<IExpression> stack = new Stack<IExpression>();

            //此处省略递归调用逻辑

            //获取最终的解析表达式：完整语法树
            IExpression expression = stack.pop();
            //递归调用获取结果
            expression.interpret(context);

            //1 + 2
        }catch (Exception e){
            e.printStackTrace();
```

```
    }

}

//上下文环境类
static class Context extends HashMap {

}

//抽象表达式
interface IExpression {
    //对表达式进行解释
    Object interpret(Context context);
}

//终结符表达式
static class TerminalExpression implements IExpression {

    private Object value;

    public Object interpret(Context context) {
        //实现文法中与终结符有关的操作
        context.put("","");
        return null;
    }

}

//非终结符表达式
static class NonterminalExpression implements IExpression {
    private IExpression [] expressions;

    public NonterminalExpression(IExpression... expressions) {
        //每个非终结符表达式都会对其他表达式产生依赖
        this.expressions = expressions;
    }


    public Object interpret(Context context) {
        //进行文法处理
        context.put("","");
        return null;
    }
}
}
```

27.2　使用解释器模式解析数学表达式

下面用解释器模式来实现一个数学表达式计算器，包含加、减、乘、除运算。

首先定义抽象表达式角色 IArithmeticInterpreter 接口。

```
public interface IArithmeticInterpreter {
    int interpret();
}
```

创建终结表达式角色 Interpreter 抽象类。

```
public abstract class Interpreter implements IArithmeticInterpreter {

    protected IArithmeticInterpreter left;
    protected IArithmeticInterpreter right;

    public Interpreter(IArithmeticInterpreter left, IArithmeticInterpreter right) {
        this.left = left;
        this.right = right;
    }
}
```

然后分别创建非终结符表达式角色加、减、乘、除解释器，加法运算表达式 AddInterpreter 类的代码如下。

```
public class AddInterpreter extends Interpreter {

    public AddInterpreter(IArithmeticInterpreter left, IArithmeticInterpreter right) {
        super(left, right);
    }

    public int interpret() {
        return this.left.interpret() + this.right.interpret();
    }
}
```

减法运算表达式 SubInterpreter 类的代码如下。

```
public class SubInterpreter extends Interpreter {
    public SubInterpreter(IArithmeticInterpreter left, IArithmeticInterpreter right) {
        super(left, right);
    }

    public int interpret() {
```

```
        return this.left.interpret() - this.right.interpret();
    }
}
```

乘法运算表达式 MultiInterpreter 类的代码如下。

```
public class MultiInterpreter extends Interpreter {

    public MultiInterpreter(IArithmeticInterpreter left, IArithmeticInterpreter right){
        super(left,right);
    }

    public int interpret() {
        return this.left.interpret() * this.right.interpret();
    }

}
```

除法运算表达式 DivInterpreter 类的代码如下。

```
public class DivInterpreter extends Interpreter {

    public DivInterpreter(IArithmeticInterpreter left, IArithmeticInterpreter right){
        super(left,right);
    }

    public int interpret() {
        return this.left.interpret() / this.right.interpret();
    }

}
```

数字表达式 NumInterpreter 类的代码如下。

```
public class NumInterpreter implements IArithmeticInterpreter {
    private int value;

    public NumInterpreter(int value) {
        this.value = value;
    }


    public int interpret() {
        return this.value;
    }
}
```

接着创建计算器 GPCalculator 类。

```
public class GPCalculator {
```

```
    private Stack<IArithmeticInterpreter> stack = new Stack<IArithmeticInterpreter>();

    public GPCalculator(String expression) {
        this.parse(expression);
    }

    private void parse(String expression) {
        String [] elements = expression.split(" ");
        IArithmeticInterpreter left,right;

        for (int i = 0; i < elements.length ; i++) {
            String operator = elements[i];
            if(OperatorUtil.ifOperator(operator)){
                left = this.stack.pop();
                right = new NumInterpreter(Integer.valueOf(elements[++i]));
                System.out.println("出栈" + left.interpret() + "和" + right.interpret());
                this.stack.push(OperatorUtil.getInterpreter(left,right,operator));
                System.out.println("应用运算符：" + operator);
            }else {
                NumInterpreter numInterpreter = new NumInterpreter(Integer.valueOf(elements[i]));
                this.stack.push(numInterpreter);
                System.out.println("入栈：" + numInterpreter.interpret());
            }

        }
    }

    public int calculate() {
        return this.stack.pop().interpret();
    }
}
```

工具类 OperatorUtil 的代码如下。

```
public class OperatorUtil {

    public static boolean isOperator(String symbol) {
        return (symbol.equals("+") || symbol.equals("-") || symbol.equals("*"));
    }

    public static Interpreter getInterpreter(IArithmeticInterpreter left, IArithmeticInterpreter
    right, String symbol) {
        if (symbol.equals("+")) {
            return new AddInterpreter(left, right);
        } else if (symbol.equals("-")) {
            return new SubInterpreter(left, right);
        } else if (symbol.equals("*")) {
```

```
            return new MultiInterpreter(left, right);
        } else if (symbol.equals("/")) {
            return new DivInterpreter(left, right);
        }
        return null;
    }
}
```

最后编写客户端测试代码。

```
public static void main(String[] args) {
    System.out.println("result: " + new GPCalculator("10 + 30").calculate());
    System.out.println("result: " + new GPCalculator("10 + 30 - 20").calculate());
    System.out.println("result: " + new GPCalculator("100 * 2 + 400 * 1 + 66").calculate());
}
```

运行结果如下图所示。

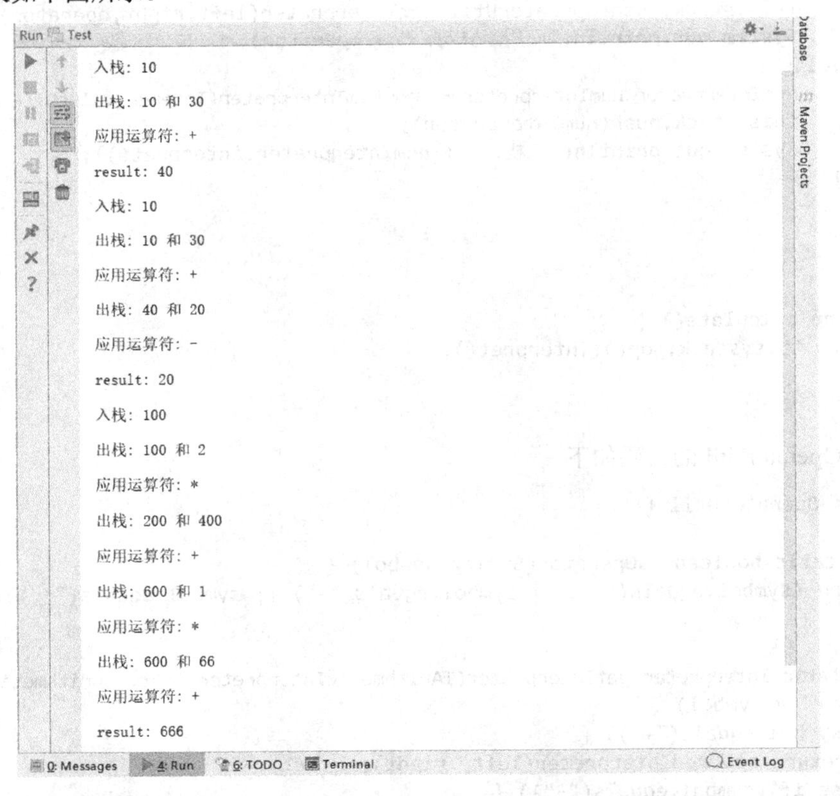

当然，上面的简易计算器还没有考虑优先级，就是从左至右依次运算的。在实际运算中，乘法和除法属于一级运算，加法和减法属于二级运算。一级运算需要优先计算。另外，我们可以通过使用括号手动调整运算的优先级。我们再优化一下代码，首先新建一个枚举类。

```
public enum OperatorEnum {
    LEFT_BRACKET("("),
    RIGHT_BRACKET(")"),
    SUB("-"),
    ADD("+"),
    MULTI("*"),
    DIV("/"),
    ;
    private String operator;

    public String getOperator() {
        return operator;
    }

    OperatorEnum(String operator) {
        this.operator = operator;
    }
}
```

然后修改 OperatorUtil 的处理逻辑，设置两个栈。

```
public class OperatorUtil {

    public static Interpreter getInterpreter(Stack<IArithmeticInterpreter> numStack, Stack<String>
operatorStack) {
        IArithmeticInterpreter right = numStack.pop();
        IArithmeticInterpreter left = numStack.pop();
        String symbol = operatorStack.pop();
        System.out.println("数字出栈: " + right.interpret() + "," + left.interpret() + ",操作
符出栈:" + symbol);
        if (symbol.equals("+")) {
            return new AddInterpreter(left, right);
        } else if (symbol.equals("-")) {
            return new SubInterpreter(left, right);
        } else if (symbol.equals("*")) {
            return new MultiInterpreter(left, right);
        } else if (symbol.equals("/")) {
            return new DivInterpreter(left, right);
        }
        return null;
    }
}
```

修改 GPCalculator 的代码。

```
public class GPCalculator {

    //数字 Stack
```

```
    private Stack<IArithmeticInterpreter> numStack = new Stack<IArithmeticInterpreter>();
    //操作符Stack
    private Stack<String> operatorStack = new Stack<String>();
    /**
     * 解析表达式
     * @param expression
     */
    public GPCalculator(String expression) {
        this.parse(expression);
    }

    private void parse(String input) {
        //对表达式去除空字符操作
        String expression = this.fromat(input);
        System.out.println("标准表达式：" + expression);
        for (String s : expression.split(" ")) {
            if (s.length() == 0){
                //如果是空格，则继续循环，什么也不操作
                continue;
            }
            //如果是加减，因为加减的优先级最低，所以这里只要遇到加减号，无论操作符栈中是什么运算符都要运算
            else if (s.equals(OperatorEnum.ADD.getOperator())
                    || s.equals(OperatorEnum.SUB.getOperator())) {
                //当栈不是空的，并且栈中最上面的一个元素是加减乘除的任意一个
                while (!operatorStack.isEmpty()
                        &&(operatorStack.peek().equals(OperatorEnum.SUB.getOperator())
                         || operatorStack.peek().equals(OperatorEnum.ADD.getOperator())
                         || operatorStack.peek().equals(OperatorEnum.MULTI.getOperator())
                         || operatorStack.peek().equals(OperatorEnum.DIV.getOperator()))) {
                    //结果存入栈中
                    numStack.push(OperatorUtil.getInterpreter(numStack,operatorStack));
                }
                //运算完后将当前的运算符入栈
                System.out.println("操作符入栈:"+s);
                operatorStack.push(s);
            }
            //当前运算符是乘除的时候，因为优先级高于加减
            //所以要判断最上面的是否是乘除，如果是乘除，则运算，否则直接入栈
            else if (s.equals(OperatorEnum.MULTI.getOperator())
                    || s.equals(OperatorEnum.DIV.getOperator())) {
                while (!operatorStack.isEmpty()&&(
                        operatorStack.peek().equals(OperatorEnum.MULTI.getOperator())
                         || operatorStack.peek().equals(OperatorEnum.DIV.getOperator()))) {
                    numStack.push(OperatorUtil.getInterpreter(numStack,operatorStack));
                }
                //将当前操作符入栈
                System.out.println("操作符入栈:"+s);
                operatorStack.push(s);
```

```
        }
        //如果是左括号，则直接入栈，什么也不用操作，trim()函数是用来去除空格的，由于上面的分割
        //操作，可能会令操作符带有空格
        else if (s.equals(OperatorEnum.LEFT_BRACKET.getOperator())) {
            System.out.println("操作符入栈:"+s);
            operatorStack.push(OperatorEnum.LEFT_BRACKET.getOperator());
        }
        //如果是右括号，则清除栈中的运算符直至左括号
        else if (s.equals(OperatorEnum.RIGHT_BRACKET.getOperator())) {
            while (!OperatorEnum.LEFT_BRACKET.getOperator().equals(operatorStack.peek())) {
                //开始运算
                numStack.push(OperatorUtil.getInterpreter(numStack,operatorStack));
            }
            //运算完之后清除左括号
            String pop = operatorStack.pop();
            System.out.println("括号运算操作完成，清除栈中右括号："+pop);
        }
        //如果是数字，则直接入数据的栈
        else {
            //将数字字符串转换成数字，然后存入栈中
            NumInterpreter numInterpreter = new NumInterpreter(Integer.valueOf(s));
            System.out.println("数字入栈："+s);
            numStack.push(numInterpreter);
        }
    }
    //最后当栈中不是空的时候继续运算，直到栈为空即可
    while (!operatorStack.isEmpty()) {
        numStack.push(OperatorUtil.getInterpreter(numStack,operatorStack));
    }
}

/**
 * 计算结果出栈
 * @return
 */
public int calculate() {
    return this.numStack.pop().interpret();
}

/**
 * 换成标准形式，便于分割
 * @param expression
 * @return
 */
private String fromat(String expression) {
    String result = "";
    for (int i = 0; i < expression.length(); i++) {
        if (expression.charAt(i) == '(' || expression.charAt(i) == ')' ||
```

```
            expression.charAt(i) == '+' || expression.charAt(i) == '-' ||
            expression.charAt(i) == '*' || expression.charAt(i) == '/')
            //在操作符与数字之间增加一个空格
            result += (" " + expression.charAt(i) + " ");
        else
            result += expression.charAt(i);
    }
    return result;
  }
}
```

此时，再来看客户端测试代码。

```
public static void main(String[] args) {
    System.out.println("result: " + new GPCalculator("10+30/((6-4)*2-2)").calculate());
  }
```

运行得到预期的结果，如下图所示。

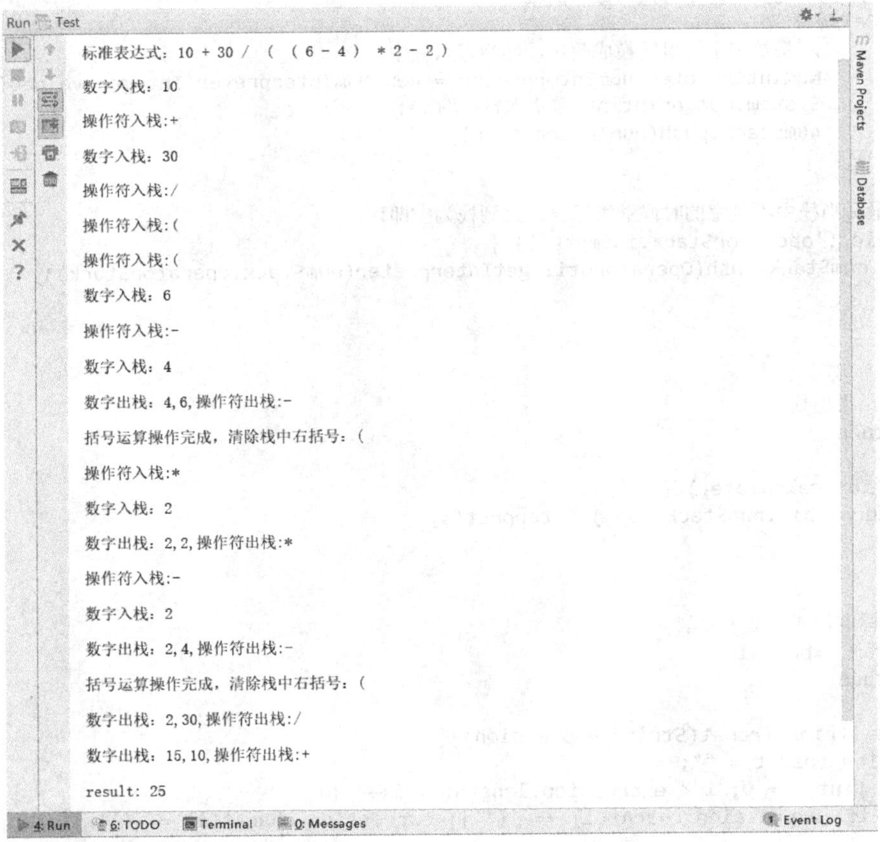

```
Run  Test
标准表达式：10 + 30 / ( ( 6 - 4 ) * 2 - 2 )
数字入栈：10
操作符入栈:+
数字入栈：30
操作符入栈:/
操作符入栈:(
操作符入栈:(
数字入栈：6
操作符入栈:-
数字入栈：4
数字出栈：4,6,操作符出栈:-
括号运算操作完成，清除栈中右括号：(
操作符入栈:*
数字入栈：2
数字出栈：2,2,操作符出栈:*
操作符入栈:-
数字入栈：2
数字出栈：2,4,操作符出栈:-
括号运算操作完成，清除栈中右括号：(
数字出栈：2,30,操作符出栈:/
数字出栈：15,10,操作符出栈:+
result: 25
4: Run  6: TODO  Terminal  0: Messages  Event Log
```

27.3　解释器模式在框架源码中的应用

27.3.1　解释器模式在 JDK 源码中的应用

先来看 JDK 源码中的 Pattern 对正则表达式的编译和解析。

```
public final class Pattern implements java.io.Serializable {
    ...
    private Pattern(String p, int f) {
        pattern = p;
        flags = f;

        if ((flags & UNICODE_CHARACTER_CLASS) != 0)
            flags |= UNICODE_CASE;


        capturingGroupCount = 1;
        localCount = 0;

        if (pattern.length() > 0) {
            compile();
        } else {
            root = new Start(lastAccept);
            matchRoot = lastAccept;
        }
    }
    ...
    public static Pattern compile(String regex) {
        return new Pattern(regex, 0);
    }
    public static Pattern compile(String regex, int flags) {
        return new Pattern(regex, flags);
    }

    ...
}
```

27.3.2　解释器模式在 Spring 源码中的应用

再来看 Spring 中的 ExpressionParser 接口。

```
public interface ExpressionParser {
```

```
    Expression parseExpression(String expressionString) throws ParseException;

    Expression parseExpression(String expressionString, ParserContext context) throws ParseException;

}
```

这里我们不深入讲解源码，通过我们前面编写的案例大致能够清楚其原理。不妨编写一段客户端代码验证一下。客户端测试代码如下。

```
public static void main(String[] args) {
    ExpressionParser parser = new SpelExpressionParser();
    Expression expression = parser.parseExpression("100 * 2 + 400 * 1 + 66");
    int result = (Integer) expression.getValue();
    System.out.println("计算结果是：" + result);
}
```

运行结果如下图所示。

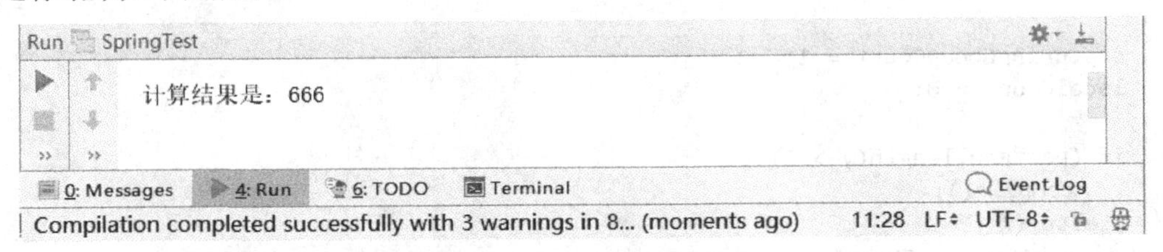

由上图可知，运行结果与预期结果是一致的。

27.4 解释器模式扩展

27.4.1 解释器模式的优点

（1）在解释器模式中，由于语法是由很多类表示的，当语法规则更改时，只需修改相应的非终结符表达式即可；当扩展语法时，只需添加相应的非终结符类即可。

（2）增加了新的解释表达式的方式。

（3）解释器模式对应的文法应当是比较简单且易于实现的，过于复杂的语法并不适合使用解释器模式。

27.4.2　解释器模式的缺点

（1）解释器模式的每个语法都要产生一个非终结符表达式，当语法规则比较复杂时，就会产生大量解释类，引起类膨胀，增加系统维护的难度。

（2）解释器模式采用递归调用方法，每个非终结符表达式都只关心与自己有关的表达式，每个表达式都需要知道最终的结果，因此完整表达式的最终结果是通过从后往前递归调用的方式获取的。当完整表达式层级较深时，解释效率下降，且出错时调试困难，因为递归迭代的层级太深。

28

第 28 章 观察者模式

28.1 观察者模式概述

28.1.1 观察者模式的定义

观察者模式（Observer Pattern）又叫作发布-订阅（Publish/Subscribe）模式、模型-视图（Model/View）模式、源-监听器（Source/Listener）模式或从属者（Dependent）模式。定义一种一对多的依赖关系，一个主题对象可被多个观察者对象同时监听，使得每当主题对象状态变化时，所有依赖它的对象都会得到通知并被自动更新，属于行为型设计模式。

> **原文**：Defines a one-to-many dependency relationship between objects so that each time an object's state changes, its dependent objects are notified and automatically updated.

观察者模式的核心是将观察者与被观察者解耦，以类似消息/广播发送的机制联动两者，使被观察者的变动能通知到感兴趣的观察者们，从而做出相应的响应。

28.1.2 观察者模式的应用场景

观察者模式在现实生活中的应用也非常广泛，比如，App 角标通知、起床闹钟设置，如下图所示，以及 GPer 生态圈消息通知、邮件通知、广播通知、桌面程序的事件响应等。

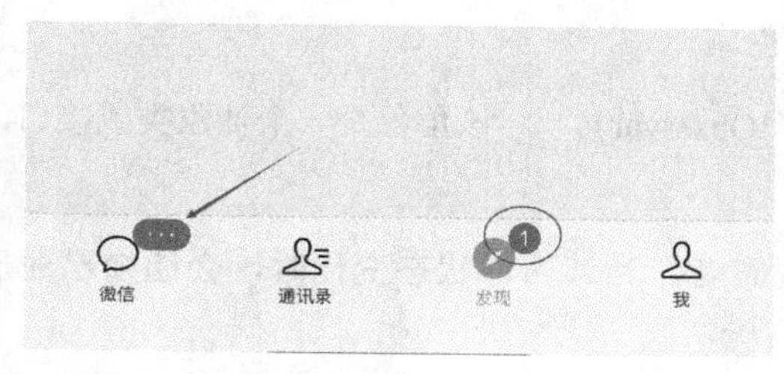

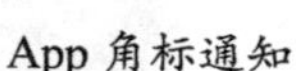
App 角标通知

起床闹钟设置

在软件系统中，当系统一方行为依赖另一方行为的变动时，可使用观察者模式松耦合联动双方，使得一方的变动可以通知到感兴趣的另一方对象，从而让另一方对象对此做出响应。观察者模式主要适用于以下应用场景。

（1）当一个抽象模型包含两方面内容，其中一方面依赖另一方面。

（2）其他一个或多个对象的变化依赖另一个对象的变化。

（3）实现类似广播机制的功能，不需要知道具体收听者，只需分发广播，系统中感兴趣的对象会自动接收该广播。

（4）多层级嵌套使用，形成一种链式触发机制，使得事件具备跨域（跨越两种观察者类型）通知。

28.1.3　观察者模式的 UML 类图

观察者模式的 UML 类图如下。

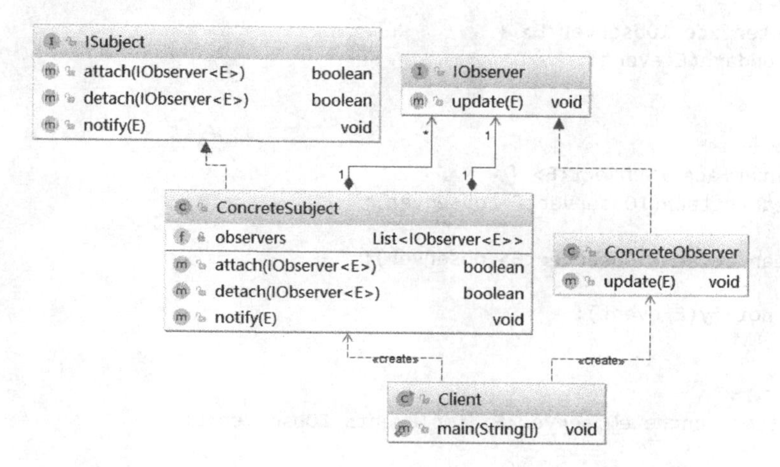

由上图可以看到，观察者模式主要包含4个角色。

（1）抽象主题（ISubject）：指被观察的对象（IObservable）。该角色是一个抽象类或接口，定义了增加、删除、通知观察者对象的方法。

（2）具体主题（ConcreteSubject）：具体被观察者，当其内部状态变化时，会通知已注册的观察者。

（3）抽象观察者（IObserver）：定义了响应通知的更新方法。

（4）具体观察者（ConcreteObserver）：当得到状态更新的通知时，会自动做出响应。

28.1.4 观察者模式的通用写法

以下是观察者模式的通用写法。

```
public class Client {

    public static void main(String[] args) {
        //被观察者
        ISubject<String> observable = new ConcreteSubject<String>();
        //观察者
        IObserver<String> observer = new ConcreteObserver<String>();
        //注册
        observable.attach(observer);
        //通知
        observable.notify("hello");
    }

    //抽象观察者
    public interface IObserver<E> {
        void update(E event);
    }

    //抽象主题者
    public interface ISubject<E> {
        boolean attach(IObserver<E> observer);

        boolean detach(IObserver<E> observer);

        void notify(E event);
    }

    //具体观察者
    static class ConcreteObserver<E> implements IObserver<E> {
```

```
        public void update(E event) {
            System.out.println("receive event: " + event);
        }
    }

    //具体主题者
    static class ConcreteSubject<E> implements ISubject<E> {
        private List<IObserver<E>> observers = new ArrayList<IObserver<E>>();

        public boolean attach(IObserver<E> observer) {
            return !this.observers.contains(observer) && this.observers.add(observer);
        }

        public boolean detach(IObserver<E> observer) {
            return this.observers.remove(observer);
        }

        public void notify(E event) {
            for (IObserver<E> observer : this.observers) {
                observer.update(event);
            }
        }
    }
}
```

28.2　使用观察者模式解决实际问题

28.2.1　基于 Java API 实现通知机制

当小伙伴们在 GPer 生态圈中提问时，如果有设置指定老师回答，则对应的老师就会收到邮件通知，这就是观察者模式的一种应用场景。有些小伙伴可能会想到 MQ、异步队列等，其实 JDK 本身就提供这样的 API。我们用代码来还原这样一个应用场景，首先创建 GPer 类。

```
/**
 * JDK 提供的一种观察者的实现方式，被观察者
 */
public class GPer extends Observable{
    private String name = "GPer 生态圈";
    private static GPer gper = null;
    private GPer(){}

    public static GPer getInstance(){
        if(null == gper){
            gper = new GPer();
        }
```

```
        return gper;
    }
    public String getName() {
        return name;
    }
    public void publishQuestion(Question question){
        System.out.println(question.getUserName() + "在" + this.name + "上提交了一个问题。");
        setChanged();
        notifyObservers(question);
    }
}
```

然后创建问题 Question 类。

```
public class Question {
    private String userName;
    private String content;

    public String getUserName() {
        return userName;
    }

    public void setUserName(String userName) {
        this.userName = userName;
    }

    public String getContent() {
        return content;
    }

    public void setContent(String content) {
        this.content = content;
    }
}
```

接着创建老师 Teacher 类。

```
public class Teacher implements Observer {

    private String name;

    public Teacher(String name) {
        this.name = name;
    }

    public void update(Observable o, Object arg) {
        GPer gper = (GPer)o;
        Question question = (Question)arg;
        System.out.println("======================");
```

```
        System.out.println(name + "老师，你好！\n" +
                "您收到了一个来自" + gper.getName() + "的提问，希望您解答。问题内容如下：\n" +
                 question.getContent() + "\n" + "提问者：" + question.getUserName());
    }
}
```

最后编写客户端测试代码。

```
    public static void main(String[] args) {
        GPer gper = GPer.getInstance();
        Teacher tom = new Teacher("Tom");
        Teacher jerry = new Teacher("Jerry");

        gper.addObserver(tom);
        gper.addObserver(jerry);

        //用户行为
        Question question = new Question();
        question.setUserName("张三");
        question.setContent("观察者模式适用于哪些场景？");

        gper.publishQuestion(question);
    }
```

运行结果如下图所示。

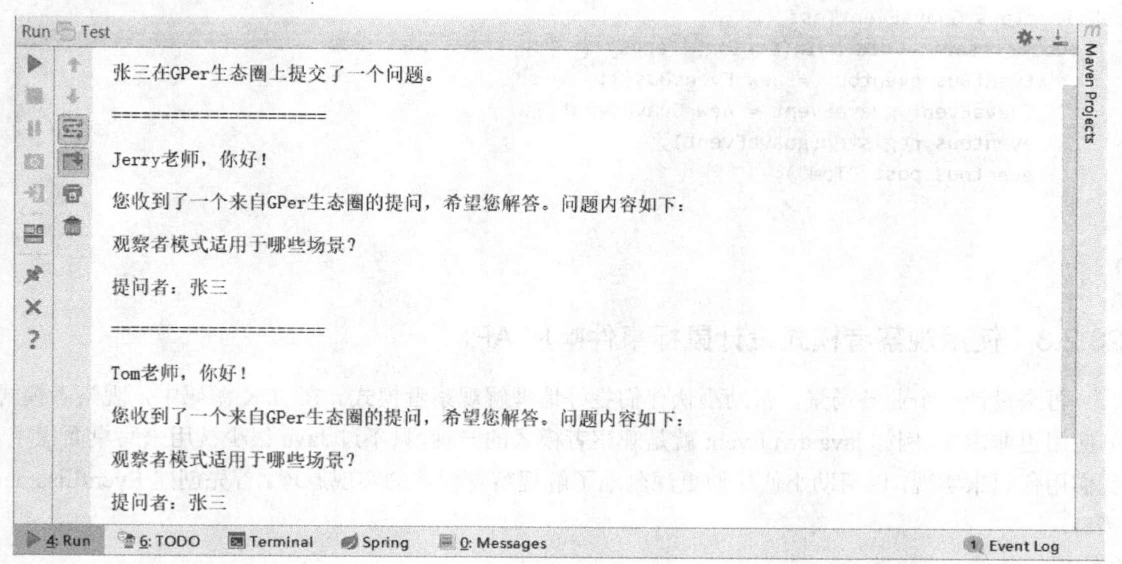

28.2.2 基于 Guava API 轻松落地观察者模式

笔者向大家推荐一个实现观察者模式的非常好用的框架，API 使用也非常简单，举个例子，首先引入 Maven 依赖包。

```
<dependency>
    <groupId>com.google.guava</groupId>
    <artifactId>guava</artifactId>
    <version>20.0</version>
</dependency>
```

然后创建侦听事件 GuavaEvent。

```
public class GuavaEvent {
    @Subscribe
    public void subscribe(String str){
        //业务逻辑
        System.out.println("执行 subscribe 方法,传入的参数是:" + str);
    }

}
```

最后编写客户端测试代码。

```
public class GuavaEventTest {
    public static void main(String[] args) {
        EventBus eventbus = new EventBus();
        GuavaEvent guavaEvent = new GuavaEvent();
        eventbus.register(guavaEvent);
        eventbus.post("Tom");
    }

}
```

28.2.3 使用观察者模式设计鼠标事件响应 API

再来设计一个业务场景，帮助小伙伴们更好地理解观察者模式。在 JDK 源码中，观察者模式的应用也非常多。例如 java.awt.Event 就是观察者模式的一种，只不过 Java 很少被用来写桌面程序。我们用代码来实现，以帮助小伙伴们更深刻地了解观察者模式的实现原理。首先创建 EventListener 接口。

```
/**
 * 观察者抽象
 */
```

```
public interface EventListener {

}
```

创建 Event 类。

```
/**
 * 标准事件源格式的定义
 */
public class Event {
    //事件源，动作是由谁发出的
    private Object source;
    //事件触发，要通知谁（观察者）
    private EventListener target;
    //观察者的回应
    private Method callback;
    //事件的名称
    private String trigger;
    //事件的触发事件
    private long time;

    public Event(EventListener target, Method callback) {
        this.target = target;
        this.callback = callback;
    }

    public Object getSource() {
        return source;
    }

    public Event setSource(Object source) {
        this.source = source;
        return this;
    }

    public String getTrigger() {
        return trigger;
    }

    public Event setTrigger(String trigger) {
        this.trigger = trigger;
        return this;
    }

    public long getTime() {
        return time;
    }
```

```
    public Event setTime(long time) {
        this.time = time;
        return this;
    }

    public Method getCallback() {
        return callback;
    }

    public EventListener getTarget() {
        return target;
    }

    @Override
    public String toString() {
        return "Event{" +
                "source=" + source +
                ", target=" + target +
                ", callback=" + callback +
                ", trigger='" + trigger + '\'' +
                ", time=" + time +
                '}';
    }
}
```

创建 EventContext 类。

```
/**
 * 被观察者的抽象
 */
public abstract class EventContext {
    protected Map<String,Event> events = new HashMap<String,Event>();

    public void addListener(String eventType, EventListener target, Method callback){
        events.put(eventType,new Event(target,callback));
    }

    public void addListener(String eventType, EventListener target){
        try {
            this.addListener(eventType, target,
target.getClass().getMethod("on"+toUpperFirstCase(eventType), Event.class));
        }catch (NoSuchMethodException e){
            return;
        }
    }
```

```
    private String toUpperFirstCase(String eventType) {
        char [] chars = eventType.toCharArray();
        chars[0] -= 32;
        return String.valueOf(chars);
    }

    private void trigger(Event event){
        event.setSource(this);
        event.setTime(System.currentTimeMillis());

        try {
            if (event.getCallback() != null) {
                //用反射调用回调函数
                event.getCallback().invoke(event.getTarget(), event);
            }
        }catch (Exception e){
            e.printStackTrace();
        }
    }

    protected void trigger(String trigger){
        if(!this.events.containsKey(trigger)){return;}
        trigger(this.events.get(trigger).setTrigger(trigger));
    }
}
```

然后创建 MouseEventType 接口。

```
public interface MouseEventType {
    //单击
    String ON_CLICK = "click";

    //双击
    String ON_DOUBLE_CLICK = "doubleClick";

    //弹起
    String ON_UP = "up";

    //按下
    String ON_DOWN = "down";

    //移动
    String ON_MOVE = "move";

    //滚动
    String ON_WHEEL = "wheel";
```

```
    //悬停
    String ON_OVER = "over";

    //失去焦点
    String ON_BLUR = "blur";

    //获得焦点
    String ON_FOCUS = "focus";
}
```

创建 Mouse 类。

```
/**
 * 具体的被观察者
 */
public class Mouse extends EventContext {

    public void click(){
        System.out.println("调用单击方法");
        this.trigger(MouseEventType.ON_CLICK);
    }

    public void doubleClick(){
        System.out.println("调用双击方法");
        this.trigger(MouseEventType.ON_DOUBLE_CLICK);
    }

    public void up(){
        System.out.println("调用弹起方法");
        this.trigger(MouseEventType.ON_UP);
    }

    public void down(){
        System.out.println("调用按下方法");
        this.trigger(MouseEventType.ON_DOWN);
    }

    public void move(){
        System.out.println("调用移动方法");
        this.trigger(MouseEventType.ON_MOVE);
    }

    public void wheel(){
        System.out.println("调用滚动方法");
```

```
        this.trigger(MouseEventType.ON_WHEEL);
    }

    public void over(){
        System.out.println("调用悬停方法");
        this.trigger(MouseEventType.ON_OVER);
    }

    public void blur(){
        System.out.println("调用获得焦点方法");
        this.trigger(MouseEventType.ON_BLUR);
    }

    public void focus(){
        System.out.println("调用失去焦点方法");
        this.trigger(MouseEventType.ON_FOCUS);
    }
}
```

创建回调方法 MouseEventLisenter 类。

```
/**
 * 观察者
 */
public class MouseEventListener implements EventListener {


    public void onClick(Event e){
        System.out.println("===========触发鼠标单击事件==========" + "\n" + e);
    }

    public void onDoubleClick(Event e){
        System.out.println("===========触发鼠标双击事件==========" + "\n" + e);
    }

    public void onUp(Event e){
        System.out.println("===========触发鼠标弹起事件==========" + "\n" + e);
    }

    public void onDown(Event e){
        System.out.println("===========触发鼠标按下事件==========" + "\n" + e);
    }

    public void onMove(Event e){
```

```
        System.out.println("===========触发鼠标移动事件==========" + "\n" + e);
    }

    public void onWheel(Event e){
        System.out.println("===========触发鼠标滚动事件==========" + "\n" + e);
    }

    public void onOver(Event e){
        System.out.println("===========触发鼠标悬停事件==========" + "\n" + e);
    }

    public void onBlur(Event e){
        System.out.println("===========触发鼠标失去焦点事件==========" + "\n" + e);
    }

    public void onFocus(Event e){
        System.out.println("===========触发鼠标获得焦点事件==========" + "\n" + e);
    }

}
```

最后编写客户端测试代码。

```
    public static void main(String[] args) {
        EventListener listener = new MouseEventListener();

        Mouse mouse = new Mouse();
        mouse.addListener(MouseEventType.ON_CLICK,listener);
        mouse.addListener(MouseEventType.ON_MOVE,listener);

        mouse.click();
        mouse.move();
    }
```

28.3 观察者模式在 Spring 源码中的应用

Spring 中的 ContextLoaderListener 实现了 ServletContextListener 接口，ServletContextListener 接口又继承了 EventListener，在 JDK 中，EventListener 有非常广泛的应用。ContextLoaderListener 的源码如下。

```
package org.springframework.web.context;

import javax.servlet.ServletContextEvent;
```

```
import javax.servlet.ServletContextListener;

public class ContextLoaderListener extends ContextLoader implements ServletContextListener {
    public ContextLoaderListener() {
    }

    public ContextLoaderListener(WebApplicationContext context) {
        super(context);
    }

    public void contextInitialized(ServletContextEvent event) {
        this.initWebApplicationContext(event.getServletContext());
    }

    public void contextDestroyed(ServletContextEvent event) {
        this.closeWebApplicationContext(event.getServletContext());
        ContextCleanupListener.cleanupAttributes(event.getServletContext());
    }
}
```

ServletContextListener 接口的源码如下。

```
package javax.servlet;
import java.util.EventListener;
public interface ServletContextListener extends EventListener {
    public void contextInitialized(ServletContextEvent sce);
    public void contextDestroyed(ServletContextEvent sce);
}
```

EventListener 接口的源码如下。

```
package java.util;
public interface EventListener {
}
```

28.4　观察者模式扩展

28.4.1　观察者模式的优点

（1）观察者和被观察者是松耦合（抽象耦合）的，符合依赖倒置原则。

（2）分离了表示层（观察者）和数据逻辑层（被观察者），并且建立了一套触发机制，使得数据的变化可以响应到多个表示层上。

（3）实现了一对多的通信机制，支持事件注册机制，支持兴趣分发机制，当被观察者触发事件时，只有感兴趣的观察者可以接收到通知。

28.4.2 观察者模式的缺点

（1）如果观察者数量过多，则事件通知会耗时较长。

（2）事件通知呈线性关系，如果其中一个观察者处理事件卡壳，则会影响后续的观察者接收该事件。

（3）如果观察者和被观察者之间存在循环依赖，则可能造成两者之间的循环调用，导致系统崩溃。

29

第 29 章 访问者模式

29.1 访问者模式概述

29.1.1 访问者模式的定义

访问者模式（Visitor Pattern）是一种将数据结构与数据操作分离的设计模式，指封装一些作用于某种数据结构中的各元素的操作，可以在不改变数据结构的前提下定义作用于这些元素的新的操作，属于行为型设计模式。

原文： Represent an operation to be performed on the elements of an object structure. Visitor lets you define a new operation without changing the classes of the elements on which it operates.

访问者模式又被称为最复杂的设计模式，并且使用频率不高，《设计模式》的作者这样评价：在大多数情况下，你不需要使用访问者模式，但是一旦需要使用它，那就是真的需要使用了。

访问者模式的基本思想是，针对系统中拥有固定类型数的对象结构（元素），在其内提供一个 accept()方法用来接收访问者对象的访问。不同的访问者对同一元素的访问内容不同，使得相同的元素集合可以产生不同的数据结果。accept()方法可以接收不同的访问者对象，然后在内部将自己（元素）转发到接收到的访问者对象的 visit()方法内。访问者内部对应类型的 visit()方法就会得到回调执行，对元素进行操作。也就是通过两次动态分发（第一次是对访问者分发 accept()方法，第二次是对元素分发 visit()方法），最终将一个具体的元素传递到一个具体的访问者。如此一来，就解耦了数据结构与数据操作，且数据操作不会改变元素状态。

访问者模式的核心是解耦数据结构与数据操作，使得对元素的操作具备优秀的扩展性。可以通过扩展不同的数据操作类型（访问者）实现对相同元素集的不同的操作。

29.1.2　访问者模式的应用场景

访问者模式在生活场景中的应用是非常多的，例如每年年底的 KPI 考核，KPI 考核标准是相对稳定的，但是参与 KPI 考核的员工可能每年都会发生变化，那么员工就是访问者。我们平时去食堂或者餐厅吃饭，餐厅的菜单和就餐方式是相对稳定的，但是去餐厅就餐的人员是每天都在发生变化的，因此就餐人员就是访问者。

当系统中存在类型数量稳定（固定）的一类数据结构时，可以通过访问者模式方便地实现对该类型所有数据结构的不同操作，而又不会对数据产生任何副作用（脏数据）。

简而言之，就是当对集合中的不同类型数据（类型数量稳定）进行多种操作时，使用访问者模式。访问者模式主要适用于以下应用场景。

（1）数据结构稳定，作用于数据结构的操作经常变化的场景。

（2）需要数据结构与数据操作分离的场景。

（3）需要对不同数据类型（元素）进行操作，而不使用分支判断具体类型的场景。

29.1.3　访问者模式的 UML 类图

访问者模式的 UML 类图如下。

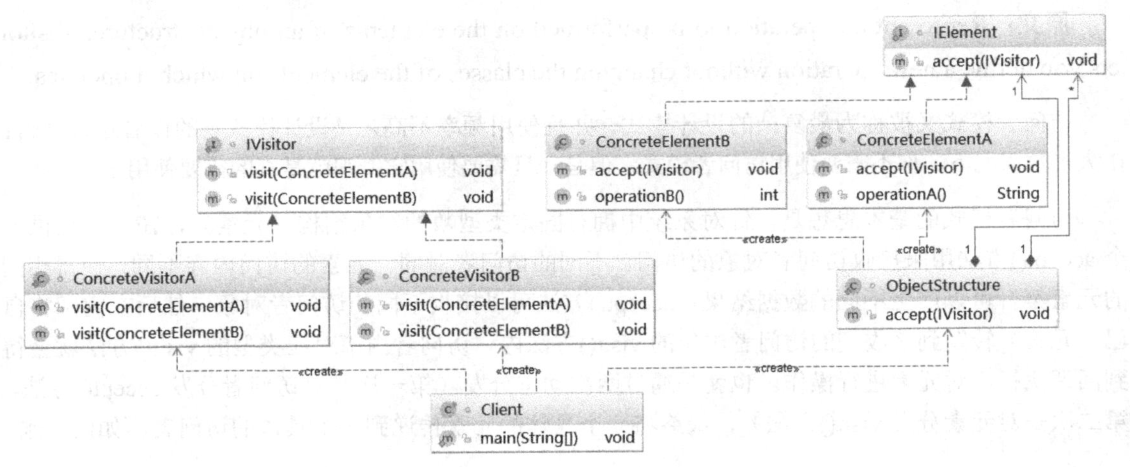

由上图可以看到，访问者模式主要包含 5 个角色。

（1）抽象访问者（IVisitor）：接口或抽象类，该类定义了一个 visit()方法用于访问每一个具体的元素，其参数就是具体的元素对象。从理论上来说，IVisitor 的方法个数与元素个数是相等的。如果元素个数经常变动，则导致 IVisitor 的方法也要进行变动，此时，该情形并不适用于访问者模式。

（2）具体访问者（ConcreteVisitor）：实现对具体元素的操作。

（3）抽象元素（IElement）：接口或抽象类，定义了一个接受访问者访问的方法 accept()，表示所有元素类型都支持被访问者访问。

（4）具体元素（ConcreteElement）：具体元素类型，提供接受访问者的具体实现。通常的实现都为 visitor.visit(this)。

（5）结构对象（ObjectStructure）：该类内部维护了元素集合，并提供方法接受访问者对该集合所有元素进行操作。

29.1.4　访问者模式的通用写法

以下是访问者模式的通用写法。

```
class Client {
    public static void main(String[] args) {
        ObjectStructure collection = new ObjectStructure();
        System.out.println("ConcreteVisitorA handle elements:");
        IVisitor visitorA = new ConcreteVisitorA();
        collection.accept(visitorA);
        System.out.println("------------------------------------");
        System.out.println("ConcreteVisitorB handle elements:");
        IVisitor visitorB = new ConcreteVisitorB();
        collection.accept(visitorB);

    }

    //抽象元素
    interface IElement {
        void accept(IVisitor visitor);
    }

    //具体元素 A
    static class ConcreteElementA implements IElement {
```

```
        public void accept(IVisitor visitor) {
            visitor.visit(this);
        }

        public String operationA() {
            return this.getClass().getSimpleName();
        }
    }

    //具体元素 B
    static class ConcreteElementB implements IElement {

        public void accept(IVisitor visitor) {
            visitor.visit(this);
        }

        public int operationB() {
            return new Random().nextInt(100);
        }
    }

    //抽象访问者
    interface IVisitor {
        void visit(ConcreteElementA element);

        void visit(ConcreteElementB element);
    }

    //具体访问者 A
    static class ConcreteVisitorA implements IVisitor {

        public void visit(ConcreteElementA element) {
            String result = element.operationA();
            System.out.println("result from " + element.getClass().getSimpleName() + ": " + result);
        }


        public void visit(ConcreteElementB element) {
            int result = element.operationB();
            System.out.println("result from " + element.getClass().getSimpleName() + ": " + result);
        }
    }

    //具体访问者 B
    static class ConcreteVisitorB implements IVisitor {

        public void visit(ConcreteElementA element) {
```

```
            String result = element.operationA();
            System.out.println("result from " + element.getClass().getSimpleName() + ": " + result);
        }

        public void visit(ConcreteElementB element) {
            int result = element.operationB();
            System.out.println("result from " + element.getClass().getSimpleName() + ": " + result);
        }
    }

    //结构对象
    static class ObjectStructure {
        private List<IElement> list = new ArrayList<IElement>();

        {
            this.list.add(new ConcreteElementA());
            this.list.add(new ConcreteElementB());
        }

        public void accept(IVisitor visitor) {
            for (IElement element : this.list) {
                element.accept(visitor);
            }
        }
    }
}
```

29.2　使用访问者模式解决实际问题

29.2.1　使用访问者模式实现 KPI 考核的场景

每到年底，管理层就要开始评定员工一年的工作绩效，员工分为工程师和经理；管理层有 CEO 和 CTO。CTO 关注工程师的代码量、经理的新产品数量；CEO 关注工程师的 KPI、经理的 KPI 及新产品数量。

由于 CEO 和 CTO 对不同的员工的关注点是不一样的，这就需要对不同的员工类型进行不同的处理。此时，访问者模式可以派上用场了，来看代码。

```
//员工基类
public abstract class Employee {

    public String name;
    public int kpi;//员工 KPI
```

```
    public Employee(String name) {
        this.name = name;
        kpi = new Random().nextInt(10);
    }
    //核心方法，接受访问者的访问
    public abstract void accept(IVisitor visitor);
}
```

Employee 类定义了员工基本信息及一个 accept()方法，accept()方法表示接受访问者的访问，由具体的子类来实现。访问者是一个接口，传入不同的实现类，可访问不同的数据。下面看工程师 Engineer 类的代码。

```
//工程师
public class Engineer extends Employee {

    public Engineer(String name) {
        super(name);
    }

    @Override
    public void accept(IVisitor visitor) {
        visitor.visit(this);
    }
    //工程师一年的代码量
    public int getCodeLines() {
        return new Random().nextInt(10 * 10000);
    }
}
```

经理 Manager 类的代码如下。

```
//经理
public class Manager extends Employee {

    public Manager(String name) {
        super(name);
    }

    @Override
    public void accept(IVisitor visitor) {
        visitor.visit(this);
    }
    //一年做的新产品数量
    public int getProducts() {
        return new Random().nextInt(10);
    }
}
```

工程师被考核的是代码量，经理被考核的是新产品数量，二者的职责不一样。也正是因为有这样的差异性，才使得访问者模式能够在这个场景下发挥作用。Employee、Engineer、Manager 3 个类型相当于数据结构，这些类型相对稳定，不会发生变化。

将这些员工添加到一个业务报表类中，公司高层可以通过该报表类的 showReport()方法查看所有员工的业绩，代码如下。

```
//员工业务报表类
public class BusinessReport {

    private List<Employee> employees = new LinkedList<Employee>();

    public BusinessReport() {
        employees.add(new Manager("经理-A"));
        employees.add(new Engineer("工程师-A"));
        employees.add(new Engineer("工程师-B"));
        employees.add(new Engineer("工程师-C"));
        employees.add(new Manager("经理-B"));
        employees.add(new Engineer("工程师-D"));
    }

    /**
     * 为访问者展示报表
     * @param visitor 公司高层，如 CEO、CTO
     */
    public void showReport(IVisitor visitor) {
        for (Employee employee : employees) {
            employee.accept(visitor);
        }
    }
}
```

下面来看访问者类型的定义，访问者声明了两个 visit()方法，分别对工程师和经理访问，代码如下。

```
public interface IVisitor {

    //访问工程师类型
    void visit(Engineer engineer);

    //访问经理类型
    void visit(Manager manager);
}
```

上面代码定义了一个 IVisitor 接口，该接口有两个 visit()方法，参数分别是 Engineer 和 Manager，也就是说对于 Engineer 和 Manager 的访问会调用两个不同的方法，以此达到差异化处理的目的。这两个访问者具体的实现类为 CEOVisitor 类和 CTOVisitor 类。首先来看 CEOVisitor 类的代码。

```
//CEO 访问者
public class CEOVisitor implements IVisitor {

    public void visit(Engineer engineer) {
        System.out.println("工程师: " + engineer.name + ", KPI: " + engineer.kpi);
    }

    public void visit(Manager manager) {
        System.out.println("经理: " + manager.name + ", KPI: " + manager.kpi +
                ", 新产品数量: " + manager.getProducts());
    }
}
```

在 CEO 的访问者中，CEO 关注工程师的 KPI、经理的 KPI 和新产品数量，通过两个 visit()方法分别进行处理。如果不使用访问者模式，只通过一个 visit()方法进行处理，则需要在这个 visit()方法中进行判断，然后分别处理，代码如下。

```
public class ReportUtil {
    public void visit(Employee employee) {
        if (employee instanceof Manager) {
            Manager manager = (Manager) employee;
            System.out.println("经理: " + manager.name + ", KPI: " + manager.kpi +
                    ", 新产品数量: " + manager.getProducts());
        } else if (employee instanceof Engineer) {
            Engineer engineer = (Engineer) employee;
            System.out.println("工程师: " + engineer.name + ", KPI: " + engineer.kpi);
        }
    }
}
```

这就导致了 if...else 逻辑的嵌套及类型的强制转换，难以扩展和维护，当类型较多时，这个 ReportUtil 就会很复杂。而使用访问者模式，通过同一个函数对不同的元素类型进行相应处理，使结构更加清晰、灵活性更高。

然后添加一个 CTO 的访问者类 CTOVisitor。

```
public class CTOVisitor implements IVisitor {

    public void visit(Engineer engineer) {
     System.out.println("工程师: " + engineer.name + ", 代码行数: " + engineer.getCodeLines());
    }
```

```
    public void visit(Manager manager) {
        System.out.println("经理: " + manager.name + ", 产品数量: " + manager.getProducts());
    }
}
```

重载的 visit()方法会对元素进行不同的操作，而通过注入不同的访问者又可以替换掉访问者的具体实现，使得对元素的操作变得更灵活，可扩展性更高，同时，消除了类型转换、if...else 等“丑陋”的代码。

客户端测试代码如下。

```
public static void main(String[] args) {
    //构建报表
    BusinessReport report = new BusinessReport();
    System.out.println("=========== CEO 看报表 ===========");
    report.showReport(new CEOVisitor());
    System.out.println("=========== CTO 看报表 ===========");
    report.showReport(new CTOVisitor());
}
```

运行结果如下图所示。

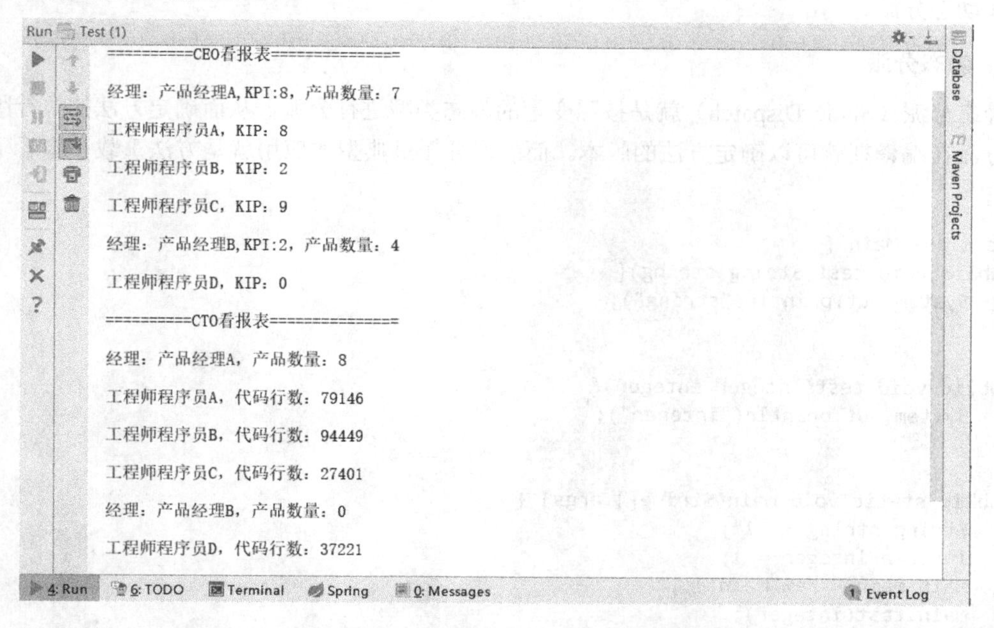

在上述案例中，Employee 扮演了 IElement 角色，Engineer 和 Manager 都是 ConcreteElement，CEOVisitor 和 CTOVisitor 都是具体的 IVisitor 对象，BusinessReport 就是 ObjectStructure。

访问者模式最大的优点就是增加访问者非常容易，从代码中可以看到，如果要增加一个访问者，则只需要新实现一个访问者接口的类，从而达到数据对象与数据操作相分离的效果。如果不使用访问者模式，而又不想对不同的元素进行不同的操作，则必定需要使用 if...else 和类型转换，这使得代码难以升级维护。

我们要根据具体情况来评估是否适合使用访问者模式。例如，对象结构是否足够稳定，是否需要经常定义新的操作，使用访问者模式是否能优化代码，而不使代码变得更复杂。

29.2.2 从静态分派到动态分派

变量被声明时的类型叫作变量的静态类型（Static Type），有些人又把静态类型叫作明显类型（Apparent Type）；而变量所引用的对象的真实类型又叫作变量的实际类型（Actual Type）。比如：

```
List list = null;
list = new ArrayList();
```

上面代码声明了一个变量 list，它的静态类型（也叫作明显类型）是 List，而它的实际类型是 ArrayList。根据对象的类型对方法进行的选择，就是分派（Dispatch）。分派又分为两种，即静态分派和动态分派。

1. 静态分派

静态分派（Static Dispatch）就是按照变量的静态类型进行分派，从而确定方法的执行版本，静态分派在编译期就可以确定方法的版本。而静态分派最典型的应用就是方法重载，来看下面的代码。

```
public class Main {
    public void test(String string){
        System.out.println("string");
    }

    public void test(Integer integer){
        System.out.println("integer");
    }

    public static void main(String[] args) {
        String string = "1";
        Integer integer = 1;
        Main main = new Main();
        main.test(integer);
        main.test(string);
    }
}
```

在静态分派判断的时候，根据多个判断依据（即参数类型和个数）判断出方法的版本，这就是多分派的概念，因为我们有一个以上的考量标准，所以 Java 是静态多分派的语言。

2. 动态分派

对于动态分派，与静态分派相反，它不是在编译期确定的方法版本，而是在运行时才能确定的。而动态分派最典型的应用就是多态的特性。举个例子，来看下面的代码。

```
interface Person{
    void test();
}
class Man implements Person{
    public void test(){
        System.out.println("男人");
    }
}
class Woman implements Person{
    public void test(){
        System.out.println("女人");
    }
}
public class Main {
    public static void main(String[] args) {
        Person man = new Man();
        Person woman = new Woman();
        man.test();
        woman.test();
    }
}
```

这段代码的输出结果为依次打印男人和女人，然而这里的 test()方法版本，无法根据 Man 和 Woman 的静态类型判断，它们的静态类型都是 Person 接口，根本无从判断。

显然，产生这样的输出结果，就是因为 test()方法的版本是在运行时判断的，这就是动态分派。

动态分派判断的方法是在运行时获取 Man 和 Woman 的实际引用类型，再确定方法的版本，而由于此时判断的依据只是实际引用类型，只有一个判断依据，所以这就是单分派的概念，这时考量标准只有一个，即变量的实际引用类型。相应地，这说明 Java 是动态单分派的语言。

29.2.3　访问者模式中的伪动态分派

通过前面的分析，我们知道 Java 是静态多分派、动态单分派的语言。Java 底层不支持动态双分派。但是通过使用设计模式，也可以在 Java 里实现伪动态双分派。在访问者模式中使用的就是伪动态双分派。所谓动态双分派就是在运行时依据两个实际类型去判断一个方法的运行行为，而访问者模式实现的手段是进行两次动态单分派来达到这个效果。

还是回到前面的 KPI 考核业务场景中，BusinessReport 类中的 showReport()方法的代码如下。

```
public void showReport(IVisitor visitor) {
    for (Employee employee : employees) {
        employee.accept(visitor);
    }
}
```

这里依据 Employee 和 IVisitor 两个实际类型决定了 showReport()方法的执行结果，从而决定了 accept()方法的动作。

accept()方法的调用过程分析如下。

（1）当调用 accept()方法时，根据 Employee 的实际类型决定是调用 Engineer 还是 Manager 的 accept()方法。

（2）这时 accept()方法的版本已经确定，假如是 Engineer，则它的 accept()方法调用下面这行代码。

```
public void accept(IVisitor visitor) {
    visitor.visit(this);
}
```

此时的 this 是 Engineer 类型，因此对应的是 IVisitor 接口的 visit(Engineer engineer)方法，此时需要再根据访问者的实际类型确定 visit()方法的版本，如此一来，就完成了动态双分派的过程。

以上过程通过两次动态双分派，第一次对 accept()方法进行动态分派，第二次对访问者的 visit()方法进行动态分派，从而达到根据两个实际类型确定一个方法的行为的效果。

而原本的做法通常是传入一个接口，直接使用该接口的方法，此为动态单分派，就像策略模式一样。在这里，showReport()方法传入的访问者接口并不是直接调用自己的 visit()方法，而是通过 Employee 的实际类型先动态分派一次，然后在分派后确定的方法版本里进行自己的动态分派。

注：这里确定 accept(IVisitor visitor)方法是由静态分派决定的，所以这个并不在此次动态双分派的范畴内，而且静态分派是在编译期完成的，所以 accept(IVisitor visitor)方法的静态分派与访问者模式的动态双分派并没有任何关系。动态双分派说到底还是动态分派，是在运行时发生的，它与静态分派有着本质上的区别，不可以说一次动态分派加一次静态分派就是动态双分派，而且访问者模式的双分派本身也另有所指。

而 this 的类型不是动态分派确定的，把它写在哪个类中，它的静态类型就是哪个类，这是在编译期就确定的，不确定的是它的实际类型，请小伙伴们也要区分开来。

29.3　访问者模式在框架源码中的应用

29.3.1　访问者模式在 JDK 源码中的应用

首先来看 JDK 的 NIO 模块下的 FileVisitor 接口，它提供了递归遍历文件树的支持。这个接口上的方法表示了遍历过程中的关键过程，允许在文件被访问、目录将被访问、目录已被访问、发生错误等过程中进行控制。换句话说，这个接口在文件被访问前、访问中和访问后，以及产生错误的时候都有相应的钩子程序进行处理。

调用 FileVisitor 中的方法，会返回访问结果的 FileVisitResult 对象值，用于决定当前操作完成后接下来该如何处理。FileVisitResult 的标准返回值存放在 FileVisitResult 枚举类型中，代码如下。

```
public interface FileVisitor<T> {

    FileVisitResult preVisitDirectory(T dir, BasicFileAttributes attrs)
        throws IOException;

    FileVisitResult visitFile(T file, BasicFileAttributes attrs)
        throws IOException;

    FileVisitResult visitFileFailed(T file, IOException exc)
        throws IOException;

    FileVisitResult postVisitDirectory(T dir, IOException exc)
        throws IOException;
}
```

（1）FileVisitResult.CONTINUE：这个访问结果表示当前的遍历过程将会继续。

（2）FileVisitResult.SKIP_SIBLINGS：这个访问结果表示当前的遍历过程将会继续，但是要忽略当前文件/目录的兄弟节点。

（3）FileVisitResult.SKIP_SUBTREE：这个访问结果表示当前的遍历过程将会继续，但是要忽略当前目录下的所有节点。

（4）FileVisitResult.TERMINATE：这个访问结果表示当前的遍历过程将会停止。

通过访问者去遍历文件树会比较方便，比如查找文件夹内符合某个条件的文件或者某一天内所创建的文件，这个类中都提供了相对应的方法。它的实现其实也非常简单，代码如下。

```
public class SimpleFileVisitor<T> implements FileVisitor<T> {
    protected SimpleFileVisitor() {
    }

    @Override
    public FileVisitResult preVisitDirectory(T dir, BasicFileAttributes attrs)
        throws IOException
    {
        Objects.requireNonNull(dir);
        Objects.requireNonNull(attrs);
        return FileVisitResult.CONTINUE;
    }

    @Override
    public FileVisitResult visitFile(T file, BasicFileAttributes attrs)
        throws IOException
    {
        Objects.requireNonNull(file);
        Objects.requireNonNull(attrs);
        return FileVisitResult.CONTINUE;
    }

    @Override
    public FileVisitResult visitFileFailed(T file, IOException exc)
        throws IOException
    {
        Objects.requireNonNull(file);
        throw exc;
    }

    @Override
    public FileVisitResult postVisitDirectory(T dir, IOException exc)
        throws IOException
    {
        Objects.requireNonNull(dir);
        if (exc != null)
            throw exc;
        return FileVisitResult.CONTINUE;
    }
}
```

29.3.2 访问者模式在 Spring 源码中的应用

再来看访问者模式在 Spring 中的应用，Spring IoC 中有个 BeanDefinitionVisitor 类，其中有一

个 visitBeanDefinition()方法，源码如下。

```
public class BeanDefinitionVisitor {

    @Nullable
    private StringValueResolver valueResolver;


    public BeanDefinitionVisitor(StringValueResolver valueResolver) {
        Assert.notNull(valueResolver, "StringValueResolver must not be null");
        this.valueResolver = valueResolver;
    }

    protected BeanDefinitionVisitor() {
    }

    public void visitBeanDefinition(BeanDefinition beanDefinition) {
        visitParentName(beanDefinition);
        visitBeanClassName(beanDefinition);
        visitFactoryBeanName(beanDefinition);
        visitFactoryMethodName(beanDefinition);
        visitScope(beanDefinition);
        if (beanDefinition.hasPropertyValues()) {
            visitPropertyValues(beanDefinition.getPropertyValues());
        }
        if (beanDefinition.hasConstructorArgumentValues()) {
            ConstructorArgumentValues cas = beanDefinition.getConstructorArgumentValues();
            visitIndexedArgumentValues(cas.getIndexedArgumentValues());
            visitGenericArgumentValues(cas.getGenericArgumentValues());
        }
    }
    ...
}
```

我们看到，在 visitBeanDefinition()方法中，访问了其他数据，比如父类的名字、自己的类名、在 IoC 容器中的名称等各种信息。

29.4　访问者模式扩展

29.4.1　访问者模式的优点

（1）解耦了数据结构与数据操作，使得操作集合可以独立变化。

（2）可以通过扩展访问者角色，实现对数据集的不同操作，程序扩展性更好。

（3）元素具体类型并非单一，访问者均可操作。

（4）各角色职责分离，符合单一职责原则。

29.4.2 访问者模式的缺点

（1）无法增加元素类型：若系统数据结构对象易于变化，经常有新的数据对象增加进来，则访问者类必须增加对应元素类型的操作，违背了开闭原则。

（2）具体元素变更困难：具体元素增加属性、删除属性等操作会导致对应的访问者类需要进行相应的修改，尤其当有大量访问者类时，修改范围太大。

（3）违背依赖倒置原则：为了达到“区别对待”，访问者角色依赖的是具体元素类型，而不是抽象。

5

第 5 篇

设计模式总结篇

30

第 30 章 专治设计模式选择困难症

30.1 设计模式到底如何落地

多年以来，敏捷（Agile）开发一直成为很多小项目的首选方案。Agile 成了一个 Good Word，和所有类似命运的词（如 Democracy、Liberty、Republic 等）相似，它的理念和实践发生了分离。所以真实世界中的软件设计水平，就笔者看到的情况而言，并没有发生质的提升。大部分实际工作中的代码还是毫无设计，如同一盘纠缠在一起的“意大利面”。

就算“天才”级的人或者说非常适合 IT 行业的人都很难去理解那种随手代码（不遵守单一职责，随意增加代码函数以实现功能的代码）在日常应用中是怎么产生的。设计原则也好，设计模式也好，DDD 也好，抽象设计都是第一步。但不得不承认，要做到抽象和设计本身就是非常困难的，有可能大部分人就不具备抽象思考问题的能力。那有没有可能在工作实践中让大家学会抽象思考呢？很难，少数人可以学会，大部分人都做不到。更别说在项目中落地，则几乎不可能。即便是某些有意设计过的代码，应用到具体的项目中，如果没有非常明确的边界和维护的手段，也会很快“腐化”，这样的例子数不胜数。

当然，一个精英（专业）开发团队可以落地软件设计。Agile 是价值观驱动的方法论的集合。软件设计并不带来当期利益，而大部分项目的领导者只关心当前版本。在这一矛盾下，当大型软件“腐化”到大部分人都觉得它需要改造一下的时候，它的改动成本已经太大了。或者，系统性能恶化到非常严重的时候，此时已经不是简单改造、优化代码就能解决的问题了，而是可能需要升级软件架构才能解决的问题。往往已在运行的系统，只要发生改动就会引入故障，带来巨大风险。几百上千人的编程习惯、捉襟见肘的人力和不确定的结果都会让这种努力变得艰难而危险。就好比肝癌患者感到疼的时候，已经是晚期了。

当然，我们也不必如此悲观。某种程度上，我们也并不希望大型项目中的所有人都是全能手。价值观正确的精英团队，往往只需要有一个程序员在编码的时候，着重关注业务拆分的正确性，关心软件设计的合理性，同时关心性能。而其他 Coder（泛指软件开发人员）只需要全心全意地关心业务。每个人都应该专注于一个点，专业分工才是效率的源泉。可运行、可监控的架构设计就是来支撑这个目标的。架构不应该只落在文档和 PPT 中，再加上统一编程规范、技能培训这些都不够。架构应该可以约束分析和编码过程，提供合适的目录结构、依赖关系、拆分方式、实现模板、非业务优化方法和某些业务、代码分析工具。这样，才可能从根本上落地软件设计、减缓代码的腐化和性能的持续恶化。不要期望所有开发人员都是精英，但是团队中的架构师或者架构团队需要承担这个责任。对于大型项目来说（数百上千人），这倒是有可能实现的。

所以，笔者在此结合个人经验，给大家重新梳理一下架构师的职责。

（1）合理地将系统拆分为组件，组件的规模应该控制在 7~10 人可以维护，组件的边界应该非常清晰，不存在调用依赖，不存在共享数据，只有消息接口。在这个基础上，做好组件内部设计是软件设计落地的重要基石。

（2）组件内要按照基础服务、通用服务、业务逻辑进行分层。大多数时候我们没有进行分层，将所有逻辑都混杂在一起，这也是一般项目的常态，因此，这也是决定程序复杂性的根源。

（3）单一职责拆分原则，做业务和逻辑拆分时要尽量保证单个接口功能简单。

（4）依赖关系管理，应该设计一套机制自动生成接口，让实现层依赖这些接口。

（5）基础服务拆分，甚至可以将基础数据库（数据结构）某些业务常用的操作封装成通用 API。

（6）非业务优化，主要是内存优化和网络资源消耗优化。

（7）准备分析工具，主要是进行代码规范检测、代码质量评价（主要是对分层结果和调用链进行评价）和业务关系分析的工具。

因此，作为架构师或者普通开发人员都应该知道，设计模式不是为每个人准备的，而是基于业务来选择设计模式，需要时就能想到它。要明白一点，技术永远为业务服务，技术只是满足业务需要的一个工具。我们需要掌握每种设计模式的应用场景、特征、优缺点，以及每种设计模式的关联关系，就能够很好地满足日常业务的需要。

需要特别声明的是，在日常应用中，设计模式从来都不是单个设计模式独立使用的。在实际应用中，通常多个设计模式混合使用，你中有我，我中有你。下图完整地描述了设计模式之间的混用关系，希望对大家有所帮助。

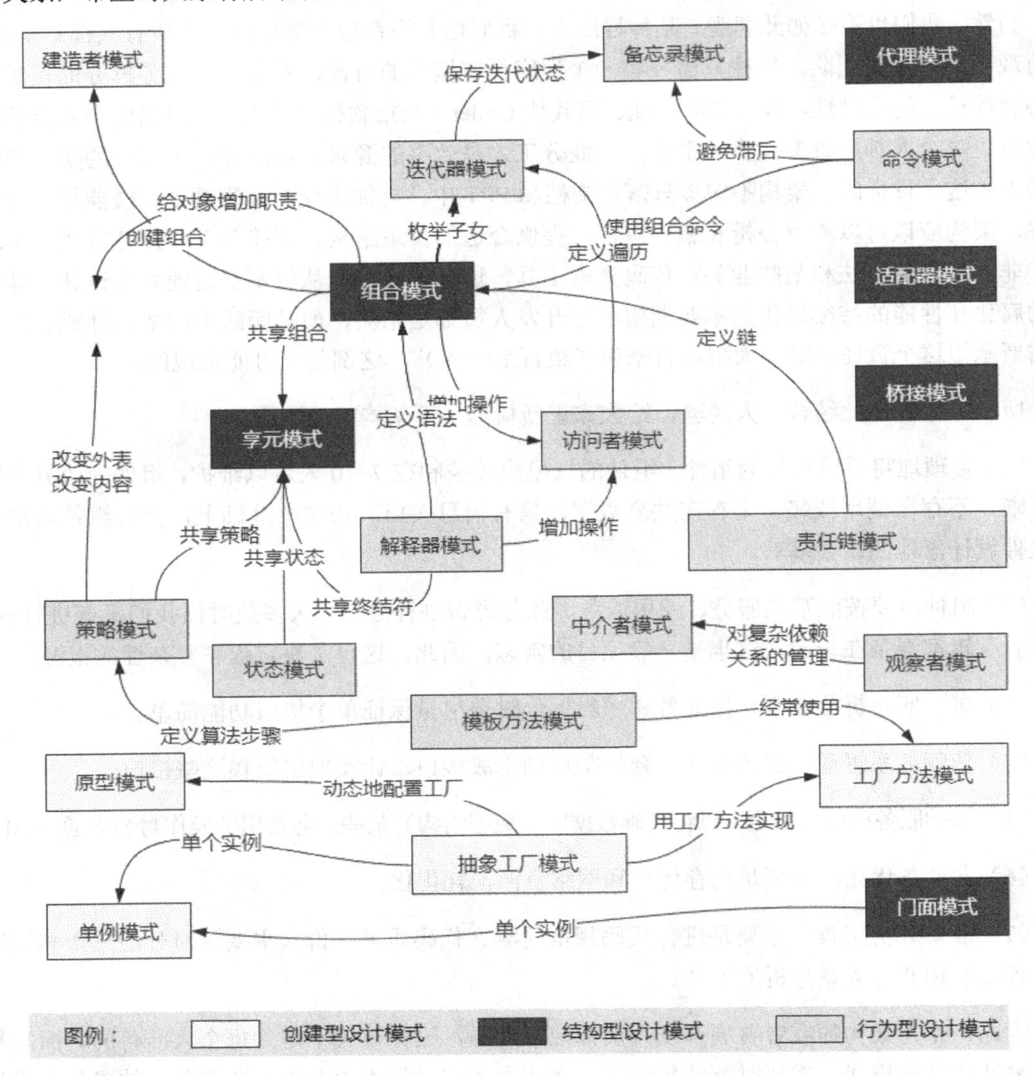

30.2　各种设计模式使用频率总结

以下是根据笔者的个人经验，对设计模式使用频率的总结，不可作为学术依据，仅供大家参考。因为设计模式的选择要依赖具体的业务场景，每个人接触的业务领域都不一样，自然设计模式的选择也会不一样。

30.2.1　创建型设计模式

如下图所示，创建型设计模式中使用频率由高到低依次为工厂方法模式、抽象工厂模式、建造者模式、单例模式、原型模式。原型模式一般都有现成的工具类，自己造轮子的情况比较少。

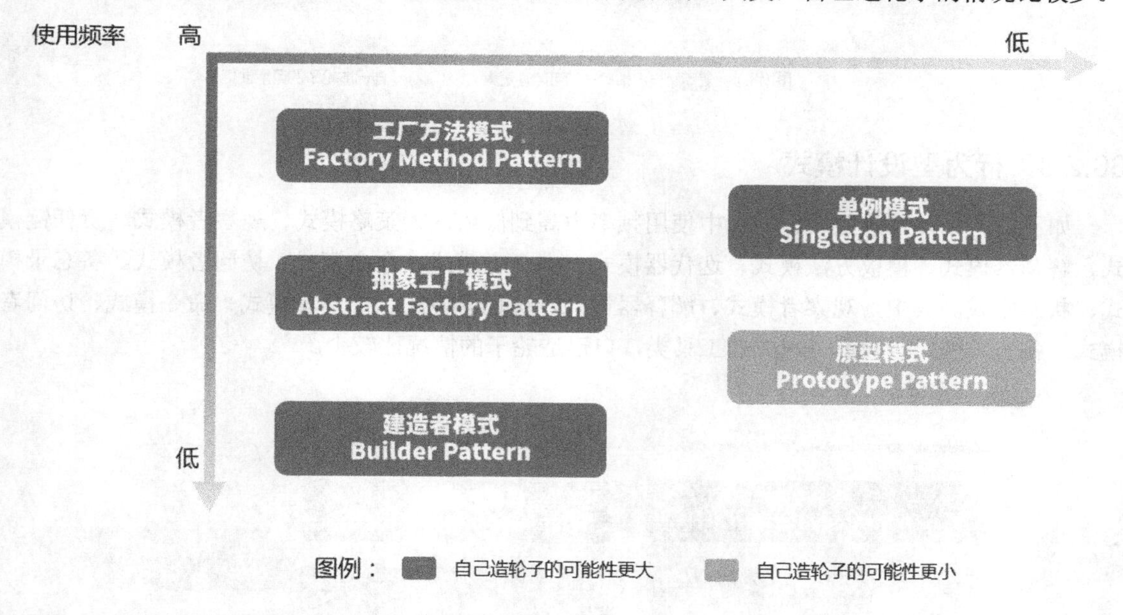

30.2.2　结构型设计模式

如下图所示，结构型设计模式中使用频率由高到低依次为适配器模式、装饰器模式、代理模式、门面模式、组合模式、享元模式、桥接模式。其中桥接模式一般都有现成的工具类，自己造轮子的情况比较少。

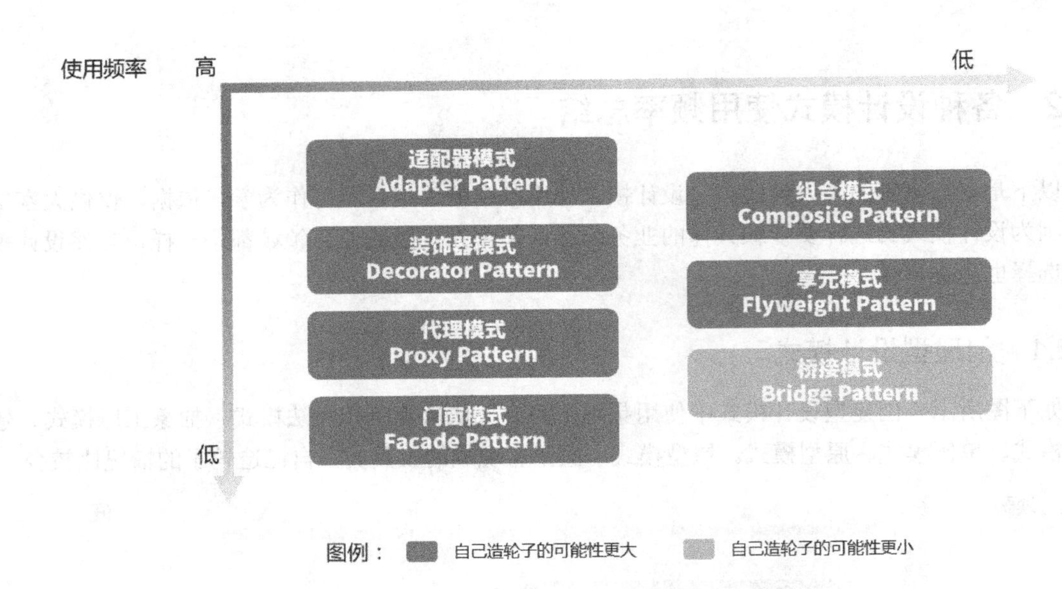

30.2.3 行为型设计模式

如下图所示，行为型设计模式中使用频率由高到低依次为策略模式、观察者模式、责任链模式、解释器模式、模板方法模式、迭代器模式、中介者模式、命令模式、访问者模式、备忘录模式、状态模式。其中，观察者模式、解释器模式、迭代器模式、中介者模式、命令模式、访问者模式、备忘录模式一般都有现成的工具类，自己造轮子的情况比较少。

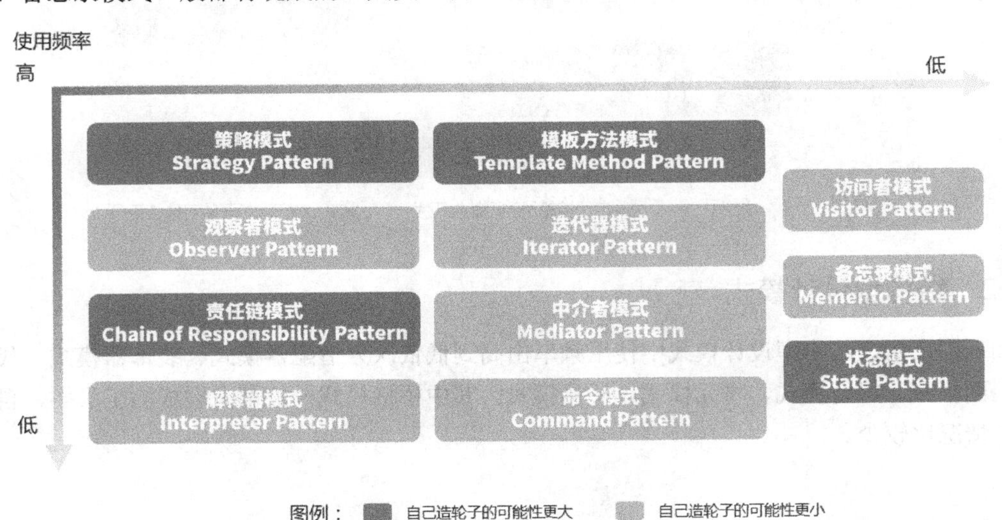

30.3　一句话归纳设计模式

各种设计模式对比及编程思想总结如下表所示。

设计模式	一句话归纳	目　的	生活案例	框架源码举例
工厂模式（Factory Pattern）	产品标准化，生产更高效	封装创建细节	实体工厂	LoggerFactory、Calender
单例模式（Singleton Pattern）	世上只有一个我	保证独一无二	CEO	BeanFactory、Runtime
原型模式（Prototype Pattern）	拔一根猴毛，吹出千万个	高效创建对象	克隆	ArrayList、PrototypeBean
建造者模式（Builder Pattern）	高配中配与低配，想选哪配就哪配	开放个性配置步骤	选配	StringBuilder、BeanDefinitionBuilder
代理模式（Proxy Pattern）	没有资源没时间，得找媒婆来帮忙	增强职责	媒婆	ProxyFactoryBean、JdkDynamicAopProxy、CglibAopProxy
门面模式（Facade Pattern）	打开一扇门，通向全世界	统一访问入口	前台	JdbcUtils、RequestFacade
装饰器模式（Decorator Pattern）	他大舅他二舅，都是他舅	灵活扩展、同宗同源	煎饼	BufferedReader、InputStream
享元模式（Flyweight Pattern）	优化资源配置，减少重复浪费	共享资源池	全国社保联网	String、Integer、ObjectPool
组合模式（Composite Pattern）	人在一起叫团伙，心在一起叫团队	统一整体和个体	组织架构树	HashMap、SqlNode
适配器模式（Adapter Pattern）	万能充电器	兼容转换	电源适配	AdvisorAdapter、HandlerAdapter
桥接模式（Bridge Pattern）	约定优于配置	不允许用继承	桥	DriverManager
委派模式（Delegate Pattern）	这个需求很简单，怎么实现我不管	只对结果负责	授权委托书	ClassLoader、BeanDefinitionParserDelegate
模板模式（Template Pattern）	流程全部标准化，需要微调请覆盖	逻辑复用	把大象装进冰箱	JdbcTemplate、HttpServlet
策略模式（Strategy Pattern）	条条大道通北京，具体哪条你来定	把选择权交给用户	选择支付方式	Comparator、InstantiationStrategy
责任链模式（Chain of Responsibility Pattern）	各人自扫门前雪，莫管他人瓦上霜	解耦处理逻辑	踢皮球	FilterChain、Pipeline
迭代器模式（Iterator Pattern）	流水线上坐一天，每个包裹扫一遍	统一对集合的访问方式	逐个检票进站	Iterator
命令模式（Command Pattern）	运筹帷幄之中，决胜千里之外	解耦请求和处理	遥控器	Runnable、TestCase

续表

设计模式	一句话归纳	目　的	生活案例	框架源码举例
状态模式（State Pattern）	状态驱动行为，行为决定状态	绑定状态和行为	订单状态跟踪	Lifecycle
备忘录（Memento Pattern）	失足不成千古恨，想重来时就重来	备份、后悔机制	草稿箱	StateManageableMessageContext
中介者（Mediator Pattern）	联系方式我给你，怎么搞定我不管	统一管理网状资源	朋友圈	Timer
解释器模式（Interpreter Pattern）	我想说“方言”，一切解释权归我	实现特定语法解析	摩斯密码	Pattern、ExpressionParser
观察者模式（Observer Pattern）	到点就通知我	解耦观察者与被观察者	闹钟	ContextLoaderListener
访问者模式（Visitor Pattern）	横看成岭侧成峰，远近高低各不同	解耦数据结构和数据操作	KPI考核	FileVisitor、BeanDefinitionVisitor

31

第 31 章 容易混淆的设计模式对比

既然都已经学到这里了，相信大家对设计模式已经有了全面透彻的了解。不管是面试也好，还是日常开发也好，相信大家都已经胸有成竹、信心满满了。但是从笔者的架构经验和教学经验总结来看，还有很多小伙伴对一些设计模式经常混淆难懂。本章内容可以说是本书最精华的部分。

本章笔者收集了很多来自学员的疑问，对各种容易混淆的设计模式进行比较，并总结整理了以下内容，希望帮助大家在以后的设计选型中能够披荆斩棘，如履平地。如果你在阅读本书之前，对设计模式较为熟悉，本章内容可以帮助你巩固加深理解。

31.1 创建型设计模式对比

31.1.1 工厂方法模式与抽象工厂模式对比

共同点	1. 都属于创建型设计模式 2. 职责相同
不同点	所创建产品的扩展程度不同：工厂方法模式只能单向维度扩展产品；而抽象工厂模式可以让产品等级结构和产品族的相互扩展，而且可以多维度扩展（至少是二维扩展）
关联性	抽象工厂很多时候不一定是接口，而是抽象类；工厂方法类一般作为抽象工厂类的子类

续表

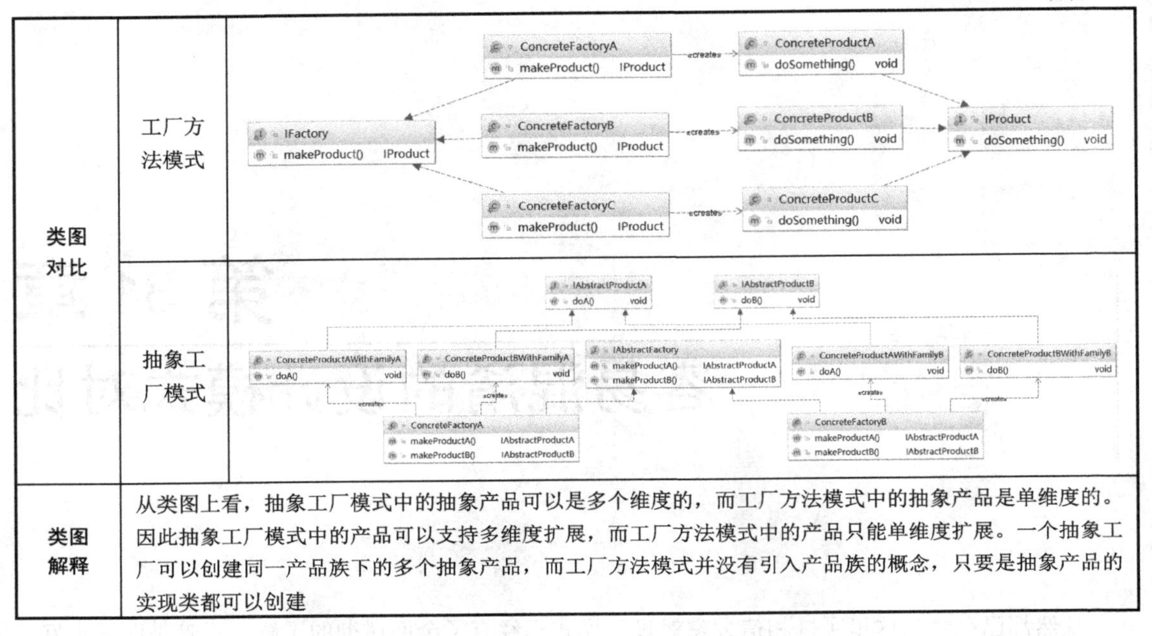

类图对比	工厂方法模式	（类图）
	抽象工厂模式	（类图）
类图解释	从类图上看，抽象工厂模式中的抽象产品可以是多个维度的，而工厂方法模式中的抽象产品是单维度的。因此抽象工厂模式中的产品可以支持多维度扩展，而工厂方法模式中的产品只能单维度扩展。一个抽象工厂可以创建同一产品族下的多个抽象产品，而工厂方法模式并没有引入产品族的概念，只要是抽象产品的实现类都可以创建	

31.1.2 简单工厂模式与单例模式对比

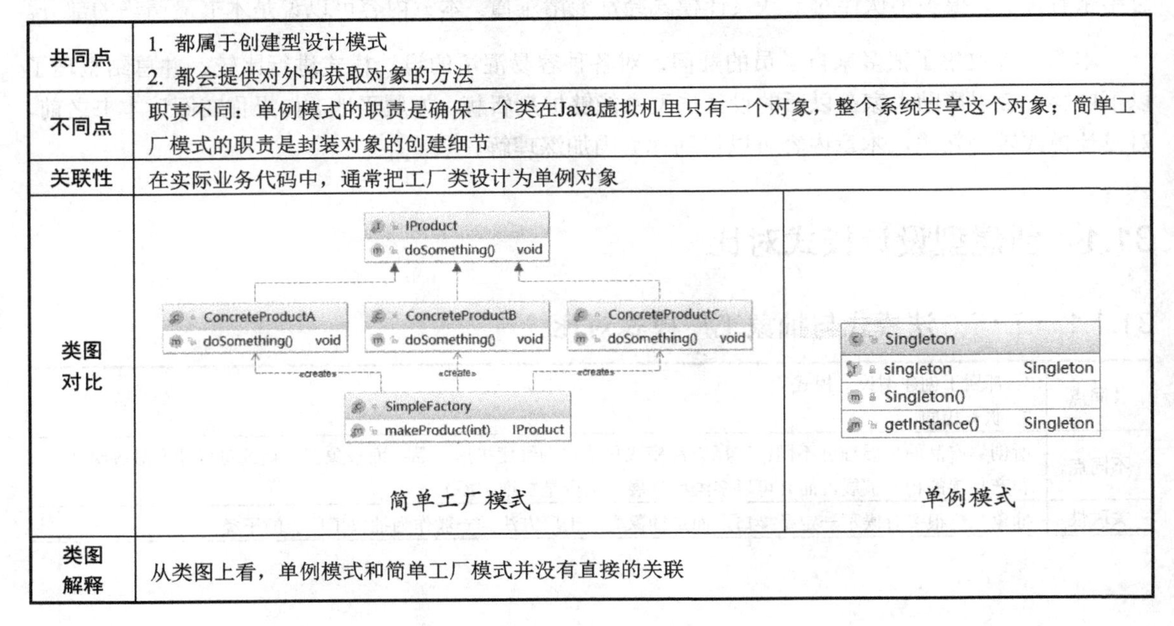

共同点	1. 都属于创建型设计模式 2. 都会提供对外的获取对象的方法	
不同点	职责不同：单例模式的职责是确保一个类在Java虚拟机里只有一个对象，整个系统共享这个对象；简单工厂模式的职责是封装对象的创建细节	
关联性	在实际业务代码中，通常把工厂类设计为单例对象	
类图对比	简单工厂模式	单例模式
类图解释	从类图上看，单例模式和简单工厂模式并没有直接的关联	

31.1.3　简单工厂模式与建造者模式对比

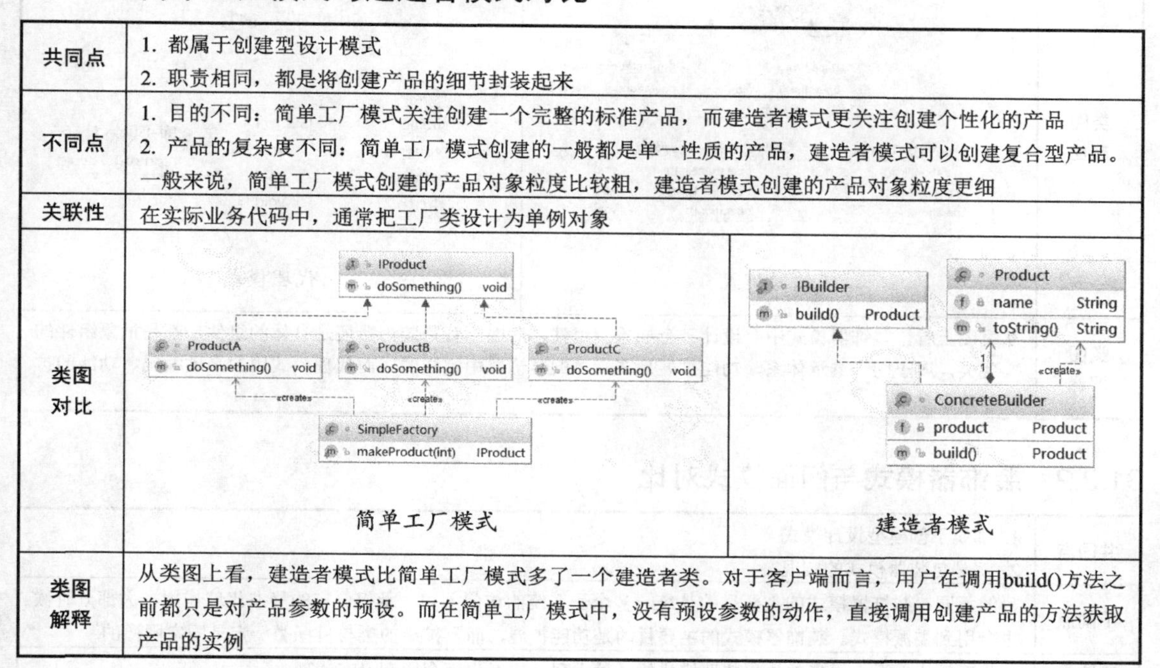

共同点	1. 都属于创建型设计模式 2. 职责相同，都是将创建产品的细节封装起来
不同点	1. 目的不同：简单工厂模式关注创建一个完整的标准产品，而建造者模式更关注创建个性化的产品 2. 产品的复杂度不同：简单工厂模式创建的一般都是单一性质的产品，建造者模式可以创建复合型产品。一般来说，简单工厂模式创建的产品对象粒度比较粗，建造者模式创建的产品对象粒度更细
关联性	在实际业务代码中，通常把工厂类设计为单例对象
类图对比	简单工厂模式 / 建造者模式
类图解释	从类图上看，建造者模式比简单工厂模式多了一个建造者类。对于客户端而言，用户在调用build()方法之前都只是对产品参数的预设。而在简单工厂模式中，没有预设参数的动作，直接调用创建产品的方法获取产品的实例

31.2　结构型设计模式对比

31.2.1　装饰器模式与代理模式对比

共同点	1. 都属于结构型设计模式 2. 都是包装器模式的实现 3. 都是为了达到功能增强的目的
不同点	1. 实现形式不同：代理模式通过组合实现功能增强和扩展，装饰器模式通过继承实现增强和扩展 2. 目的不同：代理模式着重代理的过程控制，而装饰器模式则是对类功能的加强或减弱，更注重类功能的变化 3. 可扩展程度不同：装饰器模式的代码扩展灵活度更大，代理模式的扩展相对来说依赖性更强
关联性	装饰器模式可以说是代理模式的一个特殊应用

续表

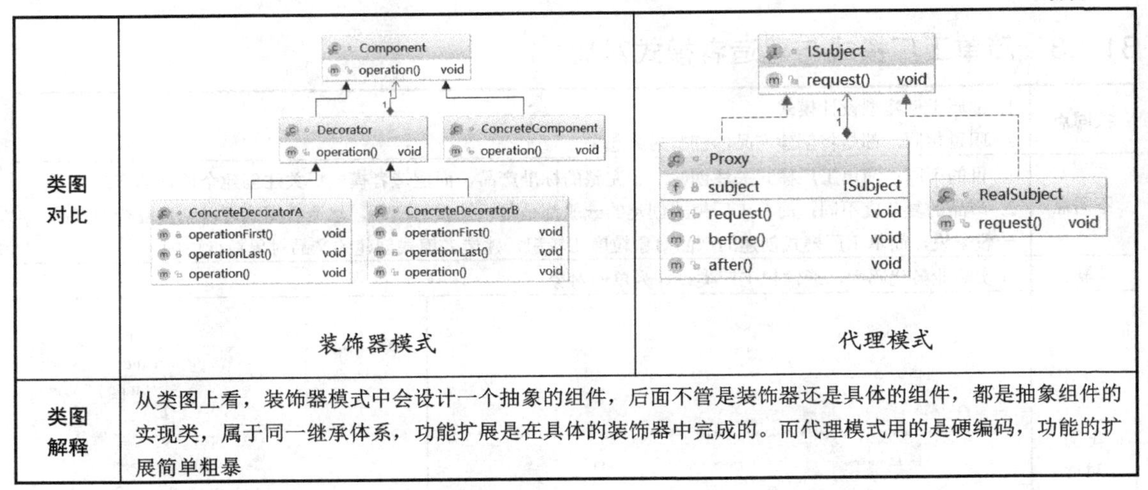

类图对比	装饰器模式	代理模式
类图解释	从类图上看，装饰器模式中会设计一个抽象的组件，后面不管是装饰器还是具体的组件，都是抽象组件的实现类，属于同一继承体系，功能扩展是在具体的装饰器中完成的。而代理模式用的是硬编码，功能的扩展简单粗暴	

31.2.2 装饰器模式与门面模式对比

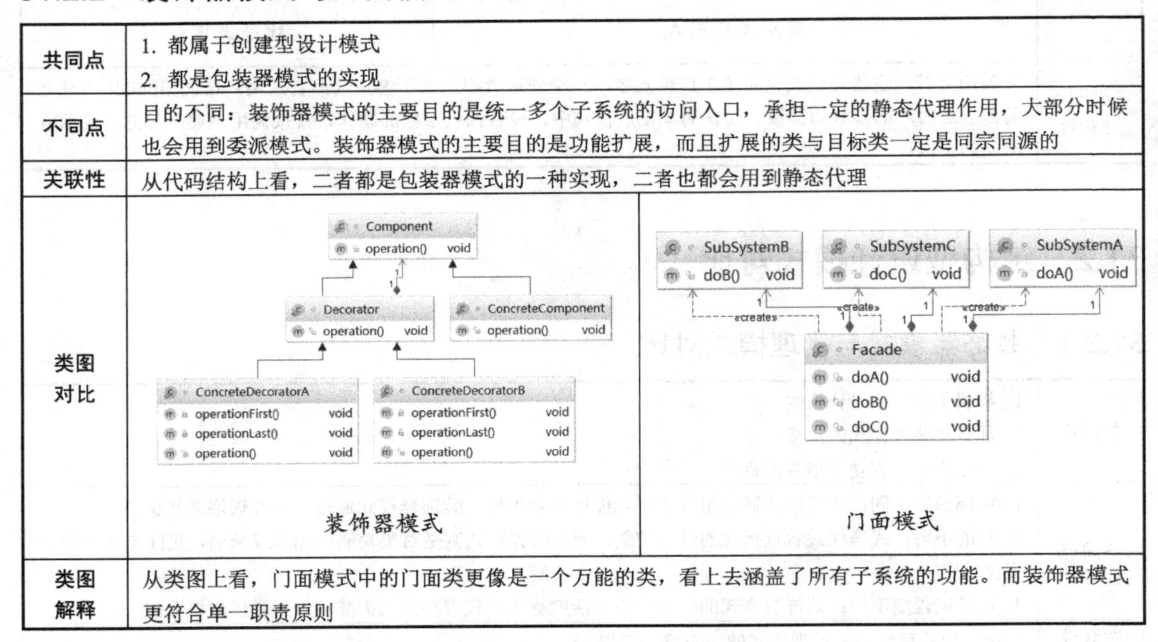

共同点	1. 都属于创建型设计模式 2. 都是包装器模式的实现	
不同点	目的不同：装饰器模式的主要目的是统一多个子系统的访问入口，承担一定的静态代理作用，大部分时候也会用到委派模式。装饰器模式的主要目的是功能扩展，而且扩展的类与目标类一定是同宗同源的	
关联性	从代码结构上看，二者都是包装器模式的一种实现，二者也都会用到静态代理	
类图对比	装饰器模式	门面模式
类图解释	从类图上看，门面模式中的门面类更像是一个万能的类，看上去涵盖了所有子系统的功能。而装饰器模式更符合单一职责原则	

31.2.3　装饰器模式与适配器模式对比

共同点	1. 都属于结构型设计模式 2. 都是包装器模式的实现
不同点	1. 代码结构不同：装饰器模式包装的都是自己的兄弟类，同宗同源；而适配器模式则是将一个非本家族的对象伪装起来 2. 设计目的不同：适配器模式的意义是将一个接口转变成另一个接口，通过改变接口来达到重复使用的目的；而装饰器模式不是要改变被装饰对象的接口，而是恰恰要保持原有的接口，但是增强原有对象的功能，或者改变原有对象的处理方式，从而提升性能
关联性	装饰器模式和适配器模式只有结构上容易混淆，在具体业务场景中还是很容易区分的
类图对比	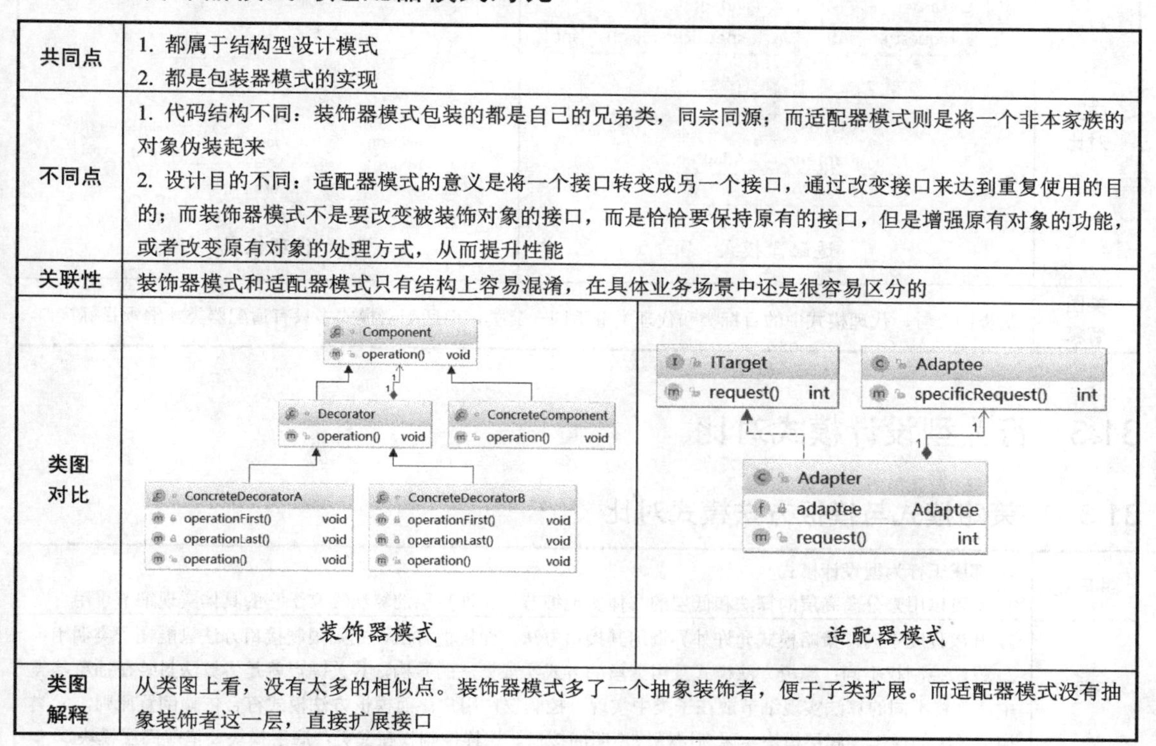装饰器模式　　适配器模式
类图解释	从类图上看，没有太多的相似点。装饰器模式多了一个抽象装饰者，便于子类扩展。而适配器模式没有抽象装饰者这一层，直接扩展接口

31.2.4　适配器模式与代理模式对比

共同点	1. 都属于结构型设计模式 2. 都是包装器模式的实现 3. 二者都起到了隐藏和保护原类的作用
不同点	目的不同：适配器模式主要解决兼容问题，会保留被适配对象已经存在的方法并继续对外开放调用；而代理模式主要是为了功能增强，目标类的方法不会直接提供给用户调用，而是调用代理类的方法获得增强后的结果
关联性	对象适配器就是静态代理的一种实现

续表

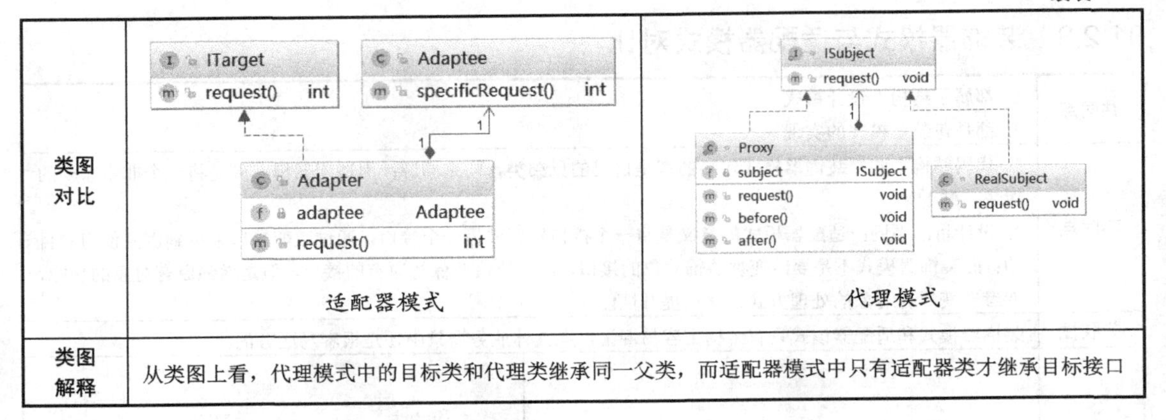

类图对比	适配器模式	代理模式
类图解释	从类图上看，代理模式中的目标类和代理类继承同一父类，而适配器模式中只有适配器类才继承目标接口	

31.3 行为型设计模式对比

31.3.1 策略模式与模板方法模式对比

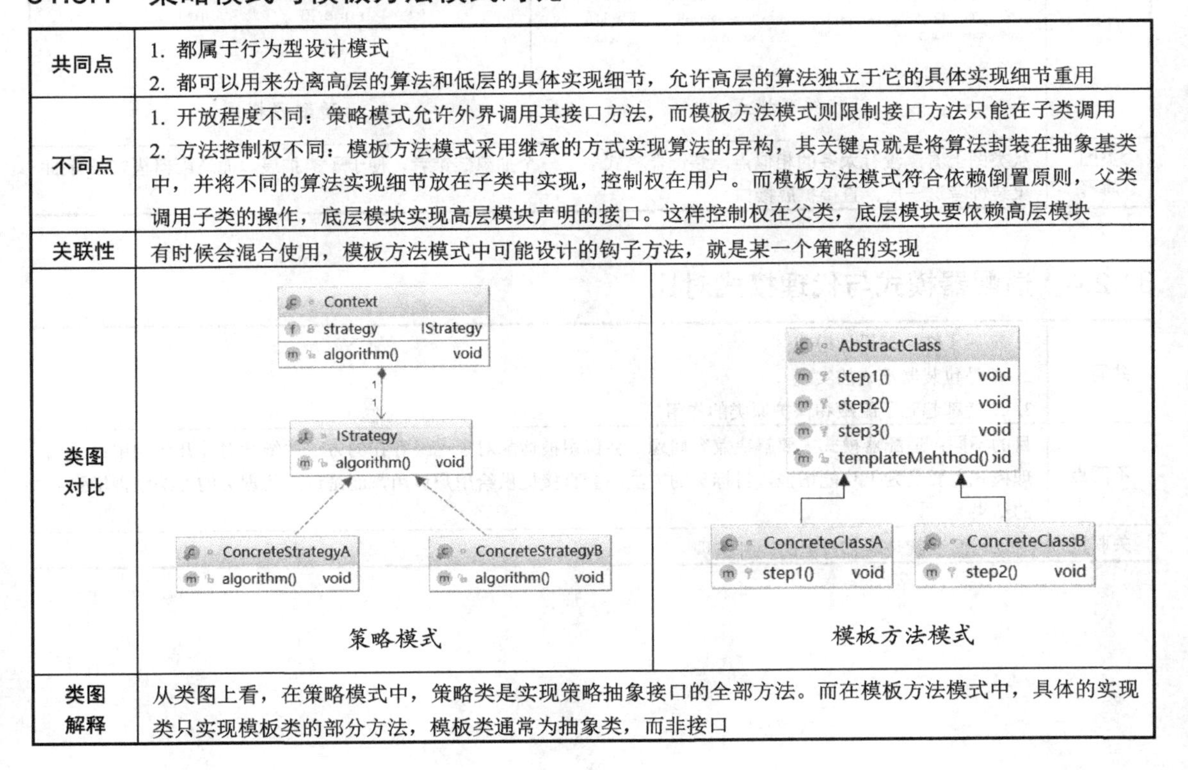

共同点	1. 都属于行为型设计模式 2. 都可以用来分离高层的算法和低层的具体实现细节，允许高层的算法独立于它的具体实现细节重用	
不同点	1. 开放程度不同：策略模式允许外界调用其接口方法，而模板方法模式则限制接口方法只能在子类调用 2. 方法控制权不同：模板方法模式采用继承的方式实现算法的异构，其关键点就是将算法封装在抽象基类中，并将不同的算法实现细节放在子类中实现，控制权在用户。而模板方法模式符合依赖倒置原则，父类调用子类的操作，底层模块实现高层模块声明的接口。这样控制权在父类，底层模块要依赖高层模块	
关联性	有时候会混合使用，模板方法模式中可能设计的钩子方法，就是某一个策略的实现	
类图对比	策略模式	模板方法模式
类图解释	从类图上看，在策略模式中，策略类是实现策略抽象接口的全部方法。而在模板方法模式中，具体的实现类只实现模板类的部分方法，模板类通常为抽象类，而非接口	

31.3.2　策略模式与命令模式对比

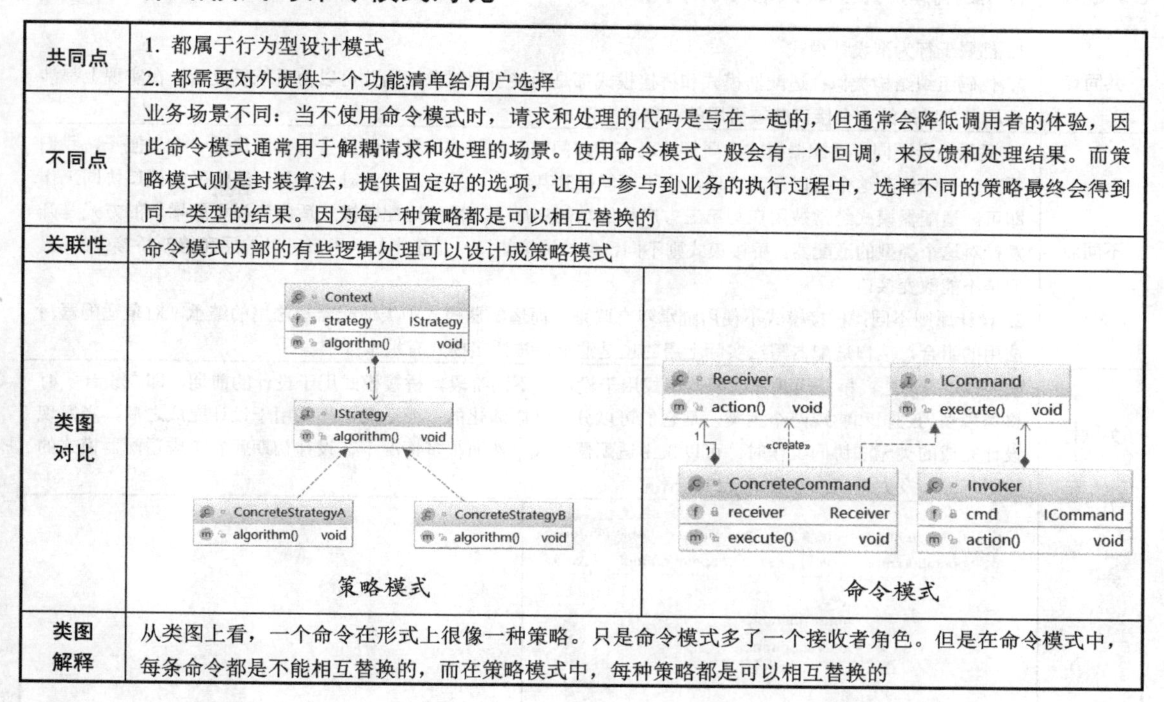

共同点	1. 都属于行为型设计模式 2. 都需要对外提供一个功能清单给用户选择
不同点	业务场景不同：当不使用命令模式时，请求和处理的代码是写在一起的，但通常会降低调用者的体验，因此命令模式通常用于解耦请求和处理的场景。使用命令模式一般会有一个回调，来反馈和处理结果。而策略模式则是封装算法，提供固定好的选项，让用户参与到业务的执行过程中，选择不同的策略最终会得到同一类型的结果。因为每一种策略都是可以相互替换的
关联性	命令模式内部的有些逻辑处理可以设计成策略模式
类图对比	策略模式 / 命令模式
类图解释	从类图上看，一个命令在形式上很像一种策略。只是命令模式多了一个接收者角色。但是在命令模式中，每条命令都是不能相互替换的，而在策略模式中，每种策略都是可以相互替换的

31.3.3　策略模式与委派模式对比

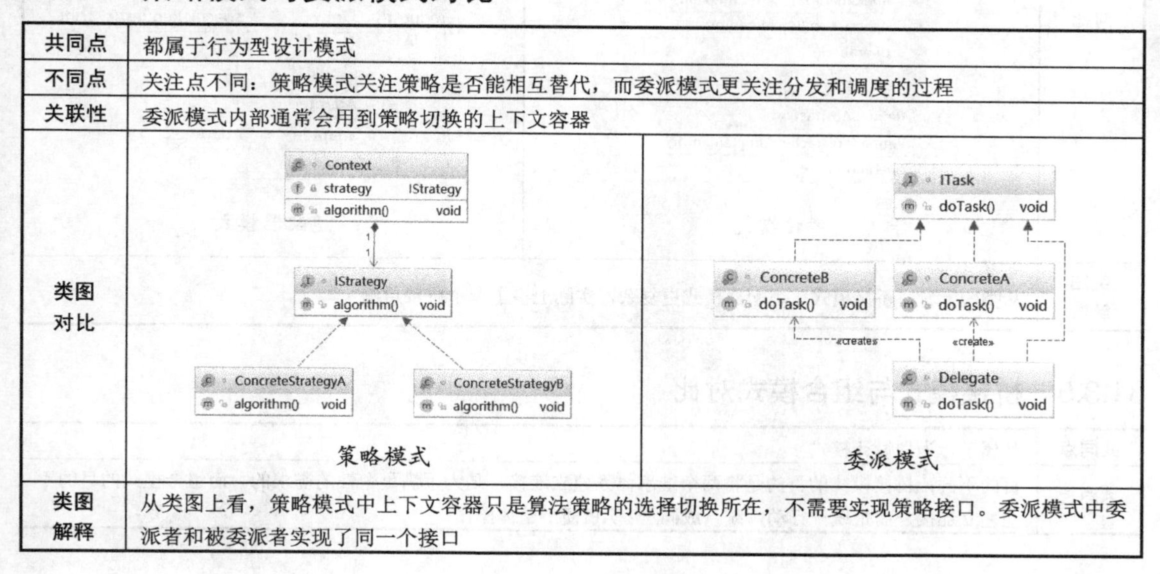

共同点	都属于行为型设计模式
不同点	关注点不同：策略模式关注策略是否能相互替代，而委派模式更关注分发和调度的过程
关联性	委派模式内部通常会用到策略切换的上下文容器
类图对比	策略模式 / 委派模式
类图解释	从类图上看，策略模式中上下文容器只是算法策略的选择切换所在，不需要实现策略接口。委派模式中委派者和被委派者实现了同一个接口

31.3.4 桥接模式与适配器模式对比

共同点	1. 都属于行为型设计模式 2. 代码组织结构类似：适配器模式和桥接模式都是间接引用对象，因此可以使系统更灵活，在实现上都涉及从自身以外的一个接口向被引用的对象发出请求
不同点	1. 适用场景不同：适配器模式主要解决已有接口间的兼容问题，被适配的接口实现像是一个黑匣子，我们不想也不能修改这个接口及其实现，也不可能控制其演化，只要相关的对象能与系统定义的接口协同工作即可。适配器模式经常被用在与第三方产品的功能集成上，采用适配器模式适应新类型的增加的方式是开发针对这个类型的适配器。桥接模式则不同，参与桥接的接口是稳定的，用户可以扩展和修改桥接中的类，但是不能改变接口 2. 设计原则不同：桥接模式不使用继承建立联系。而适配器模式中类适配器写法用的继承，对象适配器写法用的组合，接口适配器写法实际上用的也是继承，与桥接模式有根本区别
关联性	按照GoF的说法，桥接模式和适配器模式用于设计的不同阶段，桥接模式用于设计的前期，即在设计类时将类规划为逻辑和实现两个大类，是它们可以分别精心演化的；而适配器模式用于设计完成之后，当发现设计完成的类无法协同工作时，可以采用适配器模式。然而很多情况下在设计初期就要考虑适配器模式的使用，如涉及大量第三方应用接口的情况
类图对比	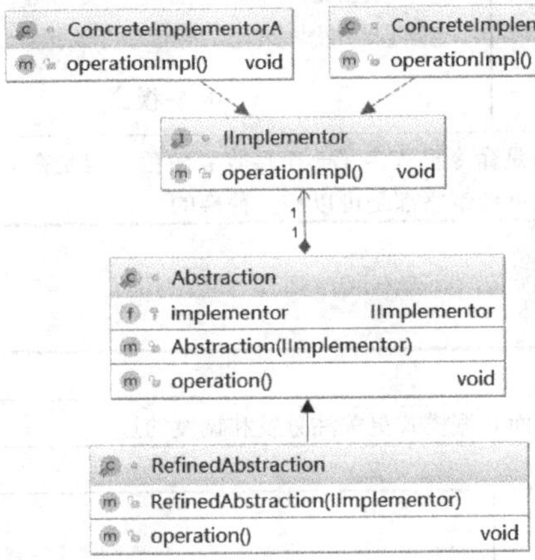 ConcreteImplementorA operationImpl() void ConcreteImplementorB operationImpl() void IImplementor operationImpl() void 1 1 Abstraction implementor IImplementor Abstraction(IImplementor) operation() void RefinedAbstraction RefinedAbstraction(IImplementor) operation() void 桥接模式 ITarget request() int Adaptee specificRequest() int 1 1 Adapter adaptee Adaptee request() int 适配器模式
类图解释	从类图上看，桥接模式比适配器模式更复杂，实际上多了一个桥接角色

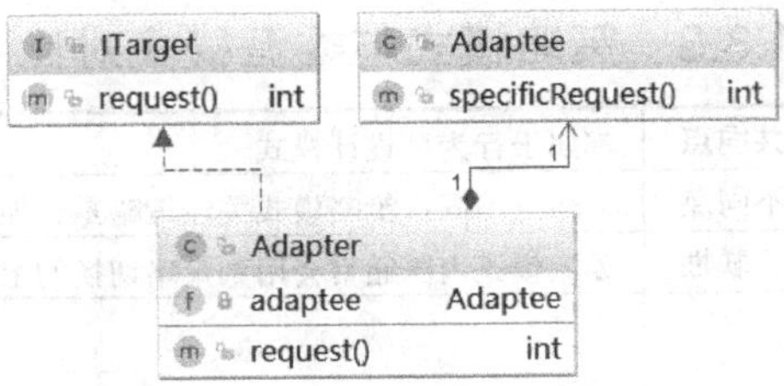

31.3.5 桥接模式与组合模式对比

共同点	都属于行为型设计模式
不同点	目的不同：桥接模式的目的是将两个继承体系建立连接，是为了满足个性的需求的。而组合模式的目的不是建立连接，而是统一行动，统一成同一套API便于整体操作

续表

关联性	桥接模式和组合模式关联性不大

类图对比

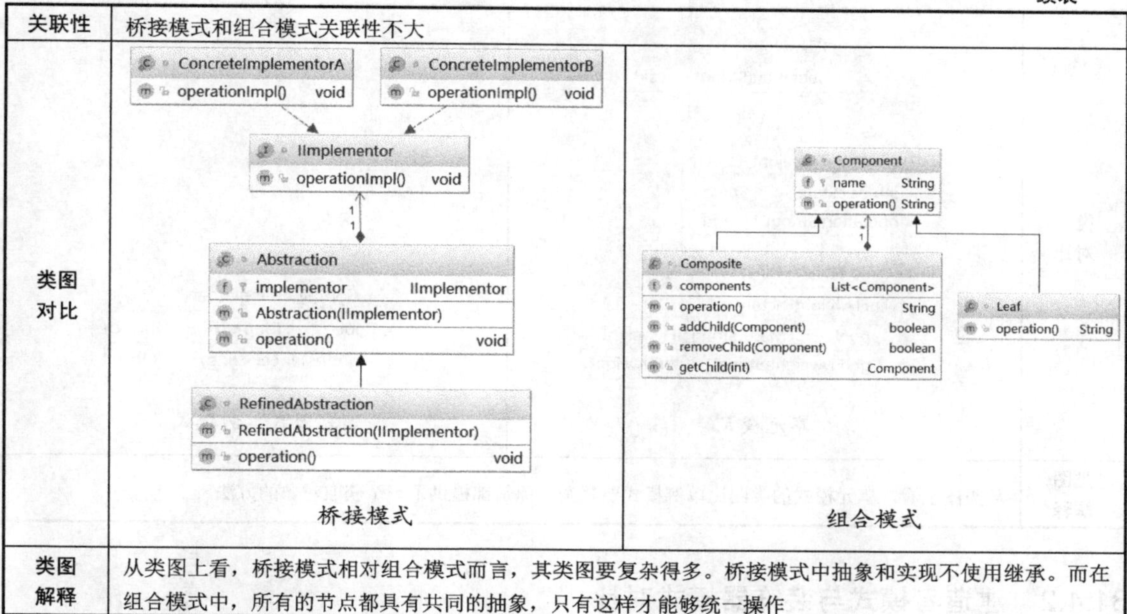

桥接模式　　　　组合模式

类图解释	从类图上看，桥接模式相对组合模式而言，其类图要复杂得多。桥接模式中抽象和实现不使用继承。而在组合模式中，所有的节点都具有共同的抽象，只有这样才能够统一操作

31.4　跨类综合对比

31.4.1　享元模式与容器式单例模式对比

共同点	1. 都设计了一个缓存对象的容器 2. 设计目的有相似之处，两者都是为了节省内存开销
不同点	1. 类型不同：享元模式属于结构型设计模式，容器式单例模式属于创建型设计模式 2. 对创建对象的控制粒度不同：享元模式可以再次创建对象，也可以取缓存对象。而单例模式则严格控制应用程序中只有一个实例对象 3. 实现形式不同：享元模式可以通过自己实现对外部的单例，也可以在需要的时候创建更多的对象；单例模式是自身控制，需要增加不属于该对象本身的逻辑
关联性	享元模式可以看成是单例模式的扩展，可以把对象池的容器设置为单例。同时，把对象池所在的类设置为单例模式的工厂。享元模式 = 单例模式 + 工厂模式 + 组合模式

续表

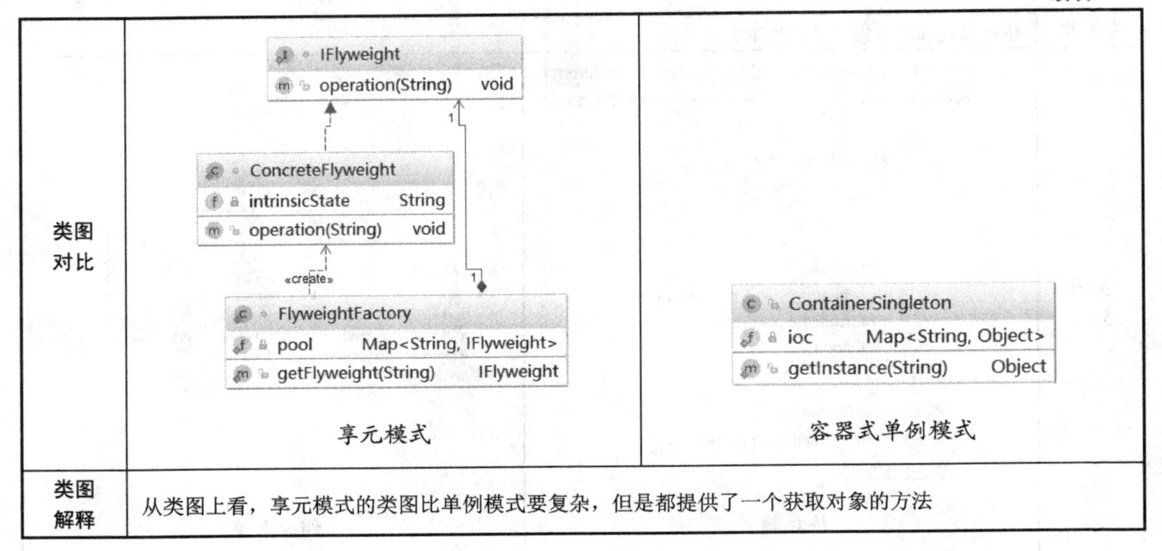

类图对比	享元模式	容器式单例模式
类图解释	从类图上看，享元模式的类图比单例模式要复杂，但是都提供了一个获取对象的方法	

31.4.2 建造者模式与装饰器模式对比

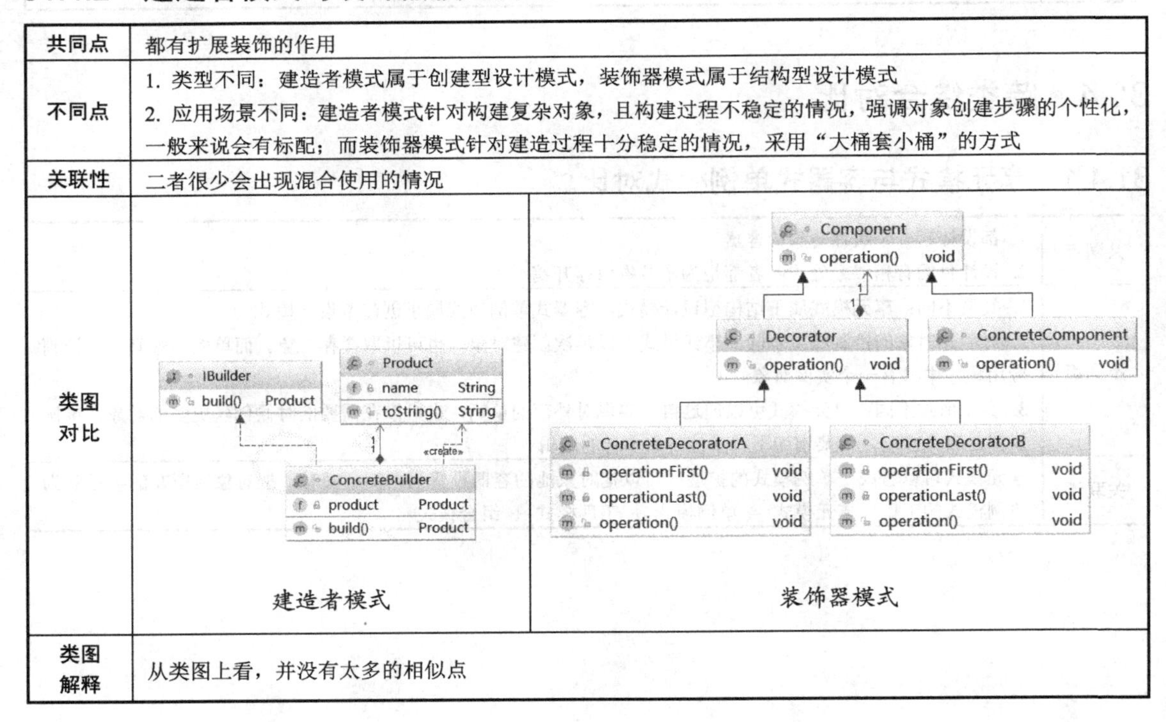

共同点	都有扩展装饰的作用	
不同点	1. 类型不同：建造者模式属于创建型设计模式，装饰器模式属于结构型设计模式 2. 应用场景不同：建造者模式针对构建复杂对象，且构建过程不稳定的情况，强调对象创建步骤的个性化，一般来说会有标配；而装饰器模式针对建造过程十分稳定的情况，采用“大桶套小桶”的方式	
关联性	二者很少会出现混合使用的情况	
类图对比	建造者模式	装饰器模式
类图解释	从类图上看，并没有太多的相似点	

31.4.3　策略模式与简单工厂模式对比

共同点	客户端调用方式相同：两者都是通过传入参数进行配置的
不同点	1. 类型不同：策略模式属于行为型设计模式，简单工厂模式属于创建型设计模式 2. 侧重点不同：简单工厂模式通过传参选择创建出需要的对象，而策略模式则通过传参配置出需要的行为算法。一个是对象创建，另一个是行为算法的替换 3. 设计逻辑不同：两者的差别很微妙，简单工厂模式直接创建具体的对象并用该对象去执行相应的动作。而策略模式设计一个上下文类，将操作给了上下文类，策略类内部没有创建具体的对象，从而实现代码的进一步封装，客户端代码并不需要知道具体的实现过程
关联性	一般来说，二者会组合使用，具体策略将由简单工厂模式来创建
类图对比	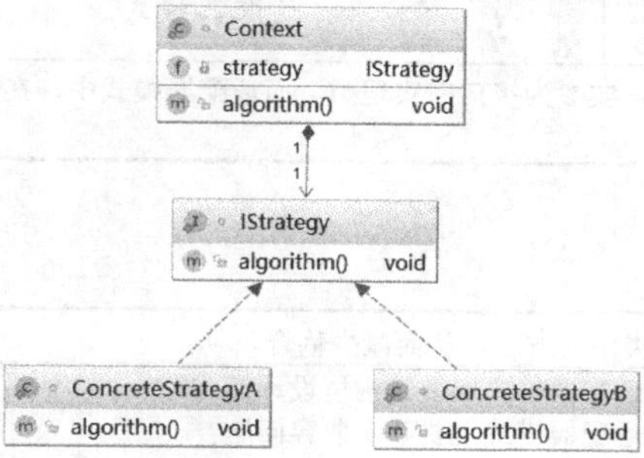 策略模式　　　　简单工厂模式
类图解释	从类图上看，策略模式和简单工厂模式非常相似，都通过多态来实现不同子类的选取，这种思想应该是从程序的整体看出来的

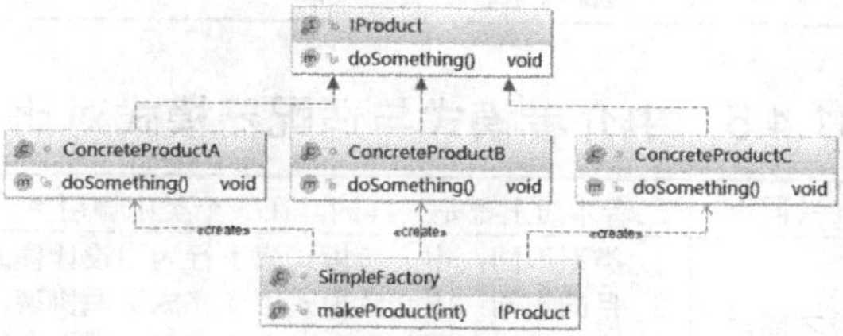

31.4.4　策略模式与适配器模式对比

共同点	都是通过找到已经存在的、运行良好的类来实现接口的
不同点	1. 类型不同：策略模式属于行为型设计模式，适配器模式属于结构型设计模式 2. 目的不同：策略模式把一系列算法封装起来，提供一个统一的接口给客户，并使这些算法可以相互间替换；而适配器模式将一个类的接口转换成客户希望的另外一个接口，从而使原本因接口不兼容不能一起工作的类可以一起工作 3. 客户单端调用方式不同：策略模式的所有策略都需要暴露出去，由客户端决定使用哪一种策略。而适配器模式是定义好接口的实现方式及内部需要引用的类，客户端直接调用适配器的方法
关联性	策略模式很多时候都和适配器模式结合使用，把具体适配器作为具体策略，用户选择不同策略从而调用不同的适配器方法

类图对比	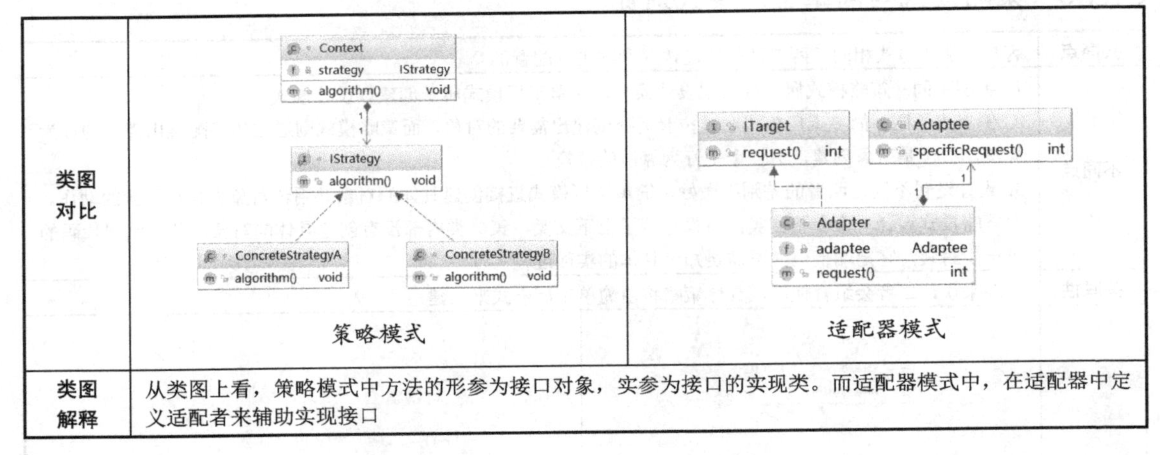
类图解释	从类图上看，策略模式中方法的形参为接口对象，实参为接口的实现类。而适配器模式中，在适配器中定义适配者来辅助实现接口

31.4.5 中介者模式与适配器模式对比

共同点	二者本质上都是一样的，在一个类中调用另一个类中的方法，从而减少耦合
不同点	1. 类型不同：中介者模式属于行为型设计模式，适配器模式属于结构型设计模式 2. 目的不同：中介者模式主要完成资源协调，而适配器模式主要解决兼容问题 3. 代码结构不同：中介者模式一定是用组合的方式实现代码复用的，所有人可能都持有中介者的引用；而适配器模式可以用继承的方式实现代码复用，也可以用组合的方式实现
关联性	二者并没有明显的关联性
类图对比	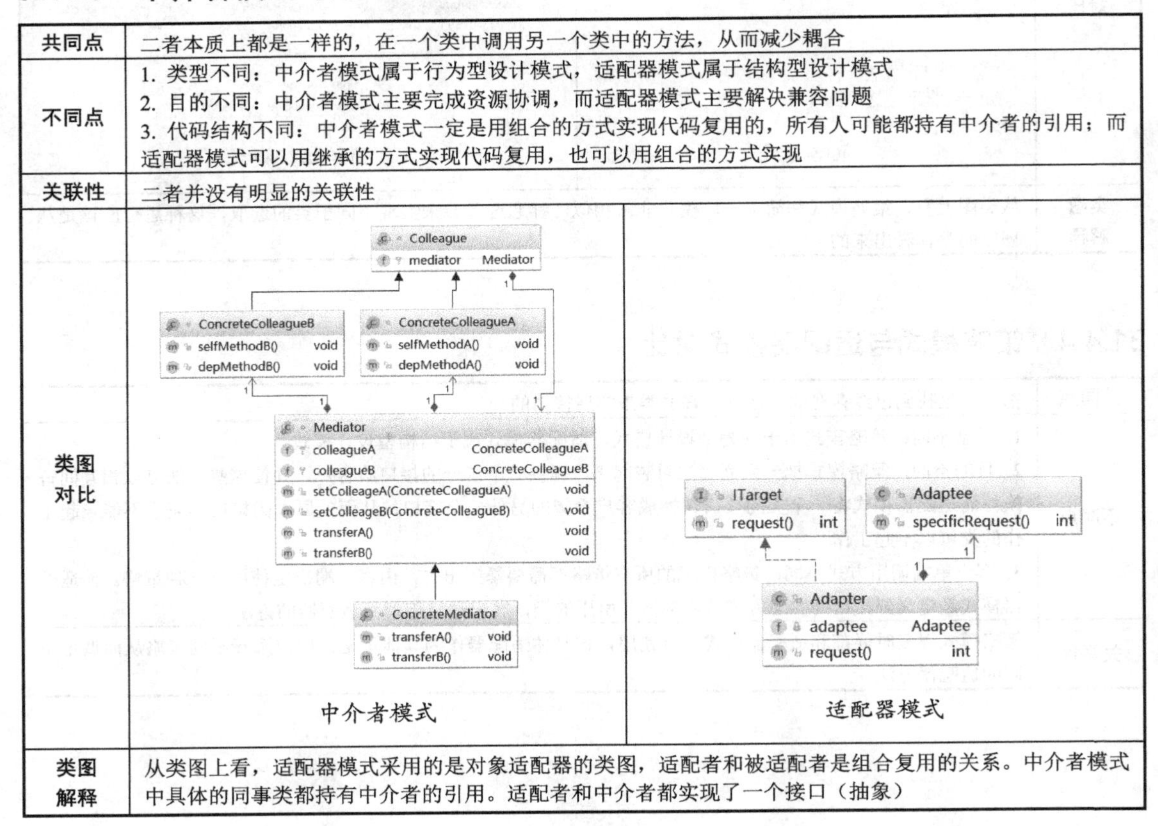
类图解释	从类图上看，适配器模式采用的是对象适配器的类图，适配者和被适配者是组合复用的关系。中介者模式中具体的同事类都持有中介者的引用。适配者和中介者都实现了一个接口（抽象）

31.4.6　中介者模式与代理模式对比

共同点	都具备保护目标对象的特性
不同点	1. 类型不同：中介者模式属于行为型设计模式，代理模式属于结构型设计模式 2. 干预程度不同：如果说代理模式是“媒婆”，那么中介者模式就是“不负责任的媒婆” 3. 职责不同：代理模式的职责是功能增强，不仅要将目标对象和代理对象建立联系，代理对象还要参与过程。而中介者模式中的中介者只负责牵线搭桥，建立联系，不参与具体的过程
关联性	中介者模式是一种面向更加复杂的对象关系的全权静态代理（委托）
类图对比	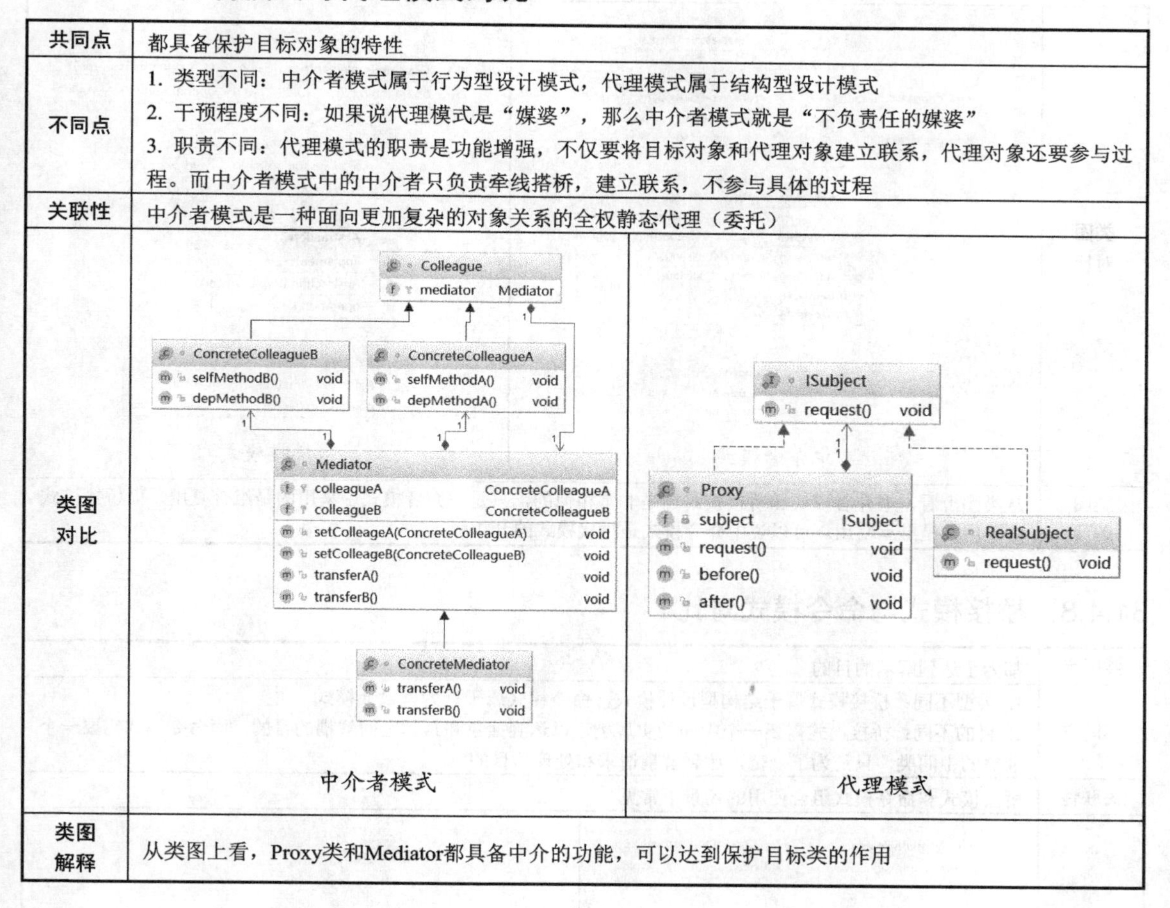
类图解释	从类图上看，Proxy类和Mediator都具备中介的功能，可以达到保护目标类的作用

31.4.7　中介者模式与桥接模式对比

共同点	都具备将两个对象建立联系的特性
不同点	1. 类型不同：中介者模式属于行为型设计模式，桥接模式属于结构型设计模式 2. 适用场景不同：桥接模式只适用于将两个维度建立连接；而中介者模式可以将多个维度建立连接
关联性	中介者模式是一种更为复杂的桥接模式，中介者可以和网状结构的对象建立连接，而桥接模式只能和两个维度的对象建立连接

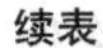
续表

<table>
<tr><td>类图对比</td><td>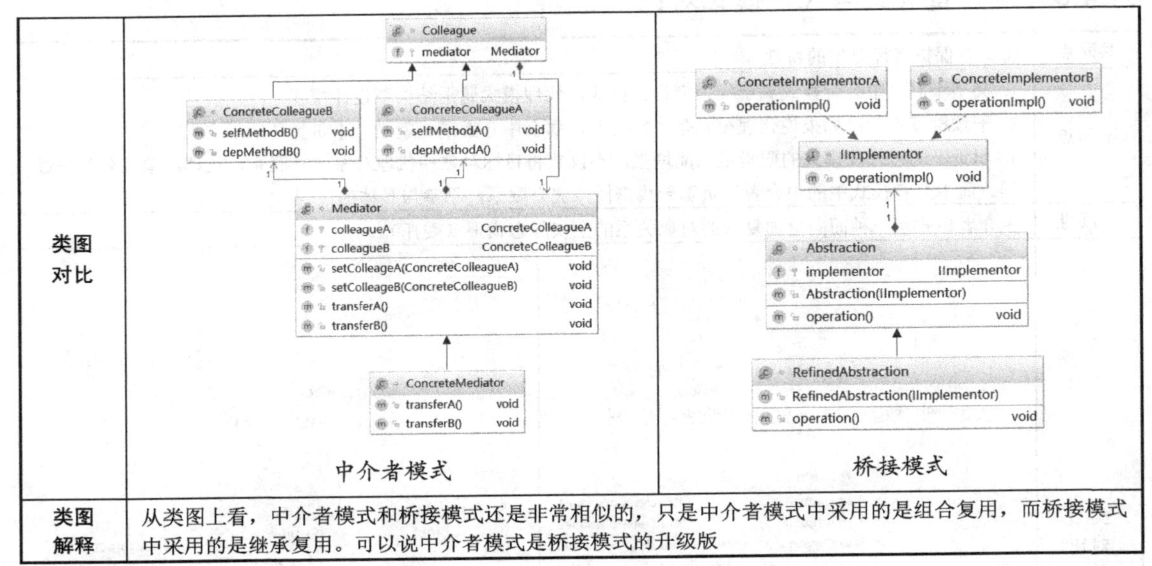

中介者模式　　　桥接模式</td></tr>
<tr><td>类图解释</td><td>从类图上看，中介者模式和桥接模式还是非常相似的，只是中介者模式中采用的是组合复用，而桥接模式中采用的是继承复用。可以说中介者模式是桥接模式的升级版</td></tr>
</table>

31.4.8 桥接模式与命令模式对比

<table>
<tr><td>共同点</td><td>都为了达到解耦的目的</td></tr>
<tr><td>不同点</td><td>1. 类型不同：桥接模式属于结构型设计模式，命令模式属于行为型设计模式
2. 目的不同：桥接模式需要一个中间的实现类，以达到抽象和具体之间解耦的目的。而命令模式需要一个抽象的中间类，只是为了规范，达到解耦请求和处理的目的</td></tr>
<tr><td>关联性</td><td>桥接模式和命令模式组合使用的场景不常见</td></tr>
<tr><td>类图对比</td><td>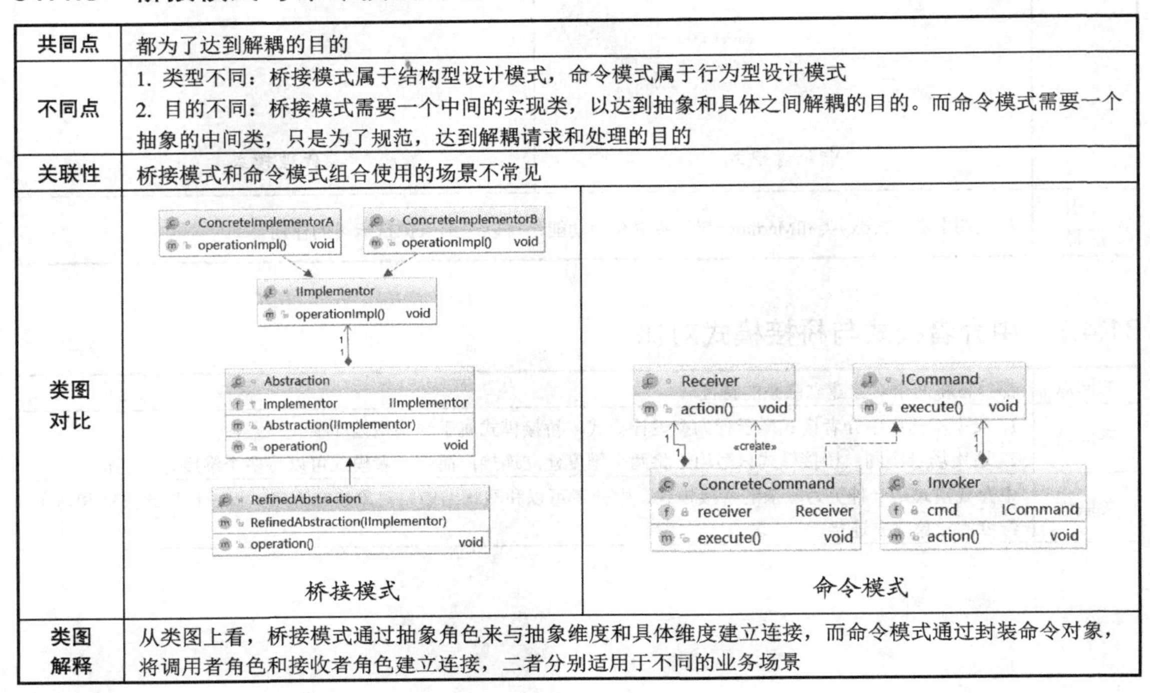

桥接模式　　　命令模式</td></tr>
<tr><td>类图解释</td><td>从类图上看，桥接模式通过抽象角色来与抽象维度和具体维度建立连接，而命令模式通过封装命令对象，将调用者角色和接收者角色建立连接，二者分别适用于不同的业务场景</td></tr>
</table>

31.4.9　委派模式与门面模式对比

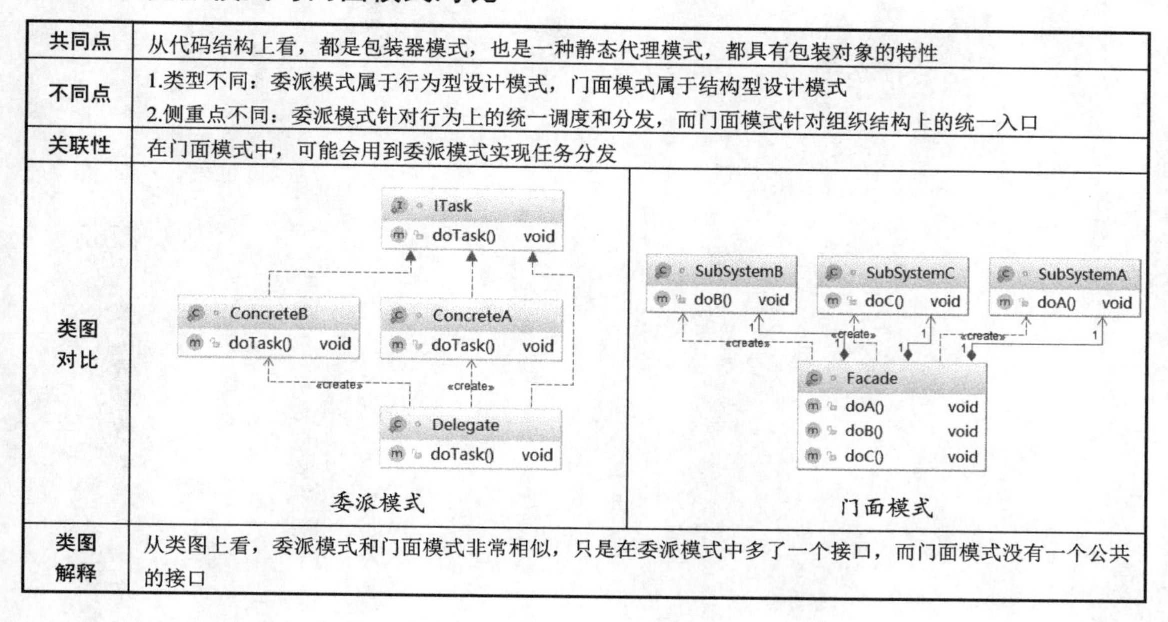

共同点	从代码结构上看，都是包装器模式，也是一种静态代理模式，都具有包装对象的特性	
不同点	1.类型不同：委派模式属于行为型设计模式，门面模式属于结构型设计模式 2.侧重点不同：委派模式针对行为上的统一调度和分发，而门面模式针对组织结构上的统一入口	
关联性	在门面模式中，可能会用到委派模式实现任务分发	
类图对比	委派模式	门面模式
类图解释	从类图上看，委派模式和门面模式非常相似，只是在委派模式中多了一个接口，而门面模式没有一个公共的接口	

31.4.10　委派模式与代理模式对比

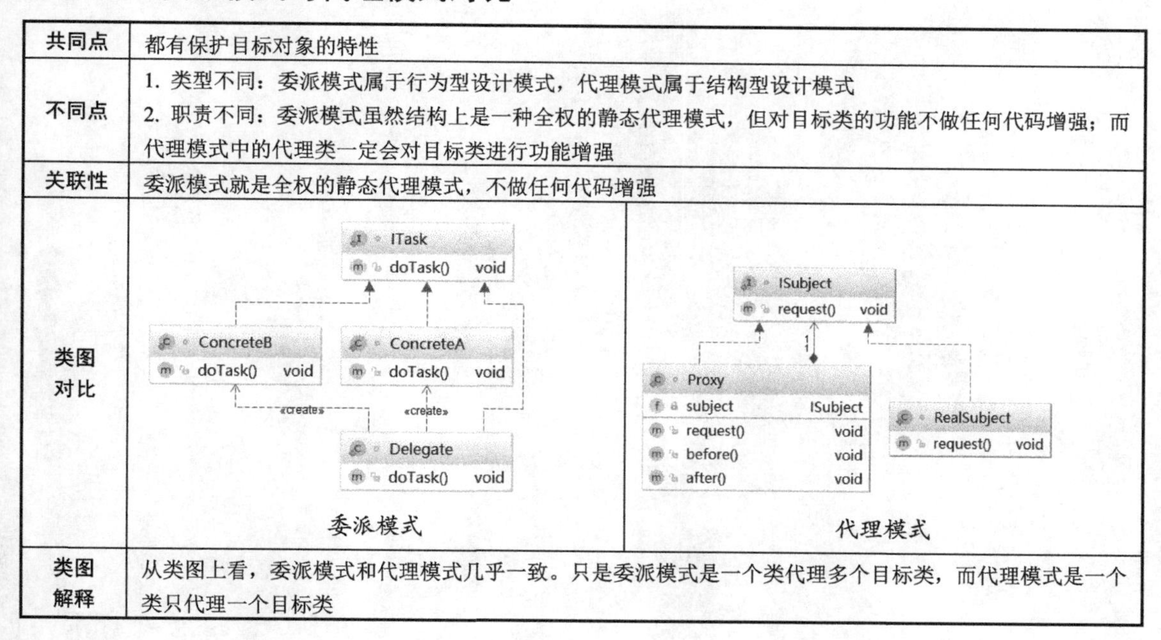

共同点	都有保护目标对象的特性	
不同点	1. 类型不同：委派模式属于行为型设计模式，代理模式属于结构型设计模式 2. 职责不同：委派模式虽然结构上是一种全权的静态代理模式，但对目标类的功能不做任何代码增强；而代理模式中的代理类一定会对目标类进行功能增强	
关联性	委派模式就是全权的静态代理模式，不做任何代码增强	
类图对比	委派模式	代理模式
类图解释	从类图上看，委派模式和代理模式几乎一致。只是委派模式是一个类代理多个目标类，而代理模式是一个类只代理一个目标类	

6

第 6 篇

架构设计扩展篇

32

第 32 章 新设计模式

32.1 对象池模式

32.1.1 对象池模式的定义

对象池模式（Object Pool Pattern）是创建型设计模式的一种，将对象预先创建并初始化后放入对象池中，对象提供者就能利用已有的对象来处理请求，减少频繁创建对象所占用的内存空间和初始化时间。

一个对象池包含一组已经初始化并且可以使用的对象，可以在有需求时创建和销毁对象。对象池的用户可以从池子中取得对象，对其进行操作处理，并在不需要时归还给池子而非直接销毁。对象池是一个特殊的工厂对象，对象池模式就是单例模式加享元模式。

32.1.2 对象池模式的应用场景

对象池模式主要适用于以下应用场景。

（1）资源受限的场景。比如，不需要可伸缩性的环境（CPU\内存等物理资源有限），CPU 性能不够强劲，内存比较紧张，垃圾收集，内存抖动会造成比较大的影响，需要提高内存管理效率，响应性比吞吐量更为重要。

（2）在内存中数量受限的对象。

（3）创建成本高的对象，可以考虑池化。

补充：常见的使用对象池模式的场景有使用 Socket 时的各种连接池、线程池、数据库连接池等。

32.1.3　对象池模式的 UML 类图

对象池模式的 UML 类图如下。

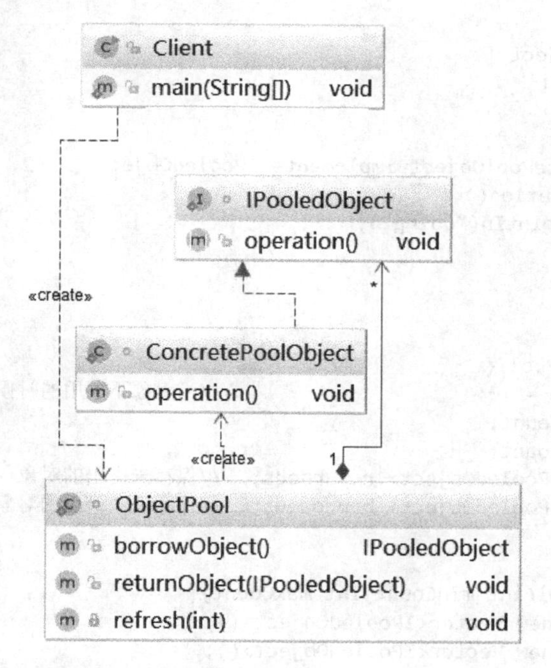

由上图可以看到，对象池模式主要包含 3 个角色。

（1）对象池（ObjectPool）：持有对象并提供取/还等方法。

（2）抽象池化对象（IPooledObject）：对池中对象的抽象。

（3）具体池化对象（ConcretePoolObject）：对池中对象的封装，封装对象的状态和一些其他信息。

32.1.4 对象池模式的通用写法

以下是对象池模式的通用写法。

```
public class Client {

    public static void main(String[] args) {
        ObjectPool pool = new ObjectPool(10,50);
        IPooledObject object = pool.borrowObject();
        object.operation();
        pool.returnObject(object);
        System.out.println();
    }

    //抽象对象
    interface IPooledObject {
        void operation();
    }
    //具体对象
    static class ConcretePoolObject implements IPooledObject {
        public void operation() {
            System.out.println("doing");
        }
    }

    //对象池
    static class ObjectPool {
        private int step = 10;                       //当对象不够用的时候，每次扩容的数量
        private int minCount;
        private int maxCount;
        private Vector<IPooledObject> returneds;  //保存未借出的对象
        private Vector<IPooledObject> borroweds;  //保存已被借出的对象

        //初始化对象池
        public ObjectPool(int minCount,int maxCount){
            borroweds = new Vector<IPooledObject>();
            returneds = new Vector<IPooledObject>();

            this.minCount = minCount;
            this.maxCount = maxCount;

            refresh(this.minCount);
        }

        //因为内部状态具备不变性，所以作为缓存的键
        public IPooledObject borrowObject() {
            IPooledObject next = null;
```

```
            if(returneds.size() > 0){
                Iterator<IPooledObject> i = returneds.iterator();
                while (i.hasNext()){
                    next = i.next();
                    returneds.remove(next);
                    borroweds.add(next);
                    return next;
                }
            }else{
                //计算出剩余可创建的对象数
                int count = (maxCount - minCount);
                //剩余可创建的数量大于单次固定创建的对象数
                //则再初始化一批固定数量的对象
                refresh(count > step ? step : count);
            }
            return next;
        }

        //不需要使用的对象被归还重复利用
        public void returnObject(IPooledObject pooledObject){
            returneds.add(pooledObject);
            if(borroweds.contains(pooledObject)){
                borroweds.remove(pooledObject);
            }
        }

        private void refresh(int count){
            for (int i = 0; i < count; i++) {
                returneds.add(new ConcretePoolObject());
            }
        }
    }
}
```

对象池模式和享元模式的最大区别在于，对象池模式中会多一个回收对象重复利用的方法。所以，对象池模式应该是享元模式更加具体的一个应用场景，相当于先将对象从对象池中借出，用完之后再还回去，以此保证有限资源的重复利用。

32.1.5　对象池模式的优缺点

1. 优点

复用池中对象，消除创建对象、回收对象所产生的内存开销、CPU 开销，以及跨网络产生的网络开销。

2. 缺点

（1）增加了分配/释放对象的开销。

（2）在并发环境中，多个线程可能同时需要获取池中对象，进而需要在堆数据结构上进行同步或者因为锁竞争而产生阻塞，这种开销要比创建销毁对象的开销高数百倍。

（3）由于池中对象的数量有限，势必成为一个可伸缩性瓶颈。

（4）很难合理设定对象池的大小，如果太小，则起不到作用；如果过大，则占用内存资源高。

32.2 规格模式

32.2.1 规格模式的定义

规格模式（Specification Pattern）可以认为是组合模式的一种扩展。很多时候程序中的某些条件决定了业务逻辑，这些条件就可以抽离出来以某种关系（与、或、非）进行组合，从而灵活地对业务逻辑进行定制。另外，在查询、过滤等应用场合中，通过预定义多个条件，然后使用这些条件的组合来处理查询或过滤，而不是使用逻辑判断语句来处理，可以简化整个实现逻辑。

这里的每个条件都是一个规格，多个规格（条件）通过串联的方式以某种逻辑关系形成一个组合式的规格。规格模式属于结构型设计模式。

32.2.2 规格模式的应用场景

规格模式主要适用于以下应用场景。

（1）验证对象，检验对象本身是否满足某些业务要求或者是否已经为实现某个业务目标做好了准备。

（2）从集合中选择符合特定业务规则的对象或对象子集。

（3）指定在创建新对象的时候必须要满足某种业务要求。

32.2.3 规格模式的 UML 类图

规格模式的 UML 类图如下。

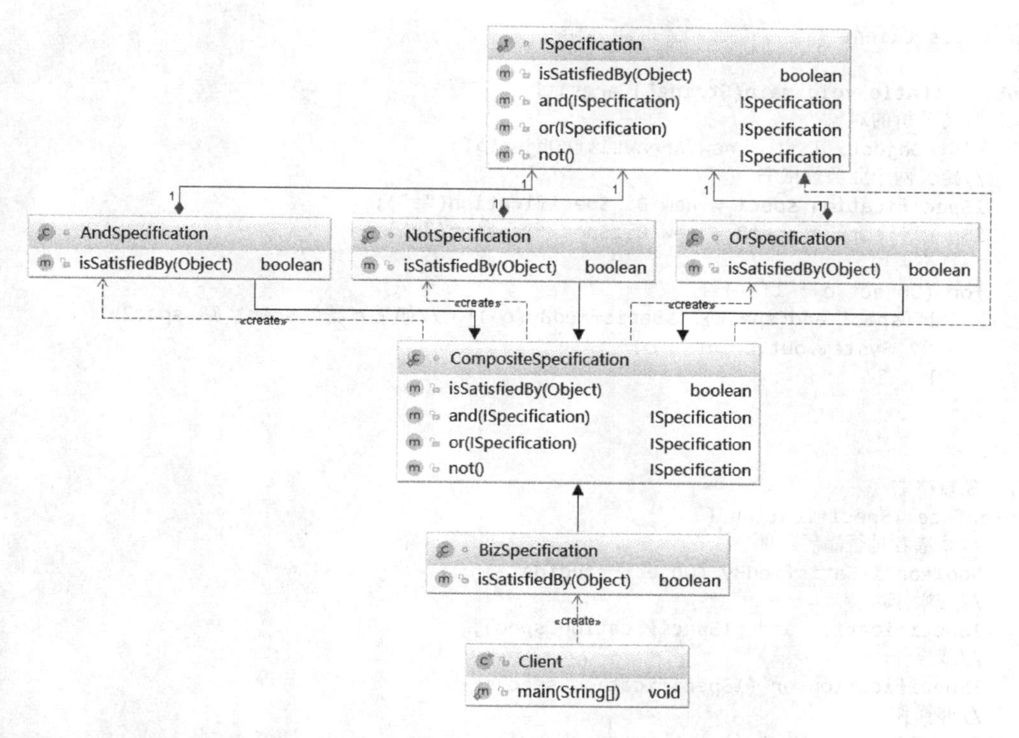

由上图可以看到，规格模式主要包含 6 个角色。

（1）抽象规格书（ISpecification）：对规格书的抽象定义。

（2）组合规格书（CompositeSpecification）：一般设计为抽象类，对规格书进行与或非操作，实现 and()、or()、not()方法，在方法中关联子类，因为子类为固定类，所以父类可以进行关联。

（3）与规格书（AndSpecification）：对规格书进行与操作，实现 isSatisfiedBy()方法。

（4）或规格书（OrSpecification）：对规格书进行或操作，实现 isSatisfiedBy()方法。

（5）非规格书（NotSpecification）：对规格书进行非操作，实现 isSatisfiedBy()方法。

（6）业务规格书（BizSpecification）：实现 isSatisfiedBy()方法，对业务进行判断，一个类为一种判断方式，可进行扩展。

32.2.4　规格模式的通用写法

以下是规格模式的通用写法。

```
public class Client {

    public static void main(String[] args) {
        //待分析的对象
        List<Object> list = new ArrayList<Object>();
        //定义两个业务规格书
        ISpecification spec1 = new BizSpecification("a");
        ISpecification spec2 = new BizSpecification("b");
        //规格调用
        for (Object o : list) {
            if(spec1.and(spec2).isSatisfiedBy(o)){  //如果o满足spec1 && spec2
                System.out.println(o);
            }
        }
    }

    //抽象规格书
    interface ISpecification {
        //候选者是否满足条件
        boolean isSatisfiedBy (Object candidate) ;
        //与操作
        ISpecification and (ISpecification spec);
        //或操作
        ISpecification or (ISpecification spec);
        //非操作
        ISpecification not ();
    }

    //组合规格书
    static abstract class CompositeSpecification implements ISpecification {
        //是否满足条件由子类实现
        public abstract boolean isSatisfiedBy (Object candidate) ;
        //与操作
        public ISpecification and (ISpecification spec) {
            return new AndSpecification(this, spec);
        }
        //或操作
        public ISpecification or(ISpecification spec) {
            return new OrSpecification(this, spec);
        }
        //非操作
        public ISpecification not() {
            return new NotSpecification(this);
        }
    }

    //与规格书
```

```
static class AndSpecification extends CompositeSpecification {
    //传递两个规格书进行与操作
    private ISpecification left;
    private ISpecification right;

    public AndSpecification(ISpecification left, ISpecification right) {
        this.left = left;
        this.right = right;
    }

    //进行与运算
    public boolean isSatisfiedBy(Object candidate) {
        return left.isSatisfiedBy(candidate) && right.isSatisfiedBy(candidate);
    }
}

static class OrSpecification extends CompositeSpecification {
    //传递两个规格书进行或操作
    private ISpecification left;
    private ISpecification right;

    public OrSpecification(ISpecification left, ISpecification right) {
        this.left= left;
        this.right = right;
    }

    //进行或运算
    public boolean isSatisfiedBy(Object candidate) {
        return left.isSatisfiedBy(candidate) || right.isSatisfiedBy(candidate);
    }
}

static class NotSpecification extends CompositeSpecification {
    //传递两个规格书进行非操作
    private ISpecification spec;

    public NotSpecification(ISpecification spec) {
        this.spec = spec;
    }

    //进行非运算
    public boolean isSatisfiedBy(Object candidate) {
        return !spec.isSatisfiedBy(candidate);
    }
}
```

```
    //业务规格书
    static class BizSpecification extends CompositeSpecification {
        //基准对象，如姓名等，也可以是int等类型
        private String obj;
        public BizSpecification(String obj) {
            this.obj = obj;
        }
        //判断是否满足要求
        public boolean isSatisfiedBy(Object candidate){
            //根据基准对象判断是否符合
            return true;
        }
    }
}
```

32.2.5 规格模式的优缺点

1. 优点

规格模式非常巧妙地实现了对象筛选功能，适合在多个对象中筛选查找，或者业务规则不适于放在任何已有实体或值对象中，而且规则变化和组合会掩盖对象的基本含义的情况。

2. 缺点

规格模式中有一个很严重的问题就是父类依赖子类，这种情景只有在非常明确不会发生变化的场景中存在，它不具备扩展性，是一种固化而不可变化的结构。一般在面向对象设计中应该尽量避免。

32.3 空对象模式

32.3.1 空对象模式的定义

空对象模式（Null Object Pattern）不属于GoF的设计模式，但是它作为一种经常出现的模式足以被视为设计模式了。其具体定义为设计一个空对象取代NULL对象实例的检查。NULL对象不是检查控制，而是反映一个不做任何动作的关系。这样的NULL对象也可以在数据不可用的时候提供默认的行为，属于行为型设计模式。

原文： Provide an object as a surrogate for the lack of an object of a given type. The Null object provides intelligent do nothing behavior, hiding the details from its collaborators.

32.3.2　空对象模式的应用场景

空对象模式适用于以下应用场景。

（1）对象实例需要一个协作实例。空对象模式不会引入协作实例，它只是使用现有的协作实例。

（2）部分协作实例不需要做任何处理。

（3）从客户端中将对象实例不存在的代码逻辑抽象出来。

32.3.3　空对象模式的 UML 类图

空对象模式的 UML 类图如下。

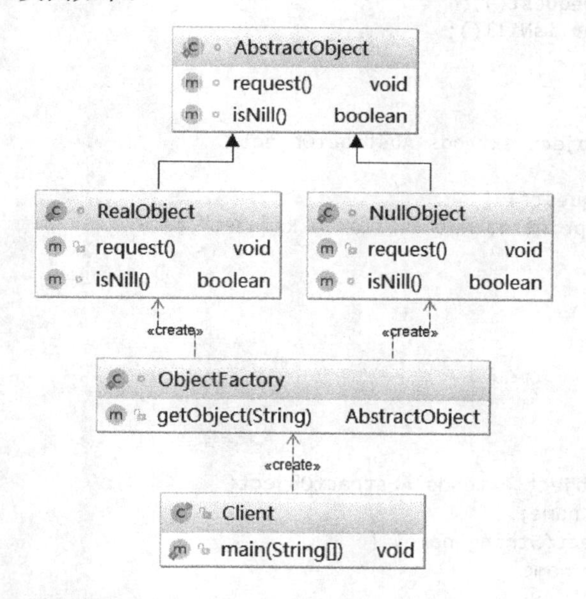

由上图可以看到，空对象模式主要包含 3 个角色。

（1）抽象对象（AbstractObject）：定义所有子类公有的行为和属性。

（2）真实对象（RealObject）：继承 AbstractObject 类，并实现所有行为。

（3）空对象（NullObject）：继承 AbstractObject 类，对父类方法和属性不做实现和赋值。

32.3.4 空对象模式的通用写法

以下是空对象模式的通用写法。

```
public class Client {

    public static void main(String[] args) {
        ObjectFactory factory = new ObjectFactory();
        System.out.println(factory.getObject("Joe").isNill());
        System.out.println(factory.getObject("Tom").isNill());
    }


    //抽象对象
    static abstract class AbstractObject{
        abstract void request();
        abstract boolean isNill();
    }

    //空对象
    static class NullObject extends AbstractObject{

        public void request() {
            System.out.println("Not Available Request");
        }

        boolean isNill() {
            return true;
        }
    }

    //真实对象
    static class RealObject extends AbstractObject{
        private String name;
        public RealObject(String name) {
            this.name = name;
        }

        public void request() {
            System.out.println("Do samething...");
        }

        boolean isNill() {
            return false;
        }
    }
```

```
    //对象工厂
    static class ObjectFactory{
        private static final String[] names = {"Tom","Mic","James"};

        public AbstractObject getObject(String name){
            for (String n : names) {
                if(n.equalsIgnoreCase(name)){
                    return new RealObject(name);
                }
            }
            return new NullObject();
        }
    }
}
```

32.3.5　空对象模式的优缺点

1. 优点

（1）它可以加强系统的稳固性，能有效地减少空指针报错对整个系统的影响，使系统更加稳定。

（2）它能够实现对空对象情况的定制化的控制，掌握处理空对象的主动权。

（3）它并不依靠 Client 来保证整个系统的稳定运行。

（4）它通过定义 isNull()对使用条件语句==null 的替换，显得更加优雅，更加易懂。

2. 缺点

每一个要返回的真实的实体都要建立一个对应的空对象模型，那样会增加类的数量。

32.4　雇工模式

32.4.1　雇工模式的定义

雇工模式（Employee Pattern）也叫作仆人模式（Servant Pattern），属于行为型设计模式，它为一组类提供通用的功能，而不需要类实现这些功能，也是命令模式的一种扩展。

32.4.2　雇工模式的应用场景

在日常开发过程中，我们可能已经接触过雇工模式，只是没有把它抽取出来，也没有汇编成册。或许大家已经看出这与命令模式非常相似，其实雇工模式是命令模式的一种简化，但更符合实际需要，更容易进入开发场景中。

32.4.3 雇工模式的 UML 类图

雇工模式的 UML 类图如下。

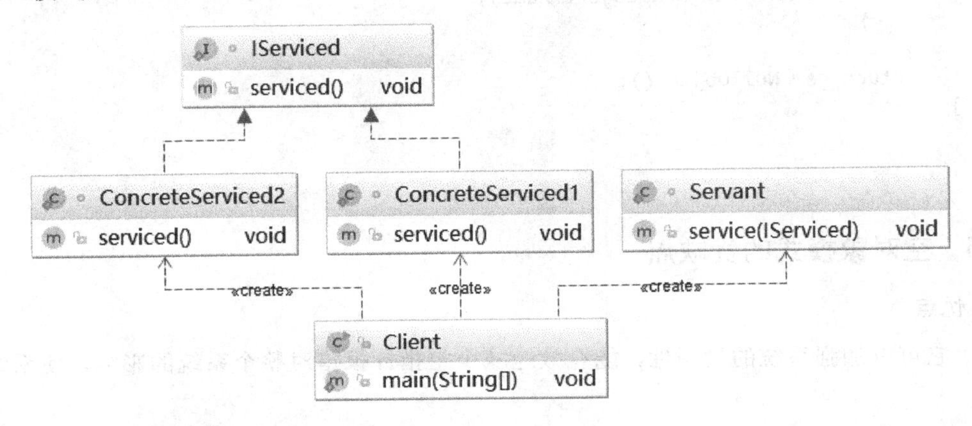

由上图可以看到，雇工模式主要包含 3 个角色。

（1）抽象服务提供者（IServiced）：用于定义服务内容的接口。

（2）具体服务提供者（ConcreteServiced）：实现所有的服务内容。

（3）雇工（Servant）：即执行者，用于执行服务。

32.4.4 雇工模式的通用写法

以下是雇工模式的通用写法。

```
public class Client {

    public static void main(String[] args) {
        Servant servant = new Servant();
        servant.service(new ConcreteServiced1());
        servant.service(new ConcreteServiced2());
    }
    /**
     * 通用功能
     */
    interface IServiced {
```

```
        //具有的特质或功能
        public void serviced();
    }

    /**
     * 具体功能
     */
    static class ConcreteServiced1 implements IServiced {
        public void serviced(){
            System.out.println("Serviced 1 doing");
        }
    }

    static class ConcreteServiced2 implements IServiced{
        public void serviced(){
            System.out.println("Serviced 2 doing");
        }
    }

    /**
     * 雇工类
     */
    static class Servant {
        //服务内容
        public void service(IServiced serviceFuture){
            serviceFuture.serviced();
        }
    }
}
```

32.4.5　雇工模式的优缺点

1. 优点

扩展性良好，可以很容易地增加雇工来执行新的任务。

2. 缺点

增加了程序的复杂度。

33

第 33 章 软件架构与设计模式

33.1 软件架构和设计模式的区别

有很多程序员经常会把软件架构和设计模式混淆，比如认为 MVC 架构是一种设计模式。实际上它们是完全不同的概念软件。软件架构通常考虑的是代码重用，而设计模式考虑的是设计重用，应用框架则介于两者之间，部分代码重用，部分设计重用，有时分析也可重用。在软件开发过程中有以下 3 种级别的重用。

（1）内部重用：即在同一应用程序中将公共使用的抽象块进行重复使用。

（2）代码重用：即将通用模块组合成库或工具集，以便在多个应用和领域都能使用。

（3）架构重用：即为专用领域提供通用的或现成的基础结构，以获得最高级别的重用性。

软件架构与设计模式虽然相似，但却有着根本的不同。设计模式是对在某种环境中反复出现的问题及解决该问题的方案的描述，它比软件架构更抽象；软件架构可以用代码表示，也能直接执行或复用，而对设计模式而言，只有实例才能用代码表示。设计模式是比软件架构更小的元素，一个软件架构中往往含有一个或多个设计模式，软件架构总是针对某一特定应用领域，但同一设计模式却可适用于各种应用。可以说，软件架构是应用程序，而设计模式是开发应用程序的具体方法。我们经常使用的软件架构有 MVC 架构、ORM 架构等。

33.2　三层架构

33.2.1　三层架构概述

通常意义上的三层架构（3-Tier Architecture）是将整个业务应用自下而上划分为数据访问层（Data Access Layer，DAL）、业务逻辑层（Business Logic Layer，BLL）和表示层（User Interface Layer，UI）。区分层次是为了体现“高聚合、低耦合”的设计理念。在软件体系架构设计中，分层式结构是最常见、也是最重要的一种结构。

33.2.2　三层架构的编程模型

三层架构的编程模型如下图所示。

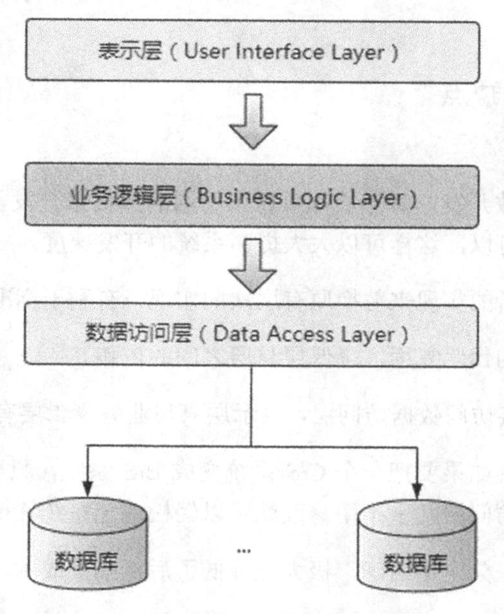

在这 3 个层次中，系统主要功能和业务逻辑都在业务逻辑层进行处理。

1. 表示层

表示层又叫作表现层，位于三层架构的最上层，与用户直接接触，主要是 B/S 信息系统中的 Web 浏览页面。作为 Web 浏览页面，表示层的主要功能是实现系统数据的传入与输出，在此过程中不需要借助逻辑判断操作就可以将数据传送到业务逻辑层进行数据处理，处理后会将结果反馈到表示层中。换句话说，表示层实现用户界面功能，将用户的需求传达和反馈，并用业务逻辑层或者 Models 进行调试，保证用户体验。

2. 业务逻辑层

业务逻辑层的功能是对具体问题进行逻辑判断与执行操作，接收到表示层的用户指令后，会连接数据访问层。业务逻辑层在三层架构中位于表示层与数据访问层的中间位置，也是表示层与数据访问层的桥梁，实现三层之间的数据连接和指令传达，可以对接收数据进行逻辑处理，实现数据的修改、获取、删除等功能，并将处理结果反馈到表示层中，实现软件功能。

3. 数据访问层

数据访问层是数据库的主要操控系统，实现数据的增加、删除、修改、查询等操作，并将操作结果反馈到业务逻辑层。在实际运行过程中，数据访问层没有逻辑判断能力，为了实现代码编写的严谨性，提高代码阅读程度，一般软件开发人员会在该层编写 SQL 语句，保证数据访问层的数据处理功能。

33.2.3 三层架构的优缺点

1. 优点

（1）有利于系统的分散开发，每一层都可以由不同的人员来开发，只要遵循接口标准，利用相同的对象模型实体类就可以，这样可以大大提高系统的开发速度。

（2）可以很容易地用新的实现来替换原有层次的实现，有利于标准化。

（3）有利于各层逻辑的代码复用，降低层与层之间的依赖。

（4）避免了表示层直接访问数据访问层，表示层只与业务逻辑层有联系，提高了数据安全性。

（5）方便系统的移植，如果要把一个 C/S 系统变成 B/S 系统，只要修改三层架构的表示层就可以。业务逻辑层和数据访问层几乎不用修改就可以轻松地把系统移植到网络上。

（6）项目结构更清楚，分工更明确，极大地降低了后期维护成本，减少了维护时间。

2. 缺点

（1）降低了系统的性能，这是不言而喻的。如果不采用分层式结构，则很多业务可以直接造访数据库，以此获取相应的数据，如今却必须通过中间层来完成。

（2）有时会导致级联的修改。这种修改尤其体现在自上而下的方向。如果在表示层中需要增加一个功能，为保证其设计符合分层式结构，则可能需要在相应的业务逻辑层和数据访问层中都增加相应的代码。

（3）增加了开发成本。

33.3　ORM 架构

33.3.1　ORM 架构概述

ORM（Object Relational Mapping，对象关系映射）是一种为了解决面向对象与关系型数据库存在的互不匹配的现象的技术。简单地说，ORM 是通过使用描述对象和数据库之间映射的元数据，将程序中的对象自动持久化到关系型数据库中。实现持久化比较简单的方案是采用硬编码方式，为每一种可能的数据库访问操作都提供单独的方法，这个操作是在三层架构中的数据访问层完成的。因此，ORM 架构就是专门为数据操作层设计的。

33.3.2　ORM 架构的编程模型

ORM 架构的编程模型如下图所示。

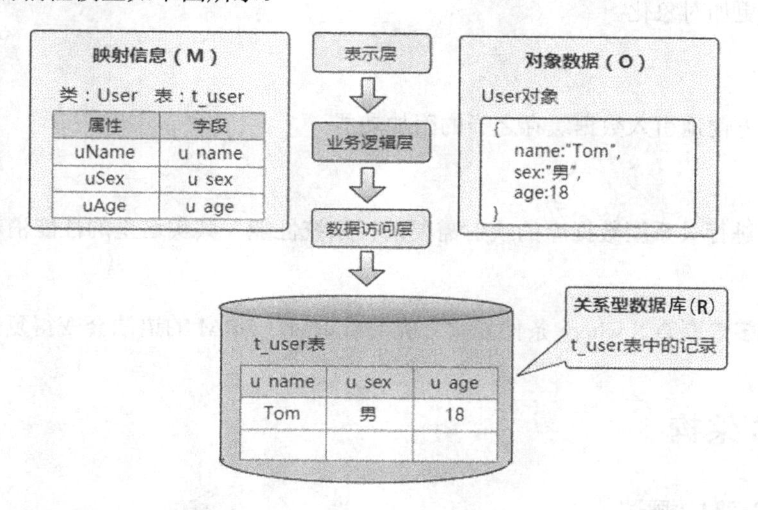

ORM 的方法论基于 4 个核心原则。

（1）简单：ORM 以最基本的形式建模数据。比如 ORM 会将 MySQL 的一张表映射成一个 Java 类（模型），表的字段就是这个类的成员变量。

（2）精确：ORM 使所有的 MySQL 数据表都按照统一的标准精确地映射成 Java 类，使系统在代码层面保持准确统一。

（3）易懂：ORM 使数据库结构文档化。比如 MySQL 数据库就被 ORM 转换成了 Java 程序员可以读懂的 Java 类，Java 程序员可以只把注意力放在他擅长的 Java 层面（当然能够熟练掌握 MySQL 更好）。

（4）易用：ORM 包含对持久化对象进行 CRUD 操作的 API，例如，创建 create()、更新 update()、保存 save()、加载 load()、查找 find()等，也就是将 SQL 查询全部封装成了编程语言中的函数，通过函数的链式组合生成最终的 SQL 语句。通过这种封装，避免了不规范、冗余、风格不统一的 SQL 语句，可以避免很多人为 Bug，方便编码风格的统一。

目前市面上采用 ORM 架构的经典框架有 MyBatis、JPA、Hibernate 等。

33.3.3 ORM 架构的优缺点

1. 优点

（1）提高开发效率，降低开发成本。

（2）使开发更加对象化。

（3）可移植。

（4）可以很方便地引入数据缓存之类的附加功能。

2. 缺点

（1）自动化进行关系型数据库的映射需要消耗系统性能。其实这里的性能消耗比较小，一般来说可以忽略。

（2）当处理多表联查、where 条件复杂之类的查询时，ORM 的语法会变得复杂。

33.4 MVC 架构

33.4.1 MVC 架构概述

MVC 的全称是 Model View Controller，是模型（Model）-视图（View）-控制器（Controller）的缩写。它指导我们用一种业务逻辑、数据、界面显示分离的方法组织代码，将业务逻辑聚集到一个部件里面，在改进和个性化定制界面及用户交互的同时，不需要重新编写业务逻辑。MVC 被独特地发展起来，用于把传统的输入、处理和输出功能映射在一个逻辑的图形化用户界面的结构中。

33.4.2　MVC 架构的编程模型

MVC 架构提供了一种对 HTML、CSS 和 JavaScript 等前端代码完全隔离的编程方式。各模块之间分工明确，职责清晰，下面对 MVC 的模块进行详细介绍。

（1）模型层用于处理应用程序数据逻辑的部分，负责在数据库中存取数据。

（2）视图层用于处理数据显示的部分，通常视图是根据模型数据来渲染的。

（3）控制层用于处理用户交互的部分，负责从视图读取数据，控制用户输入，并向模型发送数据。

三层之间主要的交互流程如下图所示。

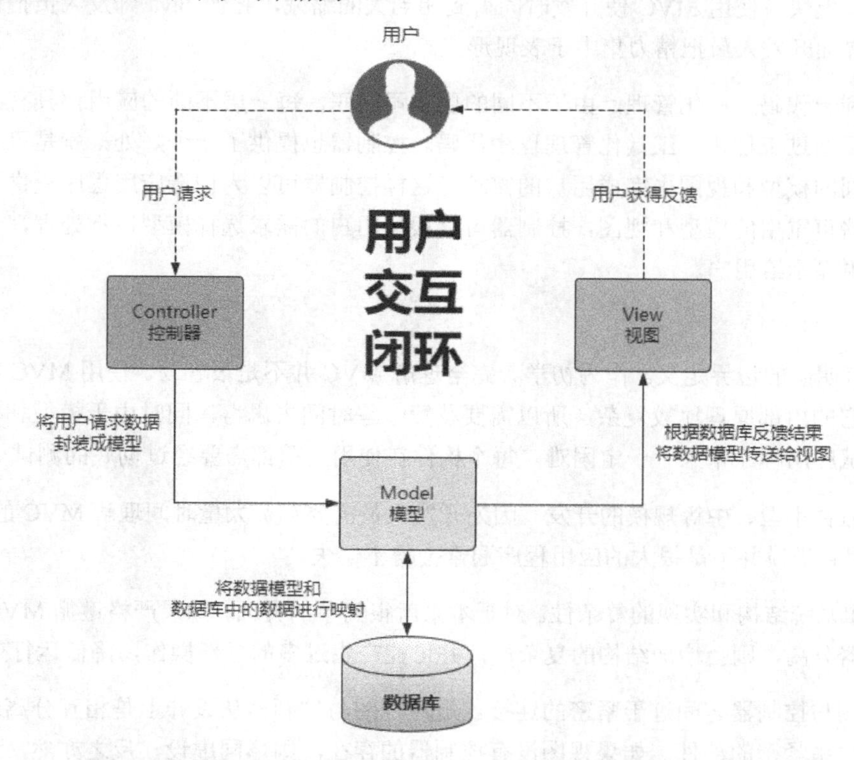

目前市面上采用 MVC 架构的主流应用框架有 Spring MVC、JSF、JFinal 等。

33.4.3 MVC 架构的优缺点

1. 优点

（1）降低耦合性。视图层和业务层分离，这样就允许更改视图层代码而不用重新编译模型和控制器代码。同样，一个应用的业务流程或者业务规则的改变只需要改动 MVC 的模型层即可。因为模型与控制器和视图相分离，所以很容易改变应用程序的数据层和业务规则。

（2）提高代码重用性。随着 IT 技术的不断升级，需要用越来越多的方式访问应用程序。MVC 模式允许使用各种不同样式的视图来访问同一个服务端的代码，因为多个视图能共享一个模型。由于模型返回的数据没有进行格式化，所以同样的构件能被不同的界面使用。

（3）软件生命周期内的维护成本降低，因为采用 MVC 使开发和维护用户接口的技术含量降低，分离视图层和业务逻辑层也使得 Web 应用更易于维护和修改。

（4）部署更快，使用 MVC 使开发时间得到相当大的缩减，它使 Java 开发人员把精力集中于业务逻辑，界面开发人员把精力集中于表现形式。

（5）有利于代码工程化管理，由于不同的层各司其职，每一层不同的应用都具有某些相同的特征，有利于通过工程化、工具化管理程序代码。控制器也提供了一个好处，就是可以使用控制器来连接不同的模型和视图去完成用户的需求，这样控制器可以为构造应用程序提供强有力的手段。给定一些可重用的模型和视图，控制器可以根据用户的需求选择模型进行处理，然后选择视图将处理结果显示给用户。

2. 缺点

（1）没有明确的边界定义。作为初学者完全理解 MVC 并不是很容易。使用 MVC 需要精心的设计，由于它的内部原理比较复杂，所以需要花费一些时间去思考。同时由于模型和视图要严格分离，给调试应用程序带来了一定困难。每个构件在使用之前都需要经过彻底的测试。

（2）不适合小型、中等规模的开发。因为开发人员需要花费大量时间理解 MVC 的设计理念，将 MVC 应用到规模并不是很大的应用程序通常会得不偿失。

（3）增加系统结构和实现的复杂性。对于本来就很简单的界面，如果严格遵循 MVC，使模型、视图与控制器分离，则会增加结构的复杂性，并可能产生过多的更新操作，降低运行效率。

（4）视图与控制器之间过于紧密的连接。虽然视图与控制器从设计上是相互分离的，但从逻辑上看却是联系紧密的部件。如果视图没有控制器的存在，则形同虚设，反之亦然，这样就使得视图和控制器的代码不能独立重用。

（5）降低了视图对模型数据的访问效率。根据模型操作接口的不同，视图可能需要多次调用才能获得足够的显示数据。对未变化数据不必要的频繁访问，会降低操作性能。

33.5 RPC 架构

33.5.1 RPC 架构概述

RPC（Remote Procedure Call，远程过程调用）是建立在 Socket 通信上的设计。在一台机器上运行的主程序，可以调用另一台机器上准备好的子程序，就像 LPC（Local Procedure Call，本地过程调用）。也就是说两台服务器 A 和 B，一个应用部署在 A 服务器上，想要调用 B 服务器上应用提供的函数/方法，由于不在同一个内存空间，不能直接调用，需要通过网络来表达调用的语义和传达调用的数据。

33.5.2 RPC 架构的编程模型

RPC 架构主要分为 3 个部分，如下图所示。

（1）服务端（Provider）：运行在服务端，提供服务接口定义与服务实现类。

（2）注册中心（Registry）：运行在服务端，负责将本地服务发布成远程服务，管理远程服务，提供给服务消费者使用。

（3）消费端（Consumer）：运行在客户端，通过远程代理对象调用远程服务。

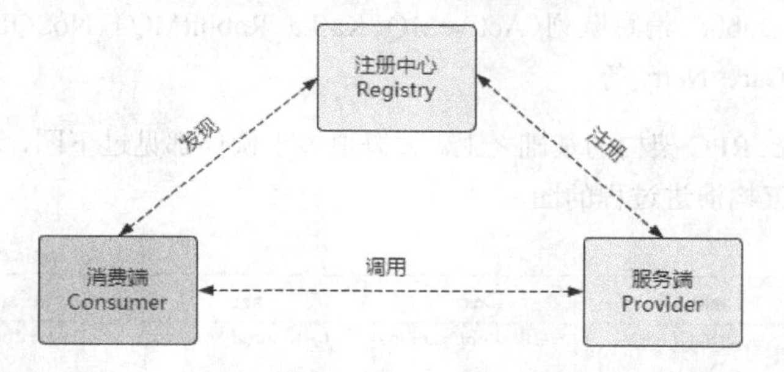

在上图中，服务端启动后主动向注册中心注册机器 IP、端口及提供的服务列表；消费端启动时由注册中心获取服务端提供方的地址列表。

目前，使用 RPC 架构的开源框架非常多，举例如下。

（1）应用级相关的服务框架：阿里的 Dubbo/Dubbox、Google GRPC、Spring Boot/Spring Cloud。

（2）远程通信相关的协议：RMI、Socket、SOAP（HTTP XML）、REST（HTTP JSON）。

（3）通信相关的框架：Mina 和 Netty。

33.5.3 RPC 架构的优缺点

1. 优点

（1）提升系统的可扩展性。

（2）提升系统的可维护性和持续交付能力。

（3）实现系统高可用。

2. 缺点

（1）一个完善的 RPC 架构的框架开发难度大。

（2）RPC 架构的框架调用成功率受限于网络状况。

（3）调用远程方法对初学者来说难度大。

33.6 未来软件架构演进之路

随着越来越多的人参与到互联网，曾经的单体应用架构越来越无法满足需求，因此，分布式集群架构出现了。也因此，分布式搭建开发成为 Web 开发者必须掌握的技能之一。这些技术包含但不限于 ZooKeeper、Dubbo、消息队列（ActiveMQ、Kafka、RabbitMQ）、NoSQL（Redis、MongoDB）、Nginx、分库分表 MyCat、Netty 等。

分布式架构建立在 RPC 架构的基础之上。大概很多小伙伴都见过下图，这是在 Dubbo 官网中找到的一张描述项目架构演进过程的图。

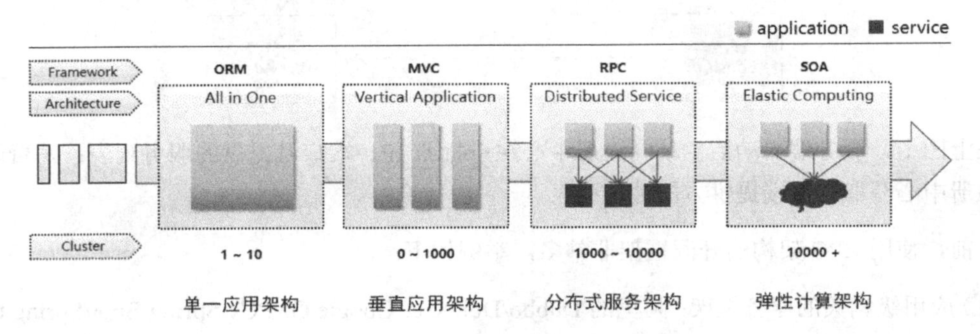

它描述了每一种架构需要的具体配置和组织形态。当网站流量很小时，只需一个应用，将所有功能都部署在一起，以减少节点部署和成本，我们通常会采用单一应用架构。之后才出现了 ORM 架构，该架构大大简化了增删改查的操作流程，提高了开发者的工作效率。

随着用户量增加，访问量逐渐增大，单一应用增加机器带来的加速度越来越小，我们需要将应用拆分成互不干扰的几个应用以提升效率，于是就出现了垂直应用架构。MVC 架构就是一种非常经典的用于加速前端页面开发的架构。

当垂直应用越来越多时，应用之间的交互不可避免，将核心业务抽取出来，作为独立的服务，逐渐形成稳定的服务中心，使前端应用能更快速地响应多变的市场需求，于是就出现了分布式服务架构。分布式架构下服务数量逐渐增加，为了提高管理效率，RPC 架构应运而生。RPC 架构用于提高业务复用及整合，在分布式服务架构下，RPC 架构是关键。

下一代架构，将会是流动计算架构占据主流。当服务越来越多时，容量的评估、小服务的资源浪费等问题逐渐明显。此时，需要增加一个调度中心，基于访问压力实时管理集群容量，提高集群利用率。SOA（Service-Oriented Architecture，面向服务的架构）架构就是用于提高集群利用率的。在资源调度和治理中心方面，SOA 架构是关键。

都已经看到这里了，相信你已经筑起了一个架构师的梦，本书将为你的架构师梦助力。本书中的内容，将是你未来技术之路的学习总纲。另外，如果你想继续深入学习 Spring 源码，推荐大家看《Spring 5 核心原理与 30 个类手写实战》；如果你想继续深入学习网络通信，推荐大家看《Netty 4 核心原理与手写 RPC 框架实战》；如果你想继续深入学习分布式微服务，推荐大家看《Spring Cloud Alibaba 微服务实战》，也欢迎大家关注“咕泡学院 Java 架构师成长丛书”的持续更新。最后，祝各位读者收获满满。

反侵权盗版声明